NITROGEN EXCRETION

This is Volume 20 in the

FISH PHYSIOLOGY series

Edited by William S. Hoar, David J. Randall, and Anthony P. Farrell

A complete list of books in this series appears at the end of the volume.

NITROGEN EXCRETION

Edited by

PATRICIA A. WRIGHT
University of Guelph
Ontario, Canada

PAUL M. ANDERSON
University of Minnesota, Duluth
Duluth, Minnesota

ACADEMIC PRESS
A Harcourt Science and Technology Company

San Diego   San Francisco   New York   Boston   London   Sydney   Tokyo

Cover photo credit: Embryo of gulf toadfish, *Opsanus beta,* 10 days after fertilization. Photo by John Barimo, University of Miami.

This book is printed on acid-free paper. ∞

Copyright © 2001 by ACADEMIC PRESS

All Rights Reserved.
No part of this publication may be reproduced or transmitted in any form or by any means, electronic or mechanical, including photocopy, recording, or any information storage and retrieval system, without permission in writing from the publisher.

Requests for permission to make copies of any part of the work should be mailed to: Permissions Department, Harcourt Inc., 6277 Sea Harbor Drive, Orlando, Florida 32887-6777

Explicit permission from Academic Press is not required to reproduce a maximum of two figures or tables from an Academic Press chapter in another scientific or research publication provided that the material has not been credited to another source and that full credit to the Academic Press chapter is given.

Academic Press
A Harcourt Science and Technology Company
525 B Street, Suite 1900, San Diego, California 92101-4495, USA
http://www.academicpress.com

Academic Press
Harcourt Place, 32 Jamestown Road, London NW1 7BY, UK
http://www.academicpress.com

Library of Congress Catalog Card Number: 2001090003

International Standard Book Number: 0-12-350444-9

PRINTED IN THE UNITED STATES OF AMERICA
01 02 03 04 05 06 QW 9 8 7 6 5 4 3 2 1

CONTENTS

CONTRIBUTORS ix
PREFACE xi

1. Evolutionary Considerations of Nitrogen Metabolism and Excretion
   Patrick J. Walsh and Thomas P. Mommsen

I.	Introduction: The Changing Roles of Ammonia and Urea	1
II.	Evolution of Urea as an Osmolyte and Excretory Product in Fish	5
III.	Uricotely and Other End Products	22
IV.	Summary and Epilogue	24
	References	26

2. Protein Synthesis
   C. G. Carter and D. F. Houlihan

I.	Introduction	31
II.	General Models of Protein Synthesis	32
III.	Protein Synthesis and Amino Acid Flux	37
IV.	Factors That Modify Protein Synthesis	57
V.	Energetic Cost of Protein Synthesis in Fish	67
VI.	Summary	67
	References	69

3. Amino Acid Metabolism
   J. S. Ballantyne

I.	Introduction	77
II.	Digestion and Uptake of Amino Acids	79

v

III.	Delivery Systems	79
IV.	Tissues	83
V.	Exercise	91
VI.	Hormonal Regulation	91
VII.	Diurnal Rhythms in Amino Acid Metabolism	93
VIII.	Diet and Starvation	94
IX.	Temperature	95
X.	Salinity	96
XI.	Anoxia	97
XII.	Summary	98
	References	99

4. Ammonia Toxicity, Tolerance, and Excretion
Y. K. Ip, S. F. Chew, and D. J. Randall

I.	Introduction	109
II.	Ammonia Toxicity to Fish	110
III.	Strategies to Reduce, Tolerate, or Avoid Ammonia Toxicity	123
IV.	Summary	139
	References	140

5. Ontogeny of Nitrogen Metabolism and Excretion
P. A. Wright and J. H. Fyhn

I.	Early Development of Fishes	149
II.	Free Amino Acids	151
III.	Nutritional Demands	164
IV.	Nitrogen End Products and Their Chemistry in Solution	169
V.	Nitrogen Metabolism	176
VI.	Nitrogen Excretion	181
VII.	Summary	185
	References	186

6. Influence of Feeding, Exercise, and Temperature on Nitrogen Metabolism and Excretion
Chris M. Wood

I.	Introduction	201
II.	Feeding and Nitrogen Metabolism	202
III.	Metabolic Fuel Usage and Nitrogen Metabolism	207
IV.	Exercise and Nitrogen Metabolism	209
V.	Temperature and Nitrogen Metabolism	216

CONTENTS

VI.	Other Nitrogen Products?	219
VII.	Excretion Mechanisms	223
VIII.	Elasmobranchs	225
IX.	Concluding Remarks	228
	References	229

7. Urea and Glutamine Synthesis: Environmental Influences on Nitrogen Excretion
Paul M. Anderson

I.	Introduction	239
II.	Carbamoyl-Phosphate Synthetase III and Nitrogen Excretion in Fish	244
III.	Glutamine Synthetase and Ammonia Detoxification	250
IV.	Urea Cycle in Elasmobranchs	253
V.	Urea Cycle in Teleosts: Adaptation to Unique Environmental Circumstances	256
VI.	Summary	269
	References	270

8. Urea Transport
P. J. Walsh and C. P. Smith

I.	Introduction	279
II.	Urea Transport in Mammalian Systems	281
III.	Physiology of Urea Transport in Fish	289
IV.	Prospects for Future Research and Evolutionary Perspectives	301
	References	302

9. Nitrogen Compounds as Osmolytes
Paul H. Yancey

I.	Introduction	309
II.	Types and Contents of Osmolytes	311
III.	Metabolism and Regulation of Osmolytes	316
IV.	Properties and Functions of Nitrogen Osmolytes	320
V.	Mechanisms of Compatibility and Counteraction: Osmolytes and Water Structure	331
VI.	Evolutionary Considerations	333
VII.	Practical Applications, Exceptions, and Unanswered Questions	334
	References	335

INDEX   343

OTHER VOLUMES IN THE FISH PHYSIOLOGY SERIES   357

CONTRIBUTORS

Numbers in parentheses indicate the pages on which the authors' contributions begin.

PAUL M. ANDERSON *(239), Department of Biochemistry and Molecular Biology, University of Minnesota, Duluth, Duluth, Minnesota 55812*

JAMES S. BALLANTYNE *(77), Department of Zoology, University of Guelph, Ontario N1G 2W1, Canada*

CHRIS G. CARTER *(31), School of Aquaculture, University of Tasmania, Launceston, Tasmania 7250, Australia*

SHIT-FUN CHEW *(109), Biology Division, Nayang Technological University, Singapore 259756, Republic of Singapore*

HANS JORGEN FYHN *(149), Institute of Zoology, University of Bergen, Bergen N-5020, Norway*

DOMINIC F. HOULIHAN *(31), Department of Zoology, University of Aberdeen, Aberdeen AB9 2TN, United Kingdom*

YUEN K. IP *(109), Department of Zoology, National University of Singapore, Singapore 0511, Republic of Singapore*

THOMAS P. MOMMSEN *(1), Department of Biochemistry and Microbiology, University of Victoria, Victoria, British Columbia V8W 3P6, Canada*

DAVE J. RANDALL *(109), Department of Zoology, University of British Columbia, Vancouver, British Columbia V6T 1Z4, Canada*

CRAIG P. SMITH *(279), School of Biological Sciences, University of Manchester, Manchester M13 9PT, United Kingdom*

PATRICK J. WALSH *(1, 279), Division of Marine Biology and Fisheries, University of Miami, Miami, Florida 33149*

CHRIS M. WOOD *(201), Department of Biology, McMaster University, Hamilton, Ontario L86 4K1, Canada*

PATRICIA A. WRIGHT *(149), Department of Zoology, University of Guelph, Ontario N1G 2W1, Canada*

PAUL H. YANCEY *(309), Biology Department, Whitman College, Walla Walla, Washington 99362*

PREFACE

Nitrogen is a component of many biological molecules in all living things—amino acids, proteins, nucleic acids, etc. Proteins and other nitrogen-containing molecules are regular and essential dietary components, and digestion and metabolism of these compounds result in excess nitrogen that must be excreted. The primary end product of the metabolic degradation of nitrogen-containing compounds is ammonia, which in fish is normally excreted directly across the gills. However, urea and perhaps other nitrogen-containing compounds, formed as direct end products of nitrogen degradation or from ammonia, are also excreted by fish under some circumstances. A series of hallmark studies by Homer Smith in the early 1930s established the fundamental aspects of nitrogen excretion in teleost and elasmobranch fishes. These studies have served as the foundation for the steady progress in our understanding of nitrogen metabolism and excretion in fish over the past 70 years.

This progress has accelerated markedly in the past decade, resulting in major advances in our understanding of virtually all aspects of nitrogen metabolism and excretion in fish, particularly in terms of molecular biology approaches to the study of these processes. Recent developments have highlighted the diversity of nitrogen excretory strategies that fish employ in adapting to a wide range of environments. These studies have provided new insights into ammonia and urea transport, the role of nitrogen metabolism in osmoregulation, ammonia toxicity, the evolution and expression of the urea cycle, environmental influences on protein synthesis, amino acid metabolism, nutrition, and the role of urea synthesis in embryonic development. This volume is a timely and comprehensive review of these topics, including highlights of relevant new directions of study.

The topic of this book was conceived by Dave Randall in 1996, the idea being to collect summaries of these new advances as well as the classical literature into one volume in the *Fish Physiology* series. The inaugural volume of this series devoted a single chapter to formation of excretory products in fish (1969, Volume 1, Chapter 5), but this is the first volume devoted entirely to nitrogen metabolism and excretion. We intentionally positioned evolutionary considerations of nitrogen metabolism and excretion as the first chapter to set the framework for the subsequent chapters. This is followed by protein metabolism and turnover and amino acid metabolism, which are reviewed along with excretion of the end products of nitrogen metabolism, i.e., ammonia, urea, and perhaps other compounds. Nitrogen excretion is reviewed in the context of evolutionary relationships, toxicity and tolerance, adaptations, environmental influences that affect adaptations, and embryonic development. Osmoregulation and urea transport are reviewed in the context of nitrogen metabolism and excretion. The chapters also reflect the diversity of research approaches, from nutritional whole-animal studies to molecular mechanisms for gene regulation.

We commend all the authors on their outstanding contributions and adherence to the guidelines and imposed timelines, allowing us to present a volume of very up-to-date information. Finally, we feel honored to have been invited to serve as guest editors of this volume in the *Fish Physiology* series. We thank Tony Farrell for his wise council on navigating the shifting seas of book editing.

We dedicate the book to Dave Randall for his seminal and prolific contributions to the field of nitrogen excretion. We note as a particular example of his natural, creative, and insightful curiosity the expedition he organized in the late 1980s to Lake Magadi, Kenya, where the water is unusually alkaline (pH $>$ 10). This trip led to the discovery of a teleost fish with adapted physiological characteristics of nitrogen excretion that have challenged us to broaden our thinking about the patterns of nitrogen excretion in fish established by Homer Smith.

1

EVOLUTIONARY CONSIDERATIONS OF NITROGEN METABOLISM AND EXCRETION

PATRICK J. WALSH AND THOMAS P. MOMMSEN

I. Introduction: The Changing Roles of Ammonia and Urea
   A. The Primordial Role of Ammonia and Urea
   B. Overview of Reactions That Transfer and Generate Ammonia
   C. Evolution of Ammonia Transport/Excretion Mechanisms
   D. Evolution of Urea Transport/Excretion Mechanisms
   E. Additional Perspectives from Piscine Systems
II. Evolution of Urea as an Osmolyte and Excretory Product in Fish
   A. Developments in the Phylogeny of Early Vertebrates
   B. An Anadromous Lifestyle for Early Vertebrates?
   C. Criteria for Ureosmotic Adaptation
   D. Is the Ureosmotic Strategy "Inferior" to the Hypoosmotic Strategy?
   E. Adaptive Radiations of Ureogenesis and Ureotely in Bony Fishes
III. Uricotely and Other End Products
IV. Summary and Epilogue
   References

I. INTRODUCTION: THE CHANGING ROLES OF AMMONIA AND UREA

In simplified explanations of requirements within the vertebrates for the evolutionary transitions from an aquatic-based existence to terrestriality, fish are often pigeonholed as being solely ammonia excreters, that is, ammonoteles. Furthermore, in this view, urea and uric acid excretion (ureotely and uricotely) are "key inventions" in higher vertebrate evolution. In fact, many successful comparative evolutionary "experiments" with alternative modes of nitrogen metabolism and excretion took place early in the piscine lineage, as well as among invertebrates and protists. Therefore, the basic menu of patterns of nitrogen metabolism and excretion needed only to be selected from, and tinkered with, on the path to vertebrate terrestriality. Our goal in this chapter is to present an evolutionary framework of nitrogen metabolism and excretion in fish that illustrates this "preexisting" diversity. We also lightly touch on and introduce a number of topics that other authors in this volume will cover in much greater detail.

A. The Primordial Role of Ammonia and Urea

In examining the basic metabolic scope of modern-day *autotrophic* organisms, as well as the common themes in the literature on the origins of life, it is clear that nitrogen, in general, and ammonia, as the most convenient and reactive form of nitrogen, are and were desirable commodities. In most scenarios and experiments on the origins of life, urea is also one of the smallest building blocks formed in prebiotic "soups" (Miller and Orgel, 1974). At the base of the food chain, and in the early evolution of life, pathways and transport mechanisms are and were geared largely toward uptake and assimilation of nitrogen, rather than excretion of nitrogenous waste.

Although many of the toxic effects of ammonia in vertebrates are related to its specific impact on the cells of the central nervous system (Cooper and Plum, 1987), clearly there are potentially more general effects at the cellular level that could have impacted early autotrophs (e.g., pHi effects, effects on ion channels/transporters, effects on enzyme activities, etc.; see Campbell, 1991). It is likely, however, that these perturbations were rarely realized due to the limiting nature of nitrogen in autotrophic systems and due to the likelihood that intracellular ammonia concentrations probably never reached concentrations high enough to cause problems. Ammonia toxicity issues probably first arose with the onset of *heterotrophy* and the need to deal with an ammonia bolus in the form of a "meal" of a phagocytozed cell or other organic particle. Likely, at this basic cellular level in early heterotrophs, there was the need to avoid mass production of free ammonia, and/or to rapidly void ammonia from the cell, to avoid potential cellular toxicity. In this context, a review of the biochemistry of ammonia is instructive.

B. Overview of Reactions That Transfer and Generate Ammonia

As tabulated by Walsh and Henry (1991), the biochemical reactions transforming ammonia in fish and other heterotrophs have several unifying themes:

1. The number of free ammonia molecules generated in the tissues can be controlled by virtue of a heavy reliance on amino acid transaminase enzymes and on the ultimate funneling of nearly all ammonia through the glutamate dehydrogenase reaction.
2. A second feature is that nearly all ammonia-producing reactions yield NH_4^+, rather than NH_3. Because one aspect of ammonia toxicity relates

1. NITROGEN METABOLISM AND EXCRETION

to the strong tendency for NH_3 to combine with a proton (at physiological pHi) to raise pH (Campbell, 1973), metabolism appears to be already adapted to minimize these effects by releasing mostly NH_4^+.

3. Un-ionized ammonia (NH_3) is potentially only the product of a limited number of reactions; for example, when unprotonated $-NH_2$ groups, as in the amide of glutamine or asparagine, or the nitrogens of purine rings, are removed. Generally, fish do not rely heavily on purine degradation to uric acid for excretion of the bulk of their nitrogenous waste; therefore, generation of large quantities of NH_3 by these means would not appear to present a problem in fish. However, formation of muscle NH_3 by AMP catabolism has been implicated as an adaptation for postexercise pHi stabilization in fish (Mommsen and Hochachka, 1988).

In summary then, early heterotrophs, from which fishes inherited their metabolic pathways, probably minimized the threat of ammonia intoxication by a metabolic makeup limiting the amount of free ammonia in cell/body fluids.

C. Evolution of Ammonia Transport/Excretion Mechanisms

The more obvious way by which cells (including unicellular heterotrophs, or even autotrophs for that matter) can avoid ammonia toxicity is through transport out of the cell. Fortunately, ammonia as NH_3 is highly permeable through membranes, and whenever its partial pressure (because it is a dissolved gas) is higher inside the cell than outside, ammonia will simply diffuse out of the cell. Because NH_3 is in free and rapid equilibrium with NH_4^+, as soon as NH_3 exits the cell, it will be replaced from the bulk pool of NH_4^+, with the liberation of a proton. A steep outwardly directed NH_3 partial pressure gradient can be maintained by the rapid addition of a proton to NH_3 in the external water environment, provided that the pH is near neutral (see, e.g., Wright *et al.*, 1989). The liberated internal proton can be ejected by Na^+/H^+ exchange or H^+-ATPase, or other pHi stabilizing transporters. Coupled with an "infinite" dilution of the exiting ammonia by the external aquatic milieu, an apparently cost-free mode of excretion and toxicity avoidance is available to all aquatic organisms. As long as the heterotroph could tolerate the "inefficiency" of the general loss of nitrogen from the system, and could make it up by simply ingesting more, a rather workable system for detoxification would be in place.

Other authors in this volume, however, will illustrate that transport of ammonia is not always that simple (Chapter 4). By virtue of the fact that hydrated ammonium ions resemble sodium and potassium ions, ammonium can be transported by many pathways that, on the surface, seem dedicated to these other ions. Although for many species of fish these other transport modes of ammonia are secondary, in some piscine examples, simple diffusion of NH_3 takes a clear backseat to other, sometimes active, processes. It is tempting to speculate that fish inherited this excretory flexibility from the transporters of ancestral autotrophs concerned with actively *extracting* ammonium from the environment, rather than excreting it.

D. Evolution of Urea Transport/Excretion Mechanisms

This topic is covered in detail in Chapter 8, but it is of interest to comment here on how unicellular organisms handle urea transport and metabolism. Notably, the enzyme that breaks down urea, urease, is absent from animal tissues, but is present in microbes (and plants), allowing them to use urea as a metabolic substrate. An interesting example of the coupling of urea transport and urease activity has been presented in a recent study on *Helicobacter pylori,* the bacterial species associated with gastric ulcers. In this species, urease activity and an H^+-gated urea channel are closely linked and appear to enable the colonization of acidic media (Weeks *et al.,* 2000). A very low intracellular concentration of urea, due to the presence of urease activity inside the cell, allows an inwardly directed gradient for urea and simple facilitated diffusion through a specific urea channel. Although this is clearly a specialized adaptation of modern-day *H. pylori* to the human stomach, this example, and the lengthier discussion in Chapter 8, suggests that specific urea transport mechanisms evolved very early and were also among the repertoire of invertebrates and early vertebrates.

E. Additional Perspectives from Piscine Systems

Considering the basic metabolic setup of heterotrophs, and the ease with which NH_3 can be voided, it may be that ammonia intoxication is a problem not encountered by organisms until emergence onto land. Clearly invertebrates encountered situations of limited ammonia excretion (e.g., land emergence) well before vertebrates, such that the genes of early fish very likely contained "solutions" (both metabolic and transport) to the problems of ammonia toxicity. However, the diverse piscine examples illustrated by a number of authors in this volume clearly suggest that it is not only the simple lack of a dilute medium that can drive how organisms deal with nitrogenous waste. Other critical factors derive from these questions: (1) How does medium osmolality affect nitrogenous solute balance? (2) Are all aquatic media equally suitable sinks for ammonia? (3) How does the need to control permeability of other compounds affect ammonia permeability (e.g., at the chorionic membranes of embryos and at the gills of adults)? (4) How do developmental changes influence ammonia excretion? (5) How important is it to not simply "waste" nitrogen as ammonia in periods of food limitation? (6) More generally, how does ammonia excretion relate to the ecology and behavior of the species? These questions illustrate that the evolutionary pressure

1. NITROGEN METABOLISM AND EXCRETION

to be more than a standard "ammonotele" does not just relate to the availability of water, and that a number of evolutionary solutions were added by fish to the invertebrate genetic repertoire, which further aided their adaption to particular environmental and developmental circumstances.

As one important point of departure in this chapter and volume, we turn to Wright (1995) and Withers (1998) who summarized many of the "alternative" roles of urea in animals, namely, as (1) a major balancing osmolyte in selected fishes, (2) a contributor to buoyancy in marine cartilaginous fishes, (3) a non-toxic nitrogen transport form in ruminant and pseudo-ruminant mammals, (4) a major part of the urine concentration mechanism of the mammalian kidney, (5) the preferred nitrogenous waste form of selected teleosts experiencing restrictions on ammonia excretion from either an inappropriate aquatic medium or air breathing, and (6) maintenance of acid–base balance by the stoichiometric removal of HCO_3^- and NH_4^+. Many of these alternative uses of urea are decidedly "piscine" and are covered in depth by many of the authors in this volume. The most thoroughly studied and debated topics are roles 1 and 5 — the evolution of the osmoregulatory use of urea and the evolution of urea as an alternative waste product in fish. We examine these combined topics in detail in the next section.

II. EVOLUTION OF UREA AS AN OSMOLYTE AND EXCRETORY PRODUCT IN FISH

Without doubt, one of the central evolutionary questions in fish biology for nearly a century has been "How and why did the ureosmotic strategy evolve in cartilaginous fishes (and the coelacanth)?" Related questions include these: Why was urea retention not used in teleost fishes? Did early vertebrate evolution take place in saltwater, freshwater, or brackish water? Was urea synthesis "invented" or modified in the vertebrate lineage? While we do not presume to be able to completely answer these questions at present, many of the pieces of the puzzle are beginning to fall more clearly into place.

One important point of departure for our discussion is the scenario constructed by Griffith a decade ago. Griffith (1991) recognized three main requirements for ureosmotic regulation: (1) urea synthesis via the ornithine–urea cycle (O-UC); (2) urea tolerance involving biochemical and physiological adjustments; and (3) urea retention that requires renal, branchial, metabolic, and reproductive adaptations. Griffith proposed that the O-UC functioned in early anadromous (jawless) agnathans (e.g., the lineage represented by modern-day lampreys), which used the *arginase* portion of the O-UC (see below) as embryos during vitellogenin catabolism, and the *arginine biosynthesis* portion of the O-UC as ammocoete-like larvae to supplement an algal diet deficient in nitrogen. Furthermore, in Griffith's scenario, descendant gnathostomes (jawed-fishes) continued evolving toward larger eggs (with presumably less facility at eliminating ammonia) and prolonged development, such that both portions of the O-UC were expressed simultaneously early in development for ammonia detoxification. If O-UC expression remained active in the adult phase (via pedogenesis), this trait would preadapt large, sluggish, euryhaline fish for ureosmotic regulation when exposed to seawater. Griffith (1991) suggested that early marine invasions were achieved by such preadapted ureosmotic fishes, a strategy that would then favor continued selection for internal fertilization and development, and that this was a common strategy of Paleozoic marine gnathostomes. Below, we discuss evidence relevant to this scenario, particularly data published in the past decade. Before addressing the three main requirements for ureosmotic adaptation, updates on other early evolutionary aspects of fish are appropriate.

A. Developments in the Phylogeny of Early Vertebrates

The 1990s saw considerable research on the phylogenetic relationships of early vertebrates, including both DNA-based and paleontological research. Traditionally, both extant classes of agnathan fish, Myxini (hagfish) and Cephalospidomorphi (lampreys), had been placed more or less equidistant from jawed fishes, making it difficult to determine if early jawed vertebrates evolved from more marine (hagfish-like) or euryhaline (lamprey-like) forms. Although it is clear that the earliest chordate groups were marine (e.g., many ostracoderms of the Ordovician and Silurian periods (400–500 mya; Forey and Janvier, 1994), many later fossil ostracoderm groups can be found in freshwater and brackish water habitats. Furthermore, cladistic analysis of morphological traits of fossil and extant groups (Forey and Janvier, 1994), as well as 18S ribosomal RNA analyses (Stock and Whitt, 1992), places ostracoderms and modern-day agnathans closer to gnathostomes than to hagfish. Most recently, conodonts are viewed as ancestral gnathostomes, with early bone appearing in the form of teeth (Donoghue, 2000).

B. An Anadromous Lifestyle for Early Vertebrates?

While much of the early debate on vertebrate origins focused rather resolutely on either freshwater versus marine extremes, Griffith (1994) proposed a rather reasonable middle ground placing *much* early fish evolution (e.g., post-*Amphioxus*-like ancestors) in an *anadromous* existence, where most of the adult phase occurred in marine or brackish environs, while reproduction and larval phases took place in brackish to freshwater environs. From this evolutionary viewpoint, ancestors of modern-day hagfish committed very early to an exclusively marine environment, but with an ionoconforming strategy similar to marine invertebrates. In light of the phylogenetic studies above, perhaps Myxinids, in fact, had no intervening euryhaline ancestors. From this early anadromous existence, elasmobranch and coelacanth ancestors reinvaded the marine environment very early on via the ureosmotic strategy, whereas other early fish groups exploited freshwater and semiterrestrial niches (e.g., lungfish ancestors). Later, as teleost forms proliferated, some of these groups reinvaded the marine environment as well, but with an ionoregulatory strategy. Furthermore, selected elasmobranch and teleostean groups also committed to a freshwater existence.

As Griffith (1994) points out, the anadromous proposal (or in our view at the very least, a variable brackish existence for early fish) makes sense from a number of perspectives. First, shallow coastal waters would have had the highest productivity due to penetration of sunlight to the bottom and nutrient loading from land runoff. Diversity of prey items (be they detrital, plant or animal) was likely highest in this niche. Second, early life history development in freshwater may have protected larvae from extreme predation, recalling that all large invertebrate predators of the day (e.g., cephalopods) were strictly marine and that the teleostean explosion into freshwater had not yet taken place. Third, an intermediate salt concentration may have facilitated internal bone deposition, because bone deposits more readily at low levels of blood ions (Griffith, 1994), whereas mineral availability in brackish waters would still be relatively high. It is possible that minerals stored as bone *enabled* forays into freshwater before ultra-efficient ion uptake mechanisms were established. Fourth, although not an evidence per se, a brackish existence would explain why teleosts "chose" approximately 300–400 mOsm as the concentration of their internal milieu; perhaps this value is simply an integrated average of their environment to which they were isosmotic most of the time, and to which their biochemical systems adapted to function optimally. Finally, early colonization and exploitation of marine environs by elasmobranchs would explain their commitment to a ureosmotic strategy (and what we believe to be an inferior strategy for dealing with variable salinity; see below). By remaining in the euryhaline "incubator" for a longer evolutionary period, teleosts likely became more *temporally* adept at osmoregulation, contributing to their abilities to expand into freshwater and marine environs.

C. Criteria for Ureosmotic Adaptation

1. UREA SYNTHESIS

a. Pathway Characteristics. The pathways for urea production in modern-day fish are illustrated in Fig. 1. Urea can be formed by four potential routes. Three of these are catabolic in nature, namely, the γ-guanidino urea hydrolase pathway, purine (urate) degradation, and argininolysis (from dietary arginine). One pathway is anabolic—the ornithine–urea cycle. As alluded to above, the O-UC can be viewed as consisting of two parts: One part is the arginine degradation portion, consisting simply of arginase, the enzyme (#4 in Fig. 1) hydrolyzing arginine into

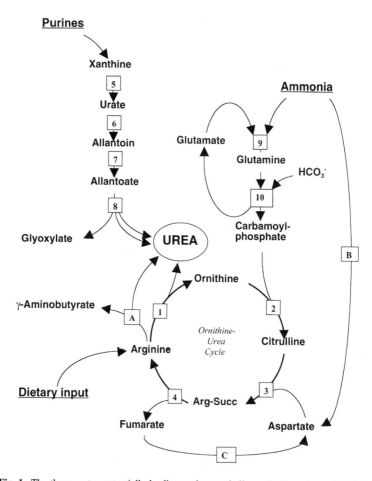

Fig. 1. The three routes potentially leading to the metabolic production of urea in fish tissues. Boxes with numbers represent individual enzymes; boxes with letters represent pathways, involving several steps. The ornithine urea cycle comprises 1, arginase; 2, ornithine-carbamoyl transferase; 3, argininosuccinate synthetase; and 4, argininosuccinate lyase. Purine degradation involves 5, xanthine oxidase; 6, urate oxidase; 7, allantoinase; and 8, allantoicase. The product glyoxylate can be recovered into glycine in the peroxisomes and mitochondria (Sakuraba et al., 1996). Glutamine synthetase (9) and carbamoyl phosphate synthetase III (10) are feeder enzymes to the O-UC, leading to the incorporation of amino acid-derived ammonia and bicarbonate into citrulline. Arginine degradation via the γ-guanidino urea hydrolase pathway (A) leads to the production of urea and γ-aminobutyrate (cf. Fig. 2). Amino acid nitrogen can also be funneled into aspartate and the urea cycle (B) via glutamate dehydrogenase and aspartate aminotransferase reactions. Fumarate from the urea cycle is recovered via parts of the Krebs cycle plus aspartate aminotransferase (C). Arg-Succ is argininosuccinate. (Adapted from Mommsen and Walsh, 1992).

ornithine and urea. The remainder of the cycle, the arginine synthetic portion, is composed of the three reactions leading to arginine synthesis from ornithine plus associated reactions. Virtually all fish express arginase and can produce a modicum of urea simply by degrading arginine obtained from dietary sources. Arginine is also the starting point for a competing route toward urea production. Interestingly, the terminal hydrolase of the route leading from arginine to urea and γ-aminobutyrate, namely, γ-guanidinobutyrate urea hydrolase, displays a higher hepatic activity than arginase—at least in the carp liver (Vellas and Serfaty, 1974), although upstream enzymes in the pathway may be limiting under physiological conditions (cf. Fig. 2). Finally, the reader should keep in mind that the impressively high activities for arginase reported in the literature are usually determined at optimized conditions (pH above 9, 1 mM Mn^2, arginine concentrations exceeding 200 mM!) that are very far removed from physiological realities: pH 7.5–8.5, micromolar concentrations of free Mn^2, and likely micromolar concentrations of arginine. Even under optimum pH conditions, the K_m values for vertebrate arginases are in the high millimolar range and thus *in vivo* arginase catalysis will be merely a fraction of test tube potential. Of the many enzymes mentioned here, arginase is likely to have the longest evolutionary history, occurs across all primary kingdoms, and hence is a suitable phylogenetic marker (Ouzounis and Kyrpides, 1994).

Independent of the actual route by which urea is derived from arginine, arginine is by no means the sole source of catabolically generated urea and the importance of uricolysis as a competing route to catabolic production of urea is under debate and deserves renewed attention. The enzymes of the urate degradation pathway are present in most fish tissues, especially the liver, and the peroxisomal allantoicase that liberates urea is quite abundant in the liver (Hayashi *et al.*, 1989). This route is often dismissed on grounds that it merely represents nucleotide (purine) turnover and thus is unlikely to experience large physiologically relevant fluctuations. However, Wright (1993) and Wilkie and coworkers (1999) noted large increases in urea excretion that were likely due to an increase in uricolytic production of urea.

b. Arginine as a Metabolic Hub. From the above discussion, it is already obvious that arginine plays a focal and versatile role in nitrogen metabolism. Also, its role goes well beyond its position as an intermediate in the urea cycle or precursor of urea, such that there were likely other evolutionary pressures on arginine metabolism. As shown in Fig. 2, this amino acid is also intricately linked to the production of a neurotransmitter (γ-aminobutyrate), an important tissue hormone (nitric oxide), a phosphagen (creatine phosphate), other amino acids (proline and glutamate), polyamines (spermine, spermidine), and not least a biogenic amine with multiple regulatory functions (agmatine) (Wu and Morris, 1998).

As an essential amino acid, especially during early development and periods

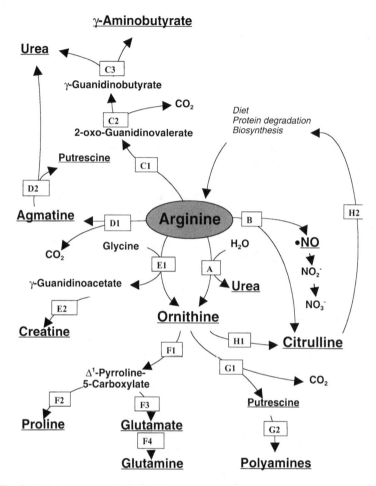

Fig. 2. Arginine as a metabolic hub. Inputs are through diet, protein degradation, and the ornithine–urea cycle (route H). Five routes of degradation can be postulated, leading to important intermediates and end products (underlined). With the exception of agmatine production, evidence has been presented for existence of all other pathways in fish tissues. The five degradation pathways are A, arginase reaction; B, synthesis of nitric oxide (•NO); C, the γ-guanidinobutyrate pathway; D, the agmatine route; and E, synthesis of creatine. Ornithine production (routes A and E) is the starting point for synthesis of other amino acids (route F) or polyamines (route G), while •NO removal can lead to production of nitrite and nitrate. Individual enzymes are A, arginase; B, nitric oxide synthase; C1, oxidative deamination of arginine; C2, 2-oxo-guanidinovalerate decarboxylase; C3, γ-guanidinobutyrate urea hydrolase; D1, arginine decarboxylase; D2, agmatine ureohydrolase; E1, arginine:glycine amidinotransferase; E2, γ-guanidinoacetate methyltransferase; F1, ornithine δ-aminotransferase; F2, Δ^1-pyrroline-5-carboxylate reductase; F3, Δ^1-pyrroline-5-carboxylate oxidase; F4, glutamine synthetase; G1, ornithine decarboxylase; G2, spermine and spermidine synthetases; H1, ornithine-carbamoylphosphate transferase; and H2, see Fig. 1.

of rapid growth, it is not surprising that arginine supplementation in the diet exerts a positive effect on growth rate in rainbow trout (*Oncorhynchus mykiss*); oddly, this is not observed for two congeners—coho (*O. kisutch*) and chinook salmon (*O. tshawytscha*) (Plisetskaya *et al.*, 1991). The dietary requirements of arginine for optimum growth of trout increase with fish size, from about 3.8% in small trout, through 4.1% in 100-g fish, up to 4.8% of diet in 300- to 500-g fish (Berge *et al.*, 1997), either indicating decreased endogenous synthesis or altered arginine demands as the fish grow larger.

Metabolically, arginine is more than an essential amino acid or a ready substrate for arginase. It is the most consistent and powerful insulinotropin known in fishes (Ince and Thorpe, 1977), and thus indirectly a positive contributor to fish growth, but, again, the role for growth promotion is blurred by the dual realities of "essential amino acid" for growth and key insulinotropin.

The transport of arginine in the mammalian liver and in other tissues in general is highly controlled and dependent on the expression of a group of cell surface proteins with properties of the amino acid transport system y^+. For instance, it is thought that the hormone-inducible arginine transporter Cat-1 is almost undetectable in normal liver to avoid equilibration of plasma arginine with the hepatocytes, thereby preventing arginine degradation by hepatic arginase. The transporter is activated under conditions of cell growth (Liu and Hatzoglou, 1998) when presumably arginine is abundant and ornithine carbon can be diverted from the O-UC into the synthesis of polyamines, which bind to nucleic acids and are essential constituents in rapidly growing cells. The level of activity of the enzyme responsible for polyamine synthesis, ornithine decarboxylase (ODC), is an excellent indicator of growth in salmonid muscle (Arndt *et al.* 1994) and is potentially an indicator of neoplasia or hyperplasia in flatfish liver (Koza *et al.*, 1993). Incidentally, fish liver has a much higher ODC activity (almost 100-fold) than mammalian liver (Calabrese *et al.*, 1993) and contains much higher titers of putrescine, while hepatic spermidine or spermine concentrations are similar to or below those found in mammalian liver (Corti *et al.*, 1987). Furthermore, the potential depletion of ornithine as an essential intermediate of arginine metabolism has yet to be addressed in fishes.

A discussion on arginine dynamics would be incomplete without mentioning extreme conditions impacting arginine availability, turnover, and metabolic fate. Female fishes, actively undergoing vitellogenesis, are an interesting case in point. Under control of estradiol, the liver increases in size by twofold or more through a mixture of hypertrophy and hyperplasia (Korsgaard and Mommsen, 1993). Under these conditions, the liver must be a net sink for large amounts of arginine, not only to support demands for arginine in the synthesis of the long amino acid backbone of vitellogenin itself, but also to support the substantial growth of liver. Because hyperplasia is involved, arginine must further serve as a precursor for polyamine biosynthesis, which appears to be an essential component of cell proliferation. In fact, putrescine and spermidine concentrations rise three- and twofold, respectively, after exposure of salmon liver to estradiol (Waters *et al.,* 1992). It would be interesting to analyze arginine flux, arginase expression, urea synthesis, and ornithine decarboxylase activity in the liver of ureogenic fish species such as toadfish (*Opsanus beta*) or the alkaline lake tilapia (*Alcolapia grahami*) under the influence of estradiol or other liver growth-promoting compounds. Is it possible that the absence (suppression?) of full urea cycle activity of most adult teleosts is linked to the fact that the female liver undergoes regular cycles producing vitellogenin? Clearly, during vitellogenesis concomitant urea production and arginine degradation in liver would compromise production or earmarking of arginine for incorporation into vitellogenin, and ornithine removal for polyamine synthesis would deprive the O-UC of its most critical intermediate. Although it is well known that estradiol, the force behind vitellogenin synthesis, regulates a number of additional genes (e.g., downregulation of albumin synthesis), its effects on urea cycle flux and O-UC enzymes remain to be explored. This topic may be especially relevant in the context of endocrine disruption (another area where humans may be impacting the evolution of nitrogen metabolism and excretion in fish; see Section IV).

c. Arginine Synthesis. It is not surprising that ureagenic fish with a consistent demand for urea production do so anabolically via expression of the O-UC. Many aspects of the expression and regulation of the O-UC are covered in other chapters, but a few highlights are relevant to our discussions in this chapter:

1. First and foremost, urea production by the O-UC is an expensive proposition, requiring 5 mol ATP for each mole of urea produced (including the production of glutamine). In fact, the metabolic cost may be even higher if essential transport phenomena for citrulline/arginine and fumarate/aspartate are taken into consideration.
2. The pathway provides for convenient entry points of nitrogen from a variety of amino acid sources, notably from the glutamate dehydrogenase "cross-road" feeding into both glutamine and aspartate, as well as the direct entry of ammonia.
3. The isozyme of carbamoyl-phosphate synthetase (enzyme #2 in Fig. 1) in fish used for urea synthesis is typically CPSase III (discussed in detail in Chapter 7). CPSase III characteristically uses glutamine (preferentially to ammonia) as a nitrogen donor, is activated allosterically by *N*-acetylglutamate, is localized in mitochondria, and is not inhibited by UTP. The ubiquitous CPSase II isozyme also utilizes glutamine preferentially, but functions primarily in pyrimidine biosynthesis, is localized in the cytosol, is not activated by *N*-acetylglutamate, and is substantially inhibited by UTP. Finally, the mitochondrial CPSase I isozyme that forms an integral part of the O-UC in mammals, utilizes ammonia preferentially to glu-

tamine, is activated by *N*-acetylglutamate, and is not inhibited by UTP (Anderson, 1995b). While the finer details of CPSase isozyme expression throughout phylogeny appear to still be unfolding (see below), the general pattern is that CPSase II is a characteristic of most organisms, CPSase III is expressed in some fish and invertebrates, and CPSase I is expressed in the lungfish and tetrapods (Anderson, 1995a,b). Thus, it is safe to say in modern-day organisms that CPSase I or III is required for full O-UC function. Indeed, Anderson and colleagues point out that measurement of activities of the correct CPSase isozyme type, and substantial ornithine-carbamoyl transferase, are typically diagnostic of a functional O-UC (Chapter 7).
4. The use of CPSase III by most piscine systems requires that the enzyme glutamine synthetase (GSase) be intimately involved with the O-UC, co-localized to the mitochondria with CPSase III (see, e.g., Anderson and Casey, 1984).

d. Evolutionary Aspects. Two main questions are raised when examining urea synthesis in light of how early elasmobranch ancestors might have embarked on the path toward the ureosmotic strategy. First, did early fish have the genes for the O-UC or was some novel invention required? We believe it is likely that these genes were already present in early fish, the main reason for this being the distribution of CPSase isozyme types in animals. CPSase II activity, the isozyme involved in pyrimidine biosynthesis, was an early invention of protists, and was modified slightly by early eukaryotes (Anderson, 1995b). It is believed that invertebrate and fish CPSase III evolved from this early eukaryotic CPSase II (see Fig. 7 in Campbell and Anderson, 1991; Hong *et al.,* 1994; see Fig. 3 in Anderson, 1995b), and while substantial expression of CPSase III is limited to organisms synthesizing large quantities of urea, this substantial CPSase III expression probably appears several times in the early metazoans. O-UC activity has been reported for land planarians (flatworms) (Campbell, 1965), which are descendants of early metazoan stock preceding the protostome–deuterostome split in animal phyla. Among protostome invertebrates, direct measurement of CPSase activity has only been demonstrated in an earthworm, and this appears to be CPSase III based on its activation by *N*-acetylglutamate (Bishop and Campbell, 1963); however, UTP inhibition and substrate preference were not extensively examined. If this enzyme is in fact a CPSase III, and its flatworm predecessor's CPSase is also a III isozyme, it is not surprising that CPSase III expression shows up again in deuterostome (fishes) lineages, and these are not likely to be independent evolutionary events. With the current advances in molecular methods being applied to CPSases, we believe it would be of interest to search for and sequence these genes from invertebrates to determine their relatedness to CPSase II in general and to CPSase III of fishes.

It is likely that the O-UC, by virtue of its high ATP usage, will divert energy away from other essential functions and systems, and will be subject to considerable evolutionary pressure and silencing by either regulatory or evolutionary (gene removal) processes when not needed. In addition, CPSases are relatively inefficient enzymes and CPSase I and III require a disproportionate amount of valuable mitochondrial real estate. It has been estimated that CPSase I alone accounts for more than 20% of rat liver mitochondrial protein (Clarke, 1976). Thus, if early fish had in fact inherited the genes for a fully functional O-UC, there may have been considerable pressure to silence them, either in complete lineages by gene deletion, or ontogenetically in other lineages by specific gene repression during development. Indeed, the expression of O-UC in muscle of the alkaline-lake tilapia, *A. grahami* (Lindley *et al.*, 1999), appears to illustrate these space limitations and selective pressures in fish. Therefore, the second question that must be answered in this discussion is "What pressures led to the retention of the genes, or activation of the genes, for the O-UC?" We believe an answer to this question lies in the dual nature of the O-UC, namely, in arginine degradation versus synthesis, as recognized by Campbell (1973), Griffith (1991), Withers (1998), and others. Clearly, all modern-day fishes possess arginase activity, and this was likely a feature of any early vertebrate because it is the first of the three enzymatic steps for degradation of arginine to glutamate. However, the arginine synthetic portion of the pathway may have been just as ubiquitous in early vertebrates if dietary protein was low, or if diets were generally unpredictable. In fact, there may have been considerable nutritional pressure alone to retain the genes for arginine biosynthesis, even if they remained inactive.

According to Griffith (1991, 1994) early freshwater environs, while attractive nurseries for embryonic and larval forms due to the lack of predation, would have been rather devoid of high-quality food. Therefore, early anadromous fishes probably had fairly yolky eggs to supply energy and building blocks for the developing fish. Because this yolk (vitellin) is rather protein rich, a substantial amount of ammonia would have been produced. As Wright and colleagues have suggested for modern-day teleost eggs (Wright *et al.*, 1995; Chadwick and Wright, 1999), either the membrane permeabilities of the eggs or a lower water circulation in the boundary layer around the eggs is conducive to ammonia buildup in the embryos (Korsgaard *et al.*, 1995); thus there is the need for detoxification, which is accomplished by the O-UC expressed briefly during early development (see Chapter 5). This need may have been a second pressure leading to the retention of genes for CPSase III and the rest of the arginine synthesis pathways. Similar pressure would have insured the persistence of GSase as a feeder enzyme to the O-UC, over and above its crucial role in the detoxification of neural ammonia.

The next step would have been simultaneous expression of arginase and the arginine synthetic portions of the full cycle. Arginine is not a particularly compatible solute for protein function (Bowlus and Somero, 1979), being rather more destabilizing to protein function than urea on a molar basis (Hochachka and Somero, 1984). Although invertebrates utilize amino acids as substantial osmolytes, only compatible solutes are used, and arginine is not a substantial osmolyte (Hochachka and Somero, 1984; Henry, 1995). Furthermore, many invertebrates possess the enzyme octopine dehydrogenase as the terminal step in glycolysis, which condenses pyruvate and arginine to yield octopine while oxidizing NADH analogously to lactate dehydrogenase, thus preventing arginine buildup during, for example, degradation of arginine phosphate energy stores during exercise or hypoxia. Of course, in light of the discussion about arginase function, the potential function of octopine dehydrogenase to sequester arginine and avoid hydrolytic attack by arginase cannot be dismissed. Because vertebrates replaced arginine phosphate with creatine phosphate (Ellington, 1989; Ellington *et al.*, 1997) as their muscle phosphagen and supplanted the need for octopine dehydrogenase, presumably the role of removing arginine excess then fell to arginase in the vertebrates. Thus, it is conceivable that there were brief times in the development of the early life-history stages of early anadromous fish that the needs to simultaneously balance arginine and ammonia concentration led to the periodic simultaneous expression of both parts of the O-UC.

It is also easy to envisage, as Griffith (1991) has suggested, both aspects of the O-UC being activated through pedomorphosis (or pedogenesis): Larvae metamorphosing into juveniles and returning to the marine environment or lineages progressively exploiting more marine niches could have been at a selective advantage by retaining osmolytes like urea. Based on the fact that modern-day lampreys fit so well into a "mold" of what an anadromous piscine ancestor might have looked like, Griffith (1991) predicted that lamprey ammocoete larvae should in fact express a fully functional O-UC. Wilkie and colleagues (1999) have subsequently examined nitrogen metabolism and excretion patterns in lampreys and found that while percentages of urea excretion can be increased, ureotely never exceeds 20%, and larval lampreys show virtually no expression of key O-UC enzymes. Of course, it is always dangerous to weigh characteristics of extant forms against predictions about ancient forms. It is possible that the modern-day lampreys represent a specialized form unlike ancestral fish.

2. UREA RETENTION

In order for urea concentrations to begin to build up in a fish migrating to more salty environs or, in an evolutionary sense, in a lineage that progressively inhabited more and more marine-like habitat, effective urea permeabilities would have had to decrease. As pointed out in Chapter 8, lipid bilayers are about an order of magnitude less permeable to urea than to water. Nonetheless, leakage of urea would have occurred, especially across respiratory surfaces. This leak of a metabolically expensive metabolite would not have been very energetically and selectively favorable. In modern-day elasmobranchs, this effective permeability decrease to urea appears to be the result of both specific secondarily active (e.g., sodium coupled) urea transporters in gills and kidney, and due to the modification of lipid composition of gills to include higher cholesterol-to-phospholipid ratios (Fines et al., 2000; also see Chapter 8). Modification of membrane lipid composition would have been a relatively straightforward adaptation, perhaps assisted by a diet high in animal tissue. However, this may have had concomitant effects on membrane fluidity (Hazel, 1995) and the function of membrane-bound proteins may have needed to be altered. As noted above and in Chapter 8, it is likely that urea transporters, even active ones, existed in early unicellular organisms; therefore, evolution of transporters with the appropriate cosolutes and polarity would not have been an untoward problem. Griffith (1991) has also pointed out that large size (and a lower surface area-to-volume ratio) might have been an adaptive, perhaps even a necessary, feature designed at lowering effective permeability.

One advantage to the choice of urea relative to ions with respect to permeability issues was pointed out by Ballantyne and Chamberlin (1988), namely, that it might equilibrate across the bilayer at a rapid enough rate throughout the various compartments of the organism without very much need for specific transporters. Indeed, specific urea transporters appear to be lacking in elasmobranch red blood cells or parenchymal hepatocytes (see Chapter 8).

Another consideration in the choice of accumulating urea as an osmolyte may have been buoyancy. Withers and colleagues (1994a,b) discovered that urea and trimethylamine oxide (TMAO) accumulated by elasmobranchs contribute substantially to positive buoyancy, in some cases at least equaling the lift provided by the "fatty" liver. Details of the mechanisms of this lift are covered by Yancey in Chapter 9. As Withers and colleagues argue, it is very likely that this lift contributes to increased overall fitness in elasmobranchs: Energy not expended in positional maintenance could be diverted to reproductive effort. However, it is conceivable that this benefit is secondary (in an historical sense) to the primary benefit of osmoregulatory balance (see below).

3. Urea Tolerance

As extensively reviewed by Ballantyne (1997), commitment to a ureosmotic strategy required some major changes in the biochemical architecture of elasmobranchs, assuming of course that comparisons to modern-day fish and other vertebrates represent a fair comparison to other protoelasmobranchs. First, there are the now well known needs to counteract the perturbing effects of urea on macromolecule structure and function by the simultaneous accumulation of stabilizing solutes (e.g., TMAO) (Hochachka and Somero, 1984, and see Chapter 9).

Second, the importance of lipids as a transportable catabolic fuel is severely limited, and they are replaced by ketone bodies. Ballantyne (1997) and Ballantyne et al. (1987) speculate that this may be due to the absence of serum albumins in elasmobranchs. Albumins transport nonesterified fatty acids in other vertebrates (both more primitive and advanced), with urea perturbation effects on albumin binding of nonesterified fatty acids being the rationale for the absence of albumins.

A related aspect of elasmobranch metabolism is the increased reliance on oxidation of amino acids. A case in point is the particularly high rate of glutamine oxidation in red muscle (Ballantyne, 1997). Ammonia is generally collected for urea synthesis by high levels of GSase in many tissues. The reliance on urea as an osmolyte is clearly related to the fact that elasmobranchs have high-protein (carnivorous) diets. Notably, although urea does appear to be the primary nitrogen compound excreted in elasmobranchs in basal (starved) conditions, it is not known if urea or ammonia is excreted following a meal (see Chapter 8).

D. Is the Ureosmotic Strategy "Inferior" to the Hypoosmotic Strategy?

In examining the characteristics of modern-day so-called "ancestral" and "derived" organisms, it is easy to fall into the trap of bestowing the labels of "primitive/poorer" and "advanced/better," respectively. Nonetheless, we should seek to explain the "fact" that the adaptive radiation of teleosts (some 25,000+ species) far exceeds that of the elasmobranchs (some 800+ species). Probably the single most important adaptation of teleosts that enables them to partition niches and speciate is the faciocranial bone structure that can be adapted to an incredibly diverse array of feeding possibilities (Romer, 1959; Bone et al., 1995). However, is it possible that the divergent osmoregulatory strategies also contribute to the differing levels of success of the two groups?

Kirschner (1993) reported a detailed energetic comparison of ureosmotic and hypoosmotic strategies, taking into account wherever possible the published costs of urea synthesis and retention and ion pumping strategies. He found the costs of osmoregulation via the two strategies to be more or less the same, comprising about 10–15% of standard metabolism. However, as Kirschner clearly recognizes, a number of uncertainties remain, and these uncertainties may likely tip the balance toward the hypoosmotic strategy being more economical from an energetic standpoint alone. For example, it is now suspected that urea retention at the elasmobranch gill is sodium coupled and this would increase ATP utilization by the Na^+,K^+-ATPase by elasmobranchs in Kirschner's calculations. Furthermore, as Kirschner has pointed out, TMAO dynamics are largely unknown. Is TMAO obtained through dietary or synthetic means? Are TMAO retention at the kidney (and gill) and its compartmentation within the animal energy-dependent processes? Furthermore, teleosts may ionoregulate at the level of the esophagus, and there are still uncertainties with respect to this process; Kirschner has predicted that better understanding of these mechanisms would tend to *decrease* the cost estimate for this group.

In addition to the likely higher costs of the ureosmotic strategy, it has other apparent disadvantages that may have ultimately contributed to diminishing the niche scope of elasmobranchs (and coelacanths). First, urea retention requires a commitment to carnivory for large amounts of nitrogen needed for urea synthesis (and either TMAO synthesis or acquisition). Second, the biochemical pathway implications illustrated by Ballantyne (1997) also channel these organisms into an altered lipid and carbohydrate metabolism that may be less compatible with herbivory and other dietary modes. Third, the need for a diet high in nitrogen, and perhaps the problems associated with potential solute loss during the high surface area-to-volume ratio period in young, further restricted the reproductive/development options of elasmobranchs. Elasmobranch species invest tremendously in the energy for egg and yolk production relative to most teleosts. They do not have the abilities of many teleost species to broadcast large numbers of eggs and larvae over a wide geographic range. Finally, although it is clear that some elasmobranchs do periodically invade freshwater, and some have permanently adapted to this condition, we predict that experiments on the dynamics of salinity adaptation will show elasmobranchs to be much slower at euryhaline adaptation than their teleostean counterparts. We know of no extensive and systematic data sets comparing the acute dynamics of elasmobranch with teleost adaptation to varying salinities (see also Chapter 8), and this is an important area for future research.

In summary, it appears as if the ureosmotic strategy has channeled elasmobranchs into a particular niche that we believe ultimately restricted their radiation. However, it is clear that, as a group, they very successfully occupy this narrow niche as top carnivores in the marine environment. As Ballantyne (1997) points out, the commitment to this mode of osmoregulation may have other benefits related to reduced neoplasia. Unfortunately, human intervention in the form of overfishing and unfortunate media imagery has radically altered the future outcome of the evolution of this group.

E. Adaptive Radiations of Ureogenesis and Ureotely in Bony Fishes

Urea production and ureotely has been used by a small (but growing) list of adult teleostean fish in adapting to specialized environmental circumstances. These uses are detailed by many authors in other chapters of this volume (e.g., Chapter 7), but we wish to highlight several here to illustrate this diversity and some evolutionary aspects. Our main question in this section is "How widespread is the O-UC in fish groups, and has it evolved within the fishes?" In our phylogenetic survey, we use primarily the classification of Nelson (1994).

1. CLASS SARCOPTERYGII (LOBE-FINNED FISHES)

a. Coelacanths. The coelacanth (*Latimeria chalumnae*), a member of the subclass Coelacanthimorpha, represents the most "primitive" bony fish (Osteichthyes) to have an extensive O-UC. It uses urea in much the same way as elasmobranchs for ureosmotic adaptation (Griffith *et al.*, 1974), although it does not appear to have the same renal urea recovery abilities as elasmobranchs (Griffith *et al.*, 1976). *L. chalumnae* shows CPSase III activity in its liver (Mommsen and Walsh, 1989). Thus, this group seemingly has not diverged much from the elasmobranch condition, and appears to be the last bony fish to use urea extensively as an osmolyte. Experimental evidence has been very difficult to come by due to the rarity of this fish, but recent discoveries of a second extant species, *L. menadoensis* (Holder *et al.*, 1999), may enable additional research.

b. Lungfish. A second early infraclass of bony fish, the Dipnoi, or lungfish, also express the O-UC. It is now believed to be the extant form most closely derived from the ancestral form, which evolved to land-dwelling amphibious vertebrates. It is not surprising that CPSase in this group appears to be type I based on kinetic characteristics, and very closely resembles that of tetrapod vertebrates (Mommsen and Walsh, 1989). However, with modern molecular techniques at hand, an effort should be made to clone this gene and verify its sequence as a CPSase I.

2. CLASS ACTINOPTERYGII (RAY-FINNED FISHES)

a. Sublass Chondrostei. The first subclass of fish representing Actinopterygii is the Chondrostei, which includes the orders Polypteriformes (e.g., the freshwater bichirs of Africa) and the Acipenseriformes (sturgeons and paddlefish). Little is known of urea in the physiology of this group, but the bichirs express a CPSase III isozyme (Mommsen and Walsh, 1989). No O-UC activity has been reported for these chondrostean fishes (e.g., sturgeons and paddlefish), and a CPSase III was not measurable (Mommsen and Walsh, 1989).

b. Subclass Neopterygii. Mommsen and Walsh (1989) initially reported expression of CPSase III in liver of a representative Neopterygii (formerly known as Holosteans) the bowfin, *Amia calva* of the order Amiiformes. This finding was later reexamined by Felskie *et al.* (1998) and found to be largely CPSase II activity. However, these authors found significant expression of CPSase III activity in the intestine and muscle of this species. However, the species does not appear to be able to estivate or produce substantial amounts of urea under stressful conditions (McKenzie and Randall, 1990).

c. Division Teleostei. Several reports have shown O-UC expression and significant ureotely in teleostean groups, although not all fit the arbitrary criterion of

50% urea excretion to be classified as "ureotelic." These reports are reviewed most exhaustively and recently by Saha and Ratha (1998) and Evans et al. (1999) (see also Chapter 7). We focus here primarily on groups for which full O-UC expression and, in particular, type of CPSase isozyme is known (see Table I) in an attempt to see if these fit any recognizable evolutionary pattern. As recently as a decade ago (see, e.g., Mommsen and Walsh, 1989), it was suspected that fish CPSases related to O-UC activity were uniformly of the CPSase III type (except for the obvious exception of the lungfish). However, recent studies showed that three teleostean species (*Heteropneustes fossilis, Clarias batrachus,* and *A. grahami*) have considerable CPSase activity with characteristics of both I and III types (Saha et al., 1997, 1999; Lindley et al., 1999). Although these are discussed in more depth in Chapter 7, an important common feature is the substantial activity with ammonia, sometimes more than with glutamine, and in two cases (*H. fossilis* and *C. batrachus*) the activities with ammonia and glutamine seem to be additive (Saha et al., 1997, 1999). It is not clear if two genes encode the expressed dual activity, or if the enzymes truly represent an intermediate between CPSase III and I. To us, the phylogenetic data presented in Table I suggest a separate origin for the "CPSase I" expressed in the teleosts compared to the true CPSase I of the lungfish and tetrapod lineage. Notably the expression of "I" is interspersed with III in many cases, and could represent an independent evolution of III into a "I" type for the expressed use of enhanced ammonia scavenging necessitated by the semiterrestrial existence, or the extreme pH of the Lake Magadi tilapia. In line with our urging above for broader CPSase sequence information from vertebrate and invertebrate lineages, searches for a true "I" isotype sequence in piscine lineages should yield some very exciting findings. As pointed out by Anderson and Walsh (Chapter 7), the liver should not be the only tissue of interest, as muscle appears to be another significant site for CPSase activity. In the Lake Magadi tilapia in particular, where requirements for CPSase activity apparently outstrip the ability to package CPSase in the liver, muscle is the major site of urea synthesis (Lindley et al., 1999). Furthermore, in *A. grahami,* the ability to use ammonia as a substrate by the CPSase is pronounced enough as to apparently obviate the need for close coupling of GSase to the O-UC, and GSase is not strongly expressed in muscle (Lindley et al., 1999).

The actual uses of ureotely within teleosts have been diverse, reaching well beyond the "expected" application for semiterrestrial existence. In the Lake Magadi tilapia, the use seems clear: to circumvent the problems associated with ammonia excretion into an alkaline environment (Randall et al., 1989). The gulf toadfish presents perhaps the most perplexing and interesting case for continued study of ureotely in teleosts. It does not appear to be associated with an intertidal (air-breathing) existence per se (Hopkins et al., 1999), and has been hypothesized to be involved in nitrogen recycling, predator avoidance (Walsh, 1997), and

Table I
Expression of Carbamoyl Phosphate Synthetase (CPSase) Substrate Preference and Phylogenetic Position in Teleostean Fish[a]

Superorder/order	Family	Species	CPSase substrate	Reference
Ostariophysi				
Cypriniformes	Cyprinidae	*Cyprinis carpio*	GLN	Felskie *et al.* (1998)
Siluriformes	Heteropneustidae	*Heteropneustes fossilis*	AMM and GLN	Saha *et al.* (1997)
	Claridae	*Clarius batrachus*	AMM and GLN	Saha *et al.* (1999)
Paracanthopterygii				
Batrachoidiformes	Batrachoididae	*Opsanus* sp.	GLN	Anderson and Walsh (1995)
Acanthopterygii				
Perciformes	Centrarchidae	*Micropterus salmoies*	GLN	Casey and Anderson (1983)
	Cichlidae	*Alcolapia grahami*	AMM and GLN	Lindley *et al.* (1999)

[a] Taxonomic groupings arranged in approximate order of primitive to advanced, based on the phylogeny of Nelson (1994).

more, the discovery that the excretion mechanism in toadfish is pulsatile (Wood *et al.*, 1995) suggests that significant ureotely may have been missed in other fish in experiments with short flux periods (e.g., a few hours). Clearly, the toadfish and the many other species using urea as adults offer exciting experimental systems. Given the findings of Wright and colleagues (1995) on the presence of the O-UC and ureotely in early developmental stages of two species which are typical ammonoteles as adults (rainbow trout and cod), it is tempting to speculate that the O-UC genes are present in most, if not all, teleosts (as proposed by Anderson, 1995a). Attempts to examine sequences and expression of CPSase III genes (and perhaps all O-UC genes and genes of urea transporters) should be expanded to many fish.

III. URICOTELY AND OTHER END PRODUCTS

The majority of this chapter (and much of the volume) is concerned with ammonia and urea, but additional nitrogenous end products should not be ignored. Uric acid excretion is usually negligible in fish, with some flexibility in response to ammonia or alkaline challenge (Wood, 1993). Although heavy reliance on uricotely is first developed in avian and reptilian lineages (Campbell, 1995), it is clear that the pathways for uric acid synthesis were well in place much earlier in the evolution of the phylum Chordata (e.g., in tunicates; Lambert *et al.*, 1998; Saffo, 1988), not least as a corollary to purine biosynthesis and degradation. Thus, it seems reasonable to caution against totally ignoring uric acid and other nitrogenous compounds as excretory/detoxification products in fish. In fact, nearly all studies of nitrogen excretion in fish measure ammonia and urea only, whereas Wood (1958) in measuring total nitrogen excretion shows that a rather significant portion of excreted nitrogen is unaccounted for by ammonia, urea, and uric acid in the few species he examined. Likely candidates include allantoin, allantoate, creatine, and creatinine. Unfortunately, only limited data are available and hence it would be premature to speculate on turnover, dynamics, and specific functions for any of these nitrogenous excretory products. Finally, nitrogenous waste products may not necessarily be voided from the fish, although how much of this retention results in distortion of the nitrogen excretion balance sheet remains to be determined. Purines are accumulated in the skin of smolting salmon (Johnston and Eales, 1967) and uric acid has been identified as an important component of fish seminal plasma (Ciereszko *et al.*, 1999).

While ammonia, urea, and uric acid may predominate and in an evolutionary sense are the most interesting excretory compounds, their production and turnover is so intricately linked to other compounds that at least a short discourse into some of these compounds seems appropriate. From a mammalian bias, glutamine and alanine usually take center stage as internal transport molecules, shuttling "fixed ammonia" and nitrogen between various tissues. Glutamine assumes a key role as a detoxification product in the brain and other tissues and as general nitrogen currency. Glutamine is also the prerequisite nitrogen-donor for the synthesis of carbamoyl phosphate for urea formation (catalyzed by CPSase III) and pyrimidine formation (catalyzed by CPSase II) (see Chapter 7). Finally, glutamine contributes two of the four nitrogen atoms of nascent purine rings. Evidence has been presented in support of glutamine's role in ammonia detoxification in fish brain, through the abundance of glutamine synthetase (e.g., Webb and Brown, 1976; Wang and Walsh, 2000) and direct evidence for substantial increases in glutamine concomitant with depletion of glutamate and ATP in the brain of ammonia-exposed fishes (Levi et al., 1974; Arillo et al., 1981; Iwata, 1988). Similarly, glutamine is being increasingly identified as a potential, general intertissue nitrogen carrier even in fishes (Jow et al., 1999; see also Chapters 3 and 4), above and beyond its specific role as a detoxification molecule specific to the central nervous system or source of nitrogen in biosynthetic reactions, but additional convincing and direct evidence for this role is needed. Considering the abundance and activity of glutamine synthetase along the digestive tract (T. P. Mommsen, unpublished observations), glutamine may also constitute an important export product of the gastrointestinal tract.

In mammals, excess muscle nitrogen is presumably carried to the liver as part of the glucose–alanine cycle and, indeed, this abundant plasma amino acid has been identified as a key carrier of nitrogen from fish muscle, at least during starvation (Leech et al., 1979; French et al., 1983). The presence of a glucose–alanine cycle is unlikely for fishes, considering the generally slow turnover of glucose (Weber and Zwingelstein, 1995) and sparing release of alanine from muscle following exhaustive exercise (Milligan, 1996). Fish, as indeterminate growers, where bulk is not an inherent deterrent to locomotion, and having an amino acid-fueled metabolism (van den Thillart, 1986), likely possess entirely different muscle amino acid requirements than land vertebrates and, hence, the interplay between muscle and liver or excretory organs will be of a different nature than in mammals. Be this as it may, flux of nitrogen from alleged transport molecules and other nitrogenous compounds need to be considered in the overall equations leading to excretory products.

Just like mammals, fishes accumulate relatively large amounts of creatine and creatine phosphate in their muscles. In fact, the total muscle creatine pools tend to be larger in fish muscle than in mammalian muscle, although the reasons for the higher retention (or higher flux?) are not obvious and probably deserve some experimental attention. In mammals, creatine turnover and homeostasis stretch over three organs: (1) The kidney for production of the arginine-derived precursor guanidino-acetate; as a common principle in metabolic pathways, it is this first step, the activity of the arginine:glycine amidinotransferase in kidney, that controls arginine flux and thus the rate of overall creatine synthesis. The enzyme is downregulated by dietary creatine and its expression is enhanced by growth hormone. (2) The liver for methylation to yield creatine. (3) Muscle as storage and functional tissue, transporting, utilizing, and turning over creatine.

For instance, starry flounder (*Platichthys stellatus*) lose almost half their white muscle creatine pool during 5 weeks of starvation (Danulat and Hochachka, 1989), most likely as urinary creatinine, which is indicative of relatively rapid turnover of creatine and insufficient supply of endogenous precursors for resynthesis (see above). Considering the white muscle mass and a pool size of almost 50 μM creatine per gram of tissue, the amount of nitrogen lost by the fish is substantial. In other aspects of creatine metabolism, fish are similar to mammals and an elasmobranch creatine transporter shows large sequence homology to the mammalian transporter (Guimbal and Kilimann, 1994). We feel that fish physiologists would be well advised to examine these alternative products in detail and attempt to put them in the general context of nitrogen turnover and excretion.

IV. SUMMARY AND EPILOGUE

The evolutionary aspects of nitrogen metabolism and excretion in fishes continue to be an exciting and fruitful area for research, hypothesis generation, and testing. The scenario proposed by Griffith (1991) can be evaluated somewhat more completely due to several significant findings of the past decade: (1) Although larval lampreys (ammocoetes) do not appear to express O-UC activity as predicted by Griffith, the O-UC occurs in embryos of teleosts that are typical ammonoteles as adults; (2) urea transport appears to be governed in fish by very specific transport molecules; (3) CPSase activities with some "I" isozyme characteristics, that is, enzymes with at least equal preference for ammonia as substrate, are found in higher teleostean fish; and (4) we are beginning to more fully understand the cost of osmoregulation in ureosmotic versus hypoosmotic regulators, and the teleostean strategy may prove to be more economical.

We believe that these findings are for the main part consistent with Griffith's (1991) hypothesis that the ornithine–urea cycle is an ancient trait in fishes, and that its presence in early piscine lineages of anadromous or brackish water ancestors led to its use in ureosmotic adaptation in descendent elasmobranch lineages. These ancestral forms likewise probably preadapted embryos of higher fish to cope with the limited exchange with the environment and avoid ammonia intoxication. These pressures clearly remain for modern-day embryos, and probably serve as strong pressure to retain the O-UC genes in fish. Selected species of teleosts have evolved to keep these genes activated in the adult phase, where ureogenesis and ureotely serve as adaptations to a semiterrestrial existence, existence in waters where ammonia excretion is impossible, and other as yet undiscovered roles. We have pointed out many research problems left to address in these contexts.

Two additional scenarios deserve attention relative to the experimental design of future research. First, the metabolic zonation of the O-UC, both within the cell, and between tissues, should be examined, particularly in light of recent findings of partial O-UC expression in muscle tissue of several species. Second, because urea-utilizing bacteria have been noted in high concentrations in the tissues of elasmobranchs (Knight *et al.*, 1988), perhaps the co-evolution of microbial systems with piscine systems is a fruitful area of research in nitrogen metabolism and excretion.

Finally, and unfortunately, humans are perhaps changing the face of the evolution of nitrogen excretion in fishes. Agricultural practices with excessive use of nitrogenous fertilizers, livestock production, industrial processes, and sewage treatment plants all contribute to nitrogen loading of the environment in different forms, including ammonia, nitrite, and nitrate. The extent of such anthropogenic nitrogen inputs is reaching critical levels in many, primarily freshwater and nearshore marine habitats, severely impacting the global nitrogen cycle (Tilman, 1999) and critically limiting the abundance, distribution, and community structure of aquatic species, including fishes (Rouse *et al.*, 1999). As evidenced numerous times in this volume, fish rely on a downhill gradient to dispose of ammonia and have to invoke some fancy finwork at considerable metabolic cost in situations where disposal of ammonia is impeded.

However, an even larger driving force might be nitrite. To their own detriment, especially some freshwater fish tend to take up nitrite from the external medium and bioaccumulate nitrite in plasma and many organs, including liver, gill, and brain (Jensen, 1995), well above toxic concentrations, largely because this anion is transported by the gill chloride transporter against a substantial concentration gradient, and the intestinal $Na^+,K^+/2Cl^-$ cotransporter can augment nitrite uptake (Grosell and Jensen, 1999). Even though the fish liver seems to have potential to convert nitrite to the less toxic nitrate (Doblander and Lackner, 1996), significant nitrite remains and some endogenous nitrite is generated through nitrate reduction. The presence of nitrite results in the formation of methemoglobin in the red blood cells rendering hemoglobin useless as an oxygen transporter (Jensen, 1995), and also participates in reactions leading to *N*-nitroso compounds that are known mutagens or carcinogens in fish models (De Flora and Arillo, 1983) and to other toxic phenomena, including the suicide binding to key P-450 detoxification enzymes (Arillo *et al.*, 1984). Given this plethora of effects of anthropogenic nitrogen, it will be especially important to have a basic understanding of the mechanisms and evolution of nitrogen metabolism and excretion in fishes.

REFERENCES

Anderson, P. M. (1995a). Urea cycle in fish: molecular and mitochondrial studies. In "Fish Physiology, Volume 14, Ionoregulation: Cellular and Molecular Approaches" (C. M. Wood and T. J. Shuttleworth, eds.), pp. 57–83. Academic Press, New York.
Anderson, P. M. (1995b). Molecular aspects of carbamoyl phosphate synthesis. In "Nitrogen Metabolism and Excretion" (P. J. Walsh and P. A. Wright, eds.), pp. 33–50. CRC Press, Boca Raton, FL.
Anderson, P. M., and Casey, C. A. (1984). Glutamine-dependent synthesis of citrulline by isolated hepatic mitochondria from *Squalus acanthias*. *J. Biol. Chem.* **259,** 456–462.
Anderson, P. M., and Walsh, P. J. (1995). Subcellular localization and biochemical properties of the enzymes of carbamoyl phosphate and urea synthesis in the Batrachoidid fishes, *Opsanus beta, Opsanus tau,* and *Porichthys notatus. J. Exp. Biol.* **198,** 755–766.
Arillo, A., Margiocco, C., Medlodia, F., Mensi, P., and Schenone, G. (1981). Ammonia toxicity mechanism in fish: studies on rainbow trout (*Salmo gairdneri* Rich.). *Ecotoxicol. Environ. Safety* **5,** 316–328.
Arillo, A., Mensi, P., and Pirozzi, G. (1984). Nitrite binding to cytochrome P-450 from liver microsomes of trout (*Salmo gairdneri* Rich.) and effects on two microsomal enzymes. *Toxicol. Lett.* **21,** 369–374.
Arndt, S. K. A., Benfey, T. J., and Cunjak, R. A. (1994). A comparison of RNA concentrations and ornithine decarboxylase activity in Atlantic salmon (*Salmo salar*) muscle tissue, with respect to specific growth rates and diel variation. *Fish Physiol. Biochem.* **13,** 463–471.
Ballantyne, J. S. (1997). Jaws: the inside story. The metabolism of elasmobranch fishes. *Comp. Biochem. Physiol.* **118B,** 703–742.
Ballantyne, J. S., and Chamberlin, M. E. (1988). Adaptation and evolution of mitochondria: osmotic and ionic considerations. *Can. J. Zool.* **66,** 1028–1035.
Ballantyne, J. S., Moyes, C. D., and Moon, T. W. (1987). Compatible and counteracting solutes and the evolution of ion and osmoregulation in fishes. *Can. J. Zool.* **65,** 1883–1888.
Berge, G. E., Lied, E., and Sveier, H. (1997). Nutrition of Atlantic salmon (*Salmo salar*): the requirement and metabolism of arginine. *Comp. Biochem. Physiol. A* **117A,** 501–509.
Bishop, S. H., and Campbell, J. W. (1963). Carbamyl phosphate synthesis in the earthworm *Lumbricus terrestris. Science* **142,** 1583–1585.
Bone, Q., Marshall, N. B., and Blaxter, J. H. S. (1995). "Biology of Fishes." Chapman and Hall, New York.
Bowlus, R. D., and Somero, G. N. (1979). Solute compatibility with enzyme function and structure: rationales for the selection of osmotic agents and end-products of anaerobic metabolism in marine invertebrates. *J. Exp. Zool.* **208,** 137–152.
Calabrese, E. J., Leonard, D. A., Baldwin, L. A., and Kostecki, P. T. (1993). Ornithine decarboxylase (ODC) activity in the liver of individual medaka (*Oryzias latipes*) of both sexes. *Ecotoxicol. Environ. Safety* **25,** 19–24.
Campbell, J. W. (1965). Arginine and urea biosynthesis in the land planaria: its significance in biochemical evolution. *Nature* **208,** 1299–1301.
Campbell, J. W. (1973). Nitrogen excretion. In "Comparative Animal Physiology" (C. L. Prosser, ed.), pp. 279–316. W. B. Saunders, Philadelphia.
Campbell, J. W. (1991). Excretory nitrogen metabolism. In "Environmental and Metabolic Animal Physiology" (C. L. Prosser, ed.), pp. 277–324. Wiley-Liss, New York.
Campbell, J. W. (1995). Excretory nitrogen metabolism in reptiles and birds. In "Nitrogen Metabolism and Excretion" (P. J. Walsh and P. A. Wright, eds.), pp. 147–178. CRC Press, Boca Raton, FL.
Campbell, J. W., and Anderson, P. M. (1991). Evolution of mitochondrial enzyme systems in fish: the mitochondrial synthesis of glutamine and citrulline. In "Biochemistry and Molecular Biology of Fishes" (P. W. Hochachka and T. P. Mommsen, eds.), Vol. I, pp. 43–76. Elsevier, Amsterdam.

Casey, C. A., and Anderson, P. M. (1983). Glutamine- and N-acetyl-glutamate-dependent carbamoyl phosphate synthetase from *Micropterus salmoides*. Purification, properties, and inhibition by glutamine analogs. *J. Biol. Chem.* **258**, 8723–8732.

Chadwick, T. D., and Wright, P. A. (1999). Nitrogen excretion and expression of urea cycle enzymes in the Atlantic cod (*Gadus morhua* L.): a comparison of early life stages with adults. *J. Exp. Biol.* **202**, 2653–2662.

Ciereszko, A., Dabrowski, K., Kucharczyk, D., Dobosz, S., Goryczko, K., and Glogowski, J. (1999). The presence of uric acid, an antioxidative substance, in fish seminal plasma. *Fish Physiol. Biochem.* **21**, 313–315.

Clarke, S. (1976). A major polypeptide component of rat liver mitochondria: carbamyl phosphate synthetase. *J. Biol. Chem.* **251**, 950–961.

Cooper, A. J. L., and Plum. F. (1987). Biochemistry and physiology of brain ammonia. *Physiol. Rev.* **67**, 440–519.

Corti, A., Astancolle, S., Davalli, P., Bacciottini, F., Casti, A., and Viviani, R. (1987). Polyamine distribution and activity of their biosynthetic enzymes in the European sea bass (*Dicentrarchus labrax* L.) compared to the rat. *Comp. Biochem. Physiol.* **88B**, 475–480.

Danulat, E., and Hochachka, P. W. (1989). Creatine turnover in the starry flounder, *Platichthys stellatus*. *Fish Physiol. Biochem.* **6**, 1–9.

De Flora, S., and Arillo, A. (1983). Mutagenic and DNA damaging activity in muscle of trout exposed *in vivo* to nitrite. *Cancer Lett.* **20**, 147–155.

Doblander, C., and Lackner, R. (1996). Metabolism and detoxification of nitrite by trout hepatocytes. *Biochim. Biophys. Acta* **1289**, 270–274.

Donoghue, P. C., Forey, P. L., and Altridge, R. J. (2000). Conodont affinity and chordate phylogeny. *Biol. Rev.* **75**, 191–251.

Ellington, W. R. (1989). Phosphocreatine represents a thermodynamic and functional improvement over other muscle phosphagens. *J. Exp. Biol.* **143**, 177–194.

Ellington, W. R., Roux, K., and Pineda, A. O. (1997). Origin of octameric creatine kinases. *FEBS Lett.* **425**, 75–78.

Evans, D. H., Claiborne, J. B., and Kormanik, G. A. (1999). Osmoregulation, acid–base regulation and nitrogen excretion. *In* "Intertidal Fishes: Life in Two Worlds" (M. H. Horn, K. L. M. Martin, and M. A. Chotkowski, eds.), pp. 79–96. Academic Press, New York.

Felskie, A. K., Anderson, P. M., and Wright, P. A. (1998). Expression and activity of carbamoyl phosphate synthetase III and ornithine urea cycle enzymes in various tissues of four fish species. *Comp. Biochem. Physiol.* **119B**, 355–364.

Fines, G. A., Ballantyne, J. S., and Wright, P. A. (2001). Active urea transport and an unusual basolateral membrane composition in the gills of a marine elasmobranch. *Am. J. Physiol.* **280**, R16–R24.

Forey, P., and Janvier, P. (1994). Evolution of the early vertebrates. *Am. Scientist* **82**, 554–565.

French, C. J., Hochachka, P. W., and Mommsen, T. P. (1983). Metabolic organization of liver during spawning migration of sockeye salmon. *Am. J. Physiol.* **245**, R827–R830.

Griffith, R. W. (1991). Guppies, toadfish, lungfish, coelacanths and frogs: a scenario for the evolution of urea retention in fishes. *Environ. Biol. Fishes* **32**, 199–218.

Griffith, R. W. (1994). The life of the first vertebrates. *BioScience* **44**, 408–417.

Griffith, R. W., Umminger, B. L., Grant, B. F., Pang, P. K. T., and Pickford, G. E. (1974). Serum composition of the coelacanth, *Latimeria chalumnae* Smith. *J. Exp. Zool.* **187**, 87–102.

Griffith, R. W., Umminger, B. L., Grant, B. F., Pang, P. K. T., Goldstein, L., and Pickford, G. E. (1976). Composition of the bladder urine of the coelacanth, *Latimeria chalumnae*. *J. Exp. Zool.* **196**, 371–380.

Grosell, M., and Jensen, F. B. (1999). NO_2^- uptake and HCO_3^- excretion in the intestine of the European flounder (*Platichthys flesus*). *J. Exp. Biol.* **202**, 2103–2110.

Guimbal, C., and Kilimann, M. W. (1994). A creatine transporter cDNA from Torpedo illustrates structure/function relationships in the GABA/noradrenaline transporter family. *J. Mol. Biol.* **241**, 317–324.

Hayashi, S., Fujiwara, S., and Noguchi, T. (1989). Degradation of uric acid in fish liver peroxisomes. Intraperoxisomal localization of hepatic allantoicase and purification of its peroxisomal membrane-bound form. *J. Biol. Chem.* **264**, 3211–3215.

Hazel, J. R. (1995). Thermal adaptation in biological membranes: is homeoviscous adaptation the explanation? *Ann. Rev. Physiol.* **57**, 19–42.

Henry, R. P. (1995). Nitrogen metabolism and excretion for cell volume regulation in invertebrates. *In* "Nitrogen Metabolism and Excretion" (P. J. Walsh and P. A. Wright, eds.), pp. 63–73. CRC Press, Boca Raton, FL.

Hochachka, P. W., and G. N. Somero. (1984). "Biochemical Adaptation." Princeton University Press, Princeton, NJ.

Holder, M. T., Erdmann, M. V., Wilcox, T. P., Caldwell, R. L., and Hillis, D. M. (1999). Two living species of coelacanths? *Proc. Natl. Acad. Sci. USA* **96**, 12616–12620.

Hong, J., Salo, W. L., Lusty, C. J., and Anderson, P. M. (1994). Carbamoyl phosphate synthetase III, an evolutionary intermediate in the transition between glutamine-dependent and ammonia-dependent carbamoyl phosphate synthetases. *J. Mol. Biol.* **243**, 131–140.

Hopkins, T. E., Serafy, J. E., and Walsh, P. J. (1997). Field studies on the ureogenic gulf toadfish, *Opsanus beta* (Goode and Bean) in a subtropical bay. II. Nitrogen excretion physiology. *J. Fish Biol.* **50**, 1271–1284.

Hopkins, T. E., Wood, C. M., and Walsh, P. J. (1999). Nitrogen metabolism and excretion in an intertidal population of the gulf toadfish (*Opsanus beta*). *Mar. Freshw. Behav. Physiol.* **33**, 21–34.

Ince, B. W., and Thorpe, A. (1977). Glucose and amino acid-stimulated insulin release *in vivo* in the European silver eel (*Anguilla anguilla* L.). *Gen. Comp. Endocrinol.* **31**, 249–256.

Iwata, K. (1988). Nitrogen metabolism in the mudskipper, *Periophthalmus cantonensis:* changes in free amino acids and related compounds in various tissues under conditions of ammonia loading, with special reference to its high ammonia tolerance. *Comp. Biochem. Physiol.* **91A**, 499–508.

Jensen, F. B. (1995). Uptake and effects of nitrite and nitrate in animals. *In* "Nitrogen Metabolism and Excretion" (P. J. Walsh and P. A. Wright, eds.), pp. 289–303. CRC Press, Boca Raton, FL.

Johnston, C. E., and Eales, J. G. (1967). Purines in the integument of the Atlantic salmon (*Salmo salar*) during parr-smolt transformation. *J. Fish. Res. Bd. Canada* **24**, 955–964.

Jow, L. Y., Chew, S. F., Lim, C. B., Anderson, P. M., and Ip, Y. K. (1999). The marble goby *Oxyeleotris marmoratus* activates hepatic glutamine synthetase and detoxifies ammonia to glutamine during air exposure. *J. Exp. Biol.* **202**, 237–245.

Kirschner, L. B. (1993). The energetics of osmotic regulation in ureotelic and hypoosmotic fishes. *J. Exp. Zool.* **267**, 19–26.

Knight, I. T., Grimes, D. J., and Colwell, R. R. (1988). Bacterial hydrolysis of urea in the tissues of carcharinid sharks. *Can. J. Fish. Aquat. Sci.* **45**, 357–360.

Korsgaard, B., and Mommsen, T. P. (1993). Gluconeogenesis in hepatocytes of immature rainbow trout (*Oncorhynchus mykiss*): control by estradiol. *Gen. Comp. Endocrinol.* **89**, 17–27.

Korsgaard, B., Mommsen, T. P., and Wright, P. A. (1995). Urea excretion in teleostean fish: adaptive relationships to environment, ontogenesis and viviparity. *In* "Nitrogen Metabolism and Excretion" (P. J. Walsh and P. A. Wright, eds.), pp. 259–287. CRC Press, Boca Raton, FL.

Koza, R. A., Moore, M. J., and Stegeman, J. J. (1993). Elevated ornithine decarboxylase activity, polyamines and cell proliferation in neoplastic and vacuolated liver cells of winter flounder (*Pleuronectes americanus*). *Carcinogenesis* **14**, 399–405.

Lambert, C. C., Lambert, G., Crundwell, G., and Kantardjieff, K. (1998). Uric acid accumulation in the solitary ascidian, *Corella inflata*. *J. Exp. Zool.* **282**, 323–331.

Leech, A. R., Goldstein, L., Cha, C. J., and Goldstein, J. M. (1979). Alanine biosynthesis during starvation in skeletal muscle of the spiny dogfish, *Squalus acanthias*. *J. Exp. Zool.* **207**, 73–80.

Levi, G., Morisi, G., Coletti, A., and Catanzaro, R. (1974). Free amino acids in fish brain: normal levels and changes upon exposure to high ammonia concentrations *in vivo,* and upon incubation of brain slices. *Comp. Biochem. Physiol.* **49A,** 623–636.

Lindley, T. E., Scheiderer, C. L., Walsh, P. J., Wood, C. M., Bergman, H. G., Bergman, A. N., Laurent, P., Wilson, P., and Anderson, P. M. (1999). Muscle as a primary site of urea cycle enzyme activity in an alkaline lake-adapted tilapia, *Oreochromis alcalicus grahami. J. Biol. Chem.* **274,** 29858–29861.

Liu, J., and Hatzoglou, M. (1998). Control of expression of the gene for the arginine transporter Cat-1 in rat liver cells by glucocorticoids and insulin. *Amino Acids* **15,** 321–337.

McKenzie, D. J., and Randall, D. J. (1990). Does *Amia calva* aestivate? *Fish Physiol. Biochem.* **8,** 147–158.

Miller, S. L., and Orgel, L. E. (1974). "The Origins of Life on Earth." Prentice Hall, Englewood Cliffs, NJ.

Milligan, C. L. (1996). Metabolic recovery from exhaustive exercise in rainbow trout. *Comp. Biochem. Physiol.* **113A,** 51–60.

Mommsen, T. P., and Hochachka, P. W. (1988). The purine nucleotide cycle as two temporally separated metabolic units—a study on trout muscle. *Metabolism* **36,** 552–556.

Mommsen, T. P., and Walsh, P. J. (1989). Evolution of urea synthesis in vertebrates: the piscine connection. *Science* **243,** 72–75.

Mommsen, T. P., and Walsh, P. J. (1992). Biochemical and environmental perspectives on nitrogen metabolism in fishes. *Experientia* **48,** 583–593.

Nelson, J. S. (1994). "Fishes of the World," 3rd ed. John Wiley & Sons, New York.

Ouzounis, C. A., and Kyrpides, N. C. (1994). On the evolution of arginases and related enzymes. *J. Mol. Evol.* **39,** 101–104.

Plisetskaya, E. M., Buchelli-Narvaez, L. I., Hardy, R. W., and Dickhoff, W. W. (1991). Effects of injected and dietary arginine on plasma insulin levels and growth of Pacific salmon and rainbow trout. *Comp. Biochem. Physiol.* **98A,** 165–170.

Randall, D. J., Wood, C. M., Perry, S. F., Bergman, H., Maloiy, G. M. O., Mommsen, T. P., and Wright, P. A. (1989). Urea excretion as a strategy for survival in a fish living in a very alkaline environment. *Nature* **337,** 165–166.

Romer, A. S. (1959). "The Vertebrate Story," 4th ed. University of Chicago Press, Chicago.

Rouse, J. D., Bishop, C. A., and Struger, J. (1999). Nitrogen pollution: an assessment of its threat to amphibian survival. *Environ. Health Perspect.* **107,** 799–803.

Saffo, M. B. (1988). Nitrogen waste or nitrogen source? Urate degradation in the renal sac of molgulid tunicates. *Biol. Bull.* **175,** 403–409.

Saha, N., and Ratha, B. K. (1998). Ureogenesis in Indian air-breathing teleosts: adaptation to environmental constraints. *Comp. Biochem. Physiol.* **120A,** 195–208.

Saha, N., Dkhar, J., Anderson, P. M., and Ratha, B. K. (1997). Carbamyl phosphate synthetases in an air-breathing teleost, *Heteropneustes fossilis. Comp. Biochem. Physiol.* **116B,** 57–63.

Saha, N., Das, L., and Dutta, S. (1999). Types of carbamyl phosphate synthetases and subcellular localization of urea cycle and related enzymes in air-breathing walking catfish, *Clarias batrachus. J. Exp. Zool.* **283,** 121–130.

Sakuraba, H., Fujiwara, S., and Noguchi, T. (1996). Metabolism of glyoxylate, the end product of purine degradation, in liver peroxisomes of fresh water fish. *Biochem. Biophys. Res. Commun.* **229,** 603–606.

Stock, D. W., and Whitt, G. S. (1992). Evidence from 18S ribosomal RNA sequences that lampreys and hagfishes form a natural group. *Science* **257,** 787–789.

Tilman, D. (1999). Global environmental impacts of agricultural expansion: the need for sustainable and efficient practices. *Proc. Natl. Acad. Sci. USA* **96,** 5995–6000.

Van den Thillart, G. (1986). Energy metabolism of swimming trout (*Salmo gairdneri*). Oxidation rates of palmitate, glucose, lactate, alanine, leucine and glutamate. *J. Comp. Physiol. B* **156 B**, 511–520.

Vellas, F., and Serfaty, A. (1974). L'ammoniaque et l'urée chez un teleostéen d'eau douce, la carpe (*Cyprinus carpio*). *J. Physiol. (Paris)* **68**, 591–614.

Walsh, P. J. (1997). Evolution and regulation of ureogenesis and ureotely in (batrachoidid) fishes. *Ann. Rev. Physiol.* **59**, 299–323.

Walsh, P. J., and R. P. Henry. (1991). Carbon dioxide and ammonia metabolism and exchange. *In* "Biochemistry and Molecular Biology of Fishes, Volume I, Phylogenetic and Biochemical Perspectives" (P. W. Hochachka and T. P. Mommsen, eds.), pp. 181–207. Elsevier, Amsterdam.

Wang, Y. S., and Walsh, P. J. (2000). High ammonia tolerance in fishes of the family Batrachoididae (toadfish and midshipmen). *Aquat. Toxicol.* **50**, 205–219.

Waters, S., Khamis, M., and von der Decken, A. (1992). Polyamines in liver and their influence on chromatin condensation after 17-beta estradiol treatment of Atlantic salmon. *Mol. Cell Biochem.* **109**, 17–24.

Webb, J. T., and Brown, G. W. (1976). Some properties and occurrence of glutamine synthetase in fish. *Comp. Biochem. Physiol B* **54B**, 171–175.

Weber, J. M., and Zwingelstein, G. (1995). Circulatory substrate fluxes and their regulation. *In* "Biochemistry and Molecular Biology of Fishes," (P. W. Hochachka and T. P. Mommsen, eds. Vol. 4), pp. 15–32. Elsevier Science, Amsterdam.

Weeks, D. L., Eskandari, S., Scott, D. R., and Sachs, G. (2000). A H^+-gated urea channel: the link between *Helicobacter pylori* urease and gastric colonization. *Science* **287**, 482–485.

Wilkie, M. P., Wang, Y., Walsh, P. J., and Youson, J. H. (1999). Urea excretion and production by the larvae of a phylogenetically ancient vertebrate: the sea lamprey (*Petromyzon marinus*). *Can. J. Zool.* **77**, 707–715.

Withers, P. C. (1998). Urea: diverse functions of a "waste" product. *Clin. Exp. Pharmacol. Physiol.* **25**, 722–727.

Withers, P. C., Morrison, G., and Guppy, M. (1994a). Buoyancy role of urea and TMAO in an elasmobranch fish, the Port Jackson shark, *Heterodontus portusjacksoni*. *Physiol. Zool.* **67**, 693–705.

Withers, P. C., Morrison, G., Hefter, G. T., and Pang, T.-S. (1994b). Role of urea and methylamines in buoyancy of elasmobranchs. *J. Exp. Biol.* **188**, 175–189.

Wood, C. M. (1993). Ammonia and urea metabolism and excretion. *In* "The Physiology of Fishes" (D. H. Evans, ed.), pp. 379–425. CRC Press, Boca Raton, FL.

Wood, C. M, Hopkins, T. E., Hogstrand, C., and Walsh, P. J. (1995). Pulsatile urea excretion in the ureagenic toadfish *Opsanus beta:* an analysis of rates and routes. *J. Exp. Biol.* **198**, 1729–1741.

Wood, J. D. (1958). Nitrogen excretion in some marine teleosts. *Can. J. Biochem. Physiol.* **38**, 1237–1242.

Wright, P. A. (1993). Nitrogen excretion and enzyme pathways for ureagenesis in freshwater tilpia (*Oreochromis niloticus*). *Physiol. Zool.* **66**, 881–901.

Wright, P. A. (1995). Nitrogen excretion: three end products, many physiological roles. *J. Exp. Biol.* **198**, 273–281.

Wright, P. A., Randall, D. J., and Perry, S. F. (1989). Fish gill water boundary layer: a site of linkage between carbon dioxide and ammonia excretion. *J. Comp. Physiol.* **158**, 627–635.

Wright, P. A., Felskie, A. K., and Anderson, P. M. (1995). Induction of ornithine-urea cycle enzymes and nitrogen metabolism and excretion in rainbow trout (*Oncorhynchus mykiss*) during early life stages. *J. Exp. Biol.* **198**, 127–135.

Wu, G., and Morris, S. M. J. (1998). Arginine metabolism: nitric oxide and beyond. *Biochem. J.* **336**, 1–17.

2

PROTEIN SYNTHESIS

C. G. CARTER AND D. F. HOULIHAN

I. Introduction
II. General Models of Protein Synthesis
   A. Mechanism of Protein Synthesis
   B. Measurement of Protein Synthesis
   C. Protein Degradation
   D. Integrated Models
III. Protein Synthesis and Amino Acid Flux
   A. Amino Acid and Protein Pools
   B. Daily Cycles of Protein Synthesis
   C. Protein Synthesis within Organs and Cells
   D. Whole-Body Protein Synthesis and Organ Integration
   E. Amino Acid Requirements and Protein Accretion
IV. Factors That Modify Protein Synthesis
   A. Environmental Factors
   B. Biotic Factors
   C. Feeding and Nutrition
V. Energetic Cost of Protein Synthesis in Fish
VI. Summary
   References

I. INTRODUCTION

Protein synthesis is fundamental to all living organisms and it has been studied intensively and at varying levels of complexity but, as Waterlow (1995) points out, a primary motivation for early interest was to help solve important human health problems such as malnutrition in infants. Many excellent recent reviews chart theoretical and practical developments in human and animal (mammalian) protein nutrition (Fuller and Garlick, 1994; Kimball et al., 1994; Waterlow, 1995, 1999; Millward, 1998; Lobley et al., 1999). The study of protein synthesis in ectotherms, particularly fish, has also received attention (Haschemeyer, 1973; Fauconneau, 1985; Houlihan, 1991; Houlihan et al., 1993a, 1995a,b). A recent motivation for measuring protein synthesis in fish has been to address global issues such as environmental change (Houlihan *et al.,* 1994; McCarthy and Houlihan, 1997; Lyndon and Houlihan, 1998) and improving aquaculture production (Carter *et al.,* 1993b; Conceicao *et al.,* 1997a; De la Higuera *et al.,* 1998). Our aim is to provide a comprehensive review of research on protein synthesis in fish, to examine data to produce simple models describing protein synthesis in terms of key variables, and to provide explanations for variations from expected or predicted rates of protein synthesis. This discussion follows a brief outline of the general models of protein synthesis. The underlying theme is to integrate information at the organismal level.

II. GENERAL MODELS OF PROTEIN SYNTHESIS

A. Mechanism of Protein Synthesis

Protein synthesis can be defined as translation: the translation of the genetic message, carried via the mRNA, into a polypeptide by the ribosome (Taylor and Brameld, 1999). Translation consists of three phases—initiation, elongation, and termination—and is accompanied by the "charging" of cytosolic tRNA molecules with all amino acids required including methionyl-tRNA that forms the initial complex with mRNA and a ribosome (Green and Noller, 1997; Taylor and Brameld, 1999). Regulation of these processes is complex and involves a number of specific proteins identified as initiation (eIF), elongation (eEF) and termination (eRF) factors. The supply of ATP, GTP, and amino acids, the number of ribosomes, their activity, and the formation of polyribosomes will also affect rates of protein synthesis (Green and Noller, 1997). The overall rate of protein synthesis of the cell, organ, or organism may, therefore, be the result of regulation at all or some of these levels as well as being regulated via transcription. Detailed accounts concerning transcription and translation and their regulation are available (Green and Noller, 1997; Hershko and Ciechanover, 1998; Taylor and Brameld, 1999).

The minimum energy requirement for the formation of one peptide bond is generally assumed to be 4 ATP equivalents made up of charging the tRNA, requiring the hydrolysis of ATP to AMP (2 ATP equivalents); entry of the aminoacyl-tRNA into the ribosome, requiring hydrolysis of GTP to GDP; and translocation of the newly formed peptidyl-tRNA, which results in hydrolysis of a further GTP to GDP. Thus, a minimum energetic cost of protein synthesis has been estimated as 40 mmol ATP equivalents per gram of protein synthesized (Reeds *et al.,* 1985). Because this considers only the cost of peptide bond formation, protein synthesis will involve further energy expenditure associated with processes such as modifications and transport of proteins. In consideration of this additional energy cost, 50 mmol ATP equivalents per gram of protein synthesized is used (Reeds et al., 1985; Houlihan et al., 1988b). Further energetic costs will be associated with protein degradation (see below). Based on these minimum costs, the importance of protein synthesis in energy expenditure of animals may be considerable and has been estimated to account for 15–25% of basal metabolic costs in endotherms (Webster, 1988), as well as in fish (Houlihan et al., 1988b; Carter et al., 1993a). In juvenile feeding fish, protein synthesis may account for as much as 42% of total energy expenditure (Houlihan et al., 1988b) and may be the most significant energy-demanding physiological process.

Rates of protein synthesis are typically expressed as fractional rates: the proportion of the protein mass of an organ or the organism synthesized per day (k_s, %/d) (Garlick et al., 1980). In the same way, rates of protein intake (k_c), protein accretion (k_g), and protein degradation (k_d) can be defined as fractional rates (Millward et al., 1975; Houlihan et al., 1995a). All the processes involved in protein synthesis and degradation are often described as protein turnover. However, the definition of protein turnover (k_t) is more complex and depends on the nutritional status of the animal: At maintenance when $k_g = 0$, then $k_t = k_s = k_d$; when $k_g > 0$, then $k_t = k_d$; when $k_g < 0$, then $k_t = k_s$ (Weisner and Zak, 1991; Houlihan et al., 1993a).

The central importance of RNA in protein synthesis is recognized by expressing RNA concentrations as the capacity for protein synthesis (C_s, mg RNA/g protein) and RNA activity (k_{RNA}, g protein synthesized/g RNA/d) (Millward et al., 1975; Sugden and Fuller, 1991). The indices of protein turnover (k_c, k_s, k_d, k_g) and of RNA concentration and activity are used as the major measurements in the following discussion.

B. Measurement of Protein Synthesis

There has been considerable debate about the relative merits of different approaches to the measurement of protein synthesis (Young et al., 1991; Wolfe, 1992; Garlick et al., 1994; Rennie et al., 1994; Waterlow, 1995). A broad division can be made into precursor and end-product methods (Waterlow, 1995). Each is based on a series of assumptions and it is necessary to consider which is appropriate for the question being asked. Not all available methods are easy to carry out with fish. Also, it is important to decide whether whole-body or tissue-specific measurements are required to answer particular questions, and methods differ between these approaches.

Early published records of protein synthesis in fish measured tracer incorporation, and data were often expressed as the accumulation of radioactivity into protein (Das and Proser, 1967; Haschemeyer, 1968). For example, a mix of radiolabeled amino acids was injected into the branchial artery of toadfish (*Opsanus tau*) and their accumulation measured in the liver (Haschemeyer, 1968). Subsequent studies by Haschemeyer and colleagues used constant infusion techniques

(Haschemeyer and Smith, 1979; Smith, 1981). For example, a radioactive amino acid was administered to large (200- to 400-g) rainbow trout (*Oncorhynchus mykiss*) over 6 h via cannula in the dorsal aorta (Smith, 1981). Mini-osmotic pumps, implanted into the peritoneal cavity of rainbow trout, maintained constant infusion of tracer for 4 weeks, and leucine flux was calculated from plasma levels (Fauconneau and Tesseraund, 1990). More recently whole proteins in which all (>99%) the nitrogen is present as the stable isotope ^{15}N have been used as tracers to calculate protein synthesis from labeling of an end product (Carter et al., 1994; Meyer-Burgdorff and Rosenow, 1995a). Details of the methods used with fish have been reviewed (Houlihan et al., 1995a,b,c); the majority of studies used a "flooding dose" of a labeled amino acid administered by injection (Garlick et al., 1980). A flooding dose can be defined as the administration, usually by injection, of an amount of amino acid that results in a large increase (flood) of all amino acid precursor pools (Garlick et al., 1980). Where injection has been impractical due to the small size of fish, larvae for example, the animals have been bathed in a flooding concentration of amino acid (Houlihan et al., 1992). The flooding dose method has been used to investigate rates of protein synthesis at the level of the organism (Houlihan et al., 1988b, 1989), organs (Pornjic et al., 1983; Houlihan et al., 1986), cellular fractions, and specific proteins (McMillan and Houlihan, 1992; Fauconneau et al., 1995).

The flooding dose method allows rates of protein synthesis in different tissues to be measured, whereas end-product methods that rely on oral administration of tracer can only provide information on whole-body rates. Stable isotopes have been shown to be reliable as tracers when using a flooding dose and this means that this method has application in situations, such as when repeat measurements are to be made, where radioactive tracers would not be considered (Carter et al., 1998; Owen et al., 1999). The end-product measurement of whole-body synthesis is useful because measurements are made over 1–3 days and, therefore, integrate the temporal fluctuations into a daily rate (Carter et al., 1994, 1998). End-point methods using a labeled meal are also advantageous because the tracer can be administered in the normal food and this can be fed without changing the daily routine and disturbing the fish (Meyer-Burgdorff and Rosenow, 1995a).

Although no single study has tried to compare estimated rates of protein synthesis from different methods with the same fish, we have constructed a table to determine whether there are differences between them (Table I). Selection of the data for inclusion in the table was based on the use of similar species, life-history stage, temperature, and feeding regimen. The data were transformed using scaling coefficients to a 100-g fish (Houlihan et al., 1995c) held at 10°C (Mathers et al., 1993). Following transformation, rates were approximately 2%/day for four experiments that used either a flooding dose of phenylalanine, containing a radioactive (^{3}H) or a stable isotope (^{15}N), or an end-product method that used ^{15}N-protein as the tracer (Table I). However, there were differences between the ex-

Table I
Comparison of Fractional Whole-Body Rates of Protein Synthesis (WBPS) in Fed Salmonids Measured Using a Variety of Methods and Tracers

Method[a] (h after feeding)	Tracer[b]	Weight (g)	Temp. (°C)	Ration (k_r) (%/day)	WBPS (k_s) (%/day)	WBPS[c] (k_s) (%/day)	Source[d]
Atlantic salmon							
FD (23 h)	^{15}N-Phe	37	12	4.7	3.3	2.1	(a)
FD (24 h)	^{3}H-Phe	180	13	2.5	2.8	2.2	(b)
Rainbow trout							
FD (3 h)	^{3}H-Phe	65	10	2.1	2.6	2.3	(c)
FD (12 h)	^{3}H-Phe	75	11	7.5	4.4	3.7	(d)
FD (1 h)	^{14}C-Leu	80	10	5.0	5.1	4.8	(e)
EP (0–24 h)	^{15}N-Protein	117	12	2.1	2.3	2.0	(f)

[a] FD, flooding dose; EP, end product.
[b] Phe, phenylalanine; Leu, leucine.
[c] Standardized to a 100-g fish (Houlihan et al., 1995c) at 10°C (Mathers et al., 1993).
[d] (a), Owen et al. (1999); (b), Carter et al. (1993b); (c), McCarthy et al. (1994); (d), Foster et al. (1991); (e), Fauconneau and Arnal (1985); (f), Carter et al. (1994).

periments in terms of the ration and the time at which measurements were made. A larger ration would be predicted to stimulate higher rates of protein synthesis (see Section IV.C). The time after feeding also has the potential to influence protein synthesis rates with some tissues showing marked temporal fluctuations (see Section III.B). The combination of high ration and measurements being made within the first 12 h of feeding probably explains the higher whole-body rates of synthesis measured in two studies (Fauconneau and Arnal, 1985; Foster et al., 1991). This comparison suggests that a variety of methods can be used to measure protein synthesis in fish. The measurement of protein synthesis is especially informative when one method is used for comparative purposes within the same experiment.

C. Protein Degradation

Protein accretion (or loss) represents the balance between protein synthesis and protein degradation. Thus, rates of protein degradation can be calculated from the more easily made measurements of protein accretion and synthesis ($k_d = k_s - k_g$) (Millward et al., 1975; Houlihan, 1991). It is recognized that this relies on short-term (hours to days) measurements of protein synthesis being compared to long-term (weeks to months) measurements of protein accretion. This equivalence will be maximized at a constant rate of protein accretion, when fish are under a stable nutritional regimen, for example, and when protein synthesis is measured over complete daily cycles to reduce the problems associated with postprandial elevations.

Individual proteins are degraded at different rates and at different sites, which makes an integrated approach to the quantification of protein degradation extremely complex. Lysosomes are the site of membrane and exogenous protein degradation, whereas endogenous proteins can be degraded in lysosomes or in the cytoplasm (Tischler, 1992). As yet, few independent methods for measuring protein degradation have been used with fish. Excretion of π-methyl-L-histidine has not been investigated but its use is likely to be associated with the problems that have led to its rejection for many other animals (Millward et al., 1983). Lysosomal and nonlysosomal proteolytic pathways, which are now thought to be both highly controlled and specific (Hershko and Ciechanover, 1998; Attaix et al., 1999), will at some stage be identified and quantified in fish. The lysosomal pathway of protein degradation has many different enzymes that break down proteins, including cathepsins exhibiting endopeptidase or exopeptidase activity. The proteasome ubiquitin pathway is the principal route of protein degradation in mammals during muscle atrophy and could be an important route for reutilization of amino acids from skeletal muscle in fish.

D. Integrated Models

This section focuses primarily on the models of protein synthesis in fish that attempt to integrate data to provide information at the level of the organism. The first model provides a description of the daily flux of amino acid nitrogen and the second model describes the relationships between indices of protein turnover over the complete range of protein intake and accretion rates.

In a simple nitrogen budget, the daily amino acid nitrogen flux derived from consumed protein is partitioned between fecal waste (fecal nitrogen), metabolic waste (nitrogenous excretion), and growth (nitrogen accretion). The model can be developed to have at its center two pools, the free amino acid and protein pools, linked by protein synthesis and protein degradation (Sugden and Fuller, 1991). The inputs to the amino acid pool are via protein intake, protein degradation and *de novo* synthesis of nonessential amino acids. Amino acids are lost from the free pool via protein synthesis, synthesis of nonprotein molecules, and oxidation to ammonia. This description of nitrogen flux has been adapted for analysis of fish protein metabolism (Carter et al., 1993a, 1995a; Houlihan et al., 1995c). For flounder (*Pleuronectes flesus*) an expanded protein-nitrogen budget was expressed as 100% C_N = (99% Z_N − 65% D_N) + 29% A_N + 10% U_N + 27% F_N, where C_N is intake, Z_N synthesis, D_N degradation, A_N ammonia excretion, U_N urea excretion, and F_N fecal loss (Carter et al., 1998). In models that do not consider protein synthesis, protein accretion (P_N) would be used, calculated as 34% P_N = 99% Z_N − 65% D_N for the example. The model also presents a comparison of the relative sizes of amino acid and protein pools and by so doing provides an explanation for differences in the protein metabolism of larval and juvenile fish (see Section III.C).

Houlihan *et al.* (1993a, 1995a,b) described a series of linear relationships between protein intake and both protein synthesis and protein accretion and between protein synthesis and protein accretion. Linear relationships between protein degradation and protein intake or protein accretion have also been demonstrated in some cases (Houlihan *et al.*, 1988b, 1989), but not others (McCarthy *et al.*, 1994; Carter *et al.*, 1998). The levels of protein intake that permit the maintenance of the total amount of whole-body protein also represent the state at which protein synthesis and degradation are equal and can be defined as the maintenance protein intake [$k_{c(m)}$ when $k_g = 0$] and has associated and equal rates of protein synthesis [$k_{s(m)}$] and degradation [$k_{d(m)}$]. There are also linear relationships between protein intake, growth, synthesis, and both the capacity for protein synthesis and the RNA activity (Carter *et al.*, 1993b; McCarthy *et al.*, 1994). Increases in protein intake seem to lead to increases in growth through increases in RNA concentration (the capacity for protein synthesis) and or RNA activity, and changes in the balance between protein synthesis and degradation. Different biotic and abiotic factors may shift linear relationships above or below the expected values (Houlihan *et al.*, 1993a).

Relationships between fractional rates indicate the efficiency of the processes and develops and links the concepts of anabolic stimulation efficiency (ASE, %: $100 \times k_s/k_c$), synthesis retention efficiency (SRE, %: $100 \times k_g/k_s$) and protein retention efficiency (PPV, %: $100 \times k_g/k_c$). The absorption of dietary amino acids stimulates organ and, consequently, whole-body protein synthesis, and the amount of protein synthesized in relation to the amount consumed is expressed as ASE. Only part of the total synthesized protein, expressed as SRE, is then retained as growth. Consequently, the efficiency with which consumed protein is retained as growth can be investigated in more detail by consideration of ASE and SRE. These three ratios were used to explore growth and individual variation in a variety of fishes (Houlihan *et al.*, 1989; Carter *et al.*, 1993a,b; McCarthy *et al.*, 1994) and provide important information for determining the optimum balance of nutrients (Section III.E).

III. PROTEIN SYNTHESIS AND AMINO ACID FLUX

A. Amino Acid and Protein Pools

Comparison of the relative size of the free amino acid and protein pools in different life-history stages of fish has shown that although juvenile fish typically have a considerably smaller whole-body free amino acid pool compared with early life-history stages, there is an allometric relationship between body weight and amino acid content (Houlihan et al., 1995c). Data ranging from fish eggs (<0.001 g) to juvenile catfish (933 g) produce weight exponents of −0.18 and −0.13 for the essential and total free amino acid concentration (μmol/g tissue), respectively (Houlihan et al., 1995c). The consequence is that for cod larvae the whole-body free amino acid pool is equivalent to 30% of the protein pool, whereas it is typically 2–3% in juvenile fish (depending on size). Similar scaling for protein synthesis and C_s may suggest that high concentrations of free amino acids as well as high RNA contents are required for high rates of protein synthesis in ectothermic animals (Houlihan et al., 1995c). However, the total free amino acid pool concentrations in liver, gastrointestinal tract (GIT), and white muscle of juvenile rainbow trout are relatively similar (35–50 μmol/g tissue) but do not reflect the large differences in protein synthesis rates between tissues (Carter et al., 1995a). The retention of synthesized protein in larval fish is of the same order of magnitude as in juvenile fish and supports the hypothesis that protein synthesis has a quantitatively similar role in protein accretion in both larval and juvenile fish (see Section IV.B).

In most studies (but see Section III.C) the synthesis of total protein in the whole fish or in particular tissues has been investigated. This can be justified in several ways. Although many thousands of proteins are synthesized in a tissue, only a relatively few make up a large proportion of the total protein synthesized so that an indication of general (whole-tissue) protein synthesis is likely to reflect synthesis of the main bulk of the cellular proteins. A 100-g rainbow trout would contain approximately 16 g of protein and the largest fraction would be accounted for by white muscle (Table II). In this species 42% of the whole-body protein is in white muscle (McCarthy et al., 1994), although other studies show it can be more than 50% (Foster et al., 1991). Myofibrillar proteins make up the largest part of the muscle and account for more than 25% of whole-body protein with a significant contribution made by myosin (Fauconneau et al., 1995). In comparison, the total protein in the liver and gastrointestinal tract (GIT) is less than the whole-body myosin content (Table II). The qualitative importance of white muscle in whole-body protein metabolism is further emphasized by the strength of the correlation between its essential amino acids content and that of the whole body ($n = 10$; $r = 0.945$; $P < 0.001$). However, there are also differences, because white muscle contains a larger proportion of essential to total amino acids than the whole body (Table II).

B. Daily Cycles of Protein Synthesis

Few studies have specifically investigated the temporal variation in rates of protein synthesis in unfed fish (diurnal cycles) or following feeding. Whole-body and tissue rates of protein synthesis are stimulated by feeding (Fig. 1). In Atlantic

Table II
Protein and Essential Amino Acid (EAA) Composition[a] of a 100-g Juvenile Rainbow Trout with a Total Protein Content of 16 g

Protein	Whole body	Muscle				Muscle fraction				Total Connective			GIT	Liver
		White	Red	Cardiac	Myofibrillar	Myosin	Actin	Ribosomal	Free pool	Collagen	Elastin			
Component weight (% BW)	100	43.7	1.7	0.09					43.7				4.8	1.6
Component protein (% organ)	16	16.0	15.2	20.4					0.055				11.3	13.8
Component protein (g)	16	6.66	0.27	0.02	4.43	0.91	0.10	1.64	0.024				0.54	0.22
Proportion of whole-body protein (%)	100	41.6	1.7	0.1	27.7	5.7	0.6	10.2	0.15	3.9	1.1		3.4	1.4
EAA(% protein)														
Arginine	6.41	4.89				3.8	7.16		4.21	8.88				
Histidine	2.96	2.19				1.3	2.93		1.41	1.09				
Isoleucine	4.34	4.93				3.5	6.70		2.62	1.34				
Leucine	7.59	8.03				7.7	7.87		6.95	2.54				
Lysine	8.49	9.37				7.2	6.75		3.75	3.43				
Methionine	2.88	2.82				2.0	2.87		1.10	2.28				
Phenylalanine	4.38	3.45				2.3	4.13		2.97	2.86				
Threonine	4.76	5.65				3.6	6.82		3.90	2.43				
Tryptophan	0.93	0.71				0.37	1.24		0.00					
Valine	5.09	6.27				4.4	5.14		4.32	2.03				
EAA/AA (%)	47.8	48.3				36.2	51.6		31.2	26.9				
Total EAA (mg/g fish)	76.5	32.3				3.3	0.5		0.1	1.7				

[a] Rainbow trout unless stated: whole-body protein content (McCarthy et al., 1994); whole-body EAA (Wilson and Cowey, 1985); muscle weights and protein content (Houlihan et al., 1986); white muscle EAA from Atlantic salmon (Carter et al., unpublished) and free AA at 24-h postfeeding (Carter et al., 1995a); total myofibrillar, myosin, actin content (Fauconneau et al., 1995); Atlantic cod myosin EAA (Skaara and Regenstein, 1990; cited by Venugopal et al., 1994); actin EAA (from Sikorski, 1994); muscle ribosomal protein (17.7%) (Von der Decken et al., 1992); connective tissue content (Love, 1957) and chum salmon collagen type I EAA (from Sikorski and Borderias, 1994); GIT and liver weights and protein content (Foster et al., 1991).

cod (*Gadus morhua*) fractional rates of protein synthesis in the gastrointestinal tract and liver peaked 6 h after feeding (Fig. 1a) and increased by 700 to 1600%, respectively (Lyndon *et al.*, 1992). The increase in whole-body protein synthesis was lower (500%) and peaked later (18 h) compared to these tissues and is likely to have reflected the pattern in white muscle synthesis (Fig. 1a). For example, in Atlantic salmon (*Salmo salar*) white muscle (Fig. 1b) protein synthesis peaked 20 h after feeding (Fauconneau *et al.*, 1989). In rainbow trout, peak rates of protein synthesis in the liver were measured soon after feeding (<2 h) but were less pronounced than for cod (McMillan and Houlihan, 1989). Differences between the relative size of the peaks probably relate to differences in feeding regimen, since the cod had not been fed for 6 days, whereas the trout were fed daily. It is clearly important to consider what part of the daily cycle of protein synthesis is of interest when devising a sampling procedure.

Peak incorporation of [^{14}C]glycine into tissue protein of unfed gulf killifish (*Fundulus grandis*) occurred at different times in muscle and liver (0 h), intestine (0 and 18 h), and the scales (12 and 18 h) and suggested diurnal cycles in protein synthesis (Negatu and Meier, 1993).

C. Protein Synthesis within Organs and Cells

Rates of protein synthesis vary between tissues and a general ranking of major tissues is usually liver > gill = gastrointestinal tract (intestine > stomach) = kidney (head kidney > kidney) >> white muscle (Fig. 2). Other organs have been studied, although less often, and rates of synthesis tend to be lower than in the tissues listed above, except in the case of the white muscle. Many factors have the potential to modify rates of protein synthesis and to differentially affect responses in different tissues (Section IV). The description here is focused on mature wild-caught fish under typical temperatures and serves to allow a comparison of the relative organ rates of protein synthesis for each species rather than across the species. This is because there were variations in the treatment of fish prior to the determination of protein synthesis, as well as in the methods used. Atlantic cod were maintained as individuals for several months and fed, although maintenance rates are given here (Houlihan *et al.*, 1988b); toadfish were unfed and measurements of protein synthesis made after 4 days (Pornjic *et al.*, 1983); icefish (*Chaenocephalus aceratus*) were maintained in aquaria and fed (Haschemeyer, 1983); female Atlantic salmon were wild caught at the start of the spawning run, when they were probably not feeding, and protein synthesis was measured 5 days after capture (Martin *et al.*, 1993).

The liver clearly has a central position in amino acid metabolism and in the synthesis and export of many proteins. Consequently, rates of protein synthesis are high as well as being extremely sensitive to feeding and nutrition and, as discussed above (Fig. 1), daily cycles of protein synthesis following feeding may be

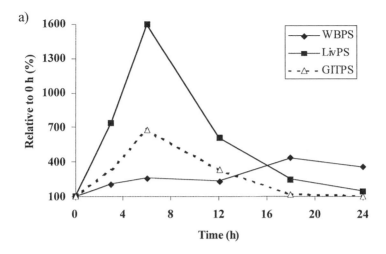
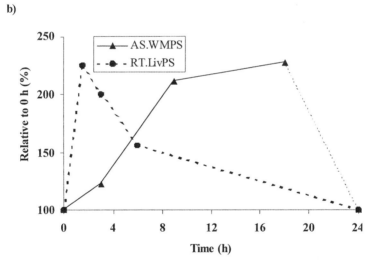

Fig. 1. The effect of feeding on the pattern of protein synthesis (% rate at 0 h) over 24 h following a meal: (a) protein synthesis in whole-body (WBPS), liver (LivPS), and GIT (GITPS) of Atlantic cod, *G. morhua* (Lyndon *et al.*, 1992); (b) protein synthesis in white muscle (WMPS) of Atlantic salmon, *S. salar* (Fauconneau *et al.*, 1989), and liver (RT.LivPS) of rainbow trout, *O. mykiss* (McMillan and Houlihan, 1989).

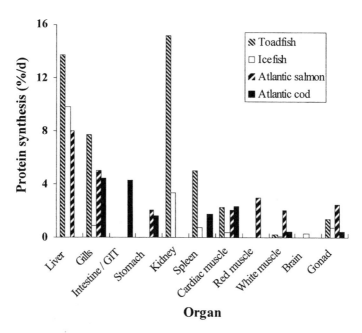

Fig. 2. Fractional rates of protein synthesis (%/day) in tissues of wild-caught fish at zero or low feed intake: toadfish (unfed), *O. tau* (Pornjic et al., 1983); icefish (fed), *C. aceratus* (Haschemeyer, 1983); salmon (unfed), *S. salar* (Martin et al., 1993); and cod, *G. morhua* (maintenance) (Houlihan et al., 1988b).

pronounced (McMillan and Houlihan, 1988, 1989, 1992; Houlihan, 1991; Martin et al., 1993). Changes in liver protein synthesis due to environmental and biotic factors are discussed in more detail in Section IV.

Fish gills are clearly identified as providing a large metabolically active interface between the fish and its environment and being involved in key functions of respiration and ionic regulation and serving as a barrier to harmful environmental agents (Lyndon and Houlihan, 1998). Under normal conditions fractional rates of protein synthesis vary between 6 and 14%/day (Table III). Generally, gill protein synthesis rates rank second or third behind the liver and kidney (Fig. 2). Intriguingly, rates of protein synthesis were different between the holobranchs of flounder and significantly higher in the first, compared to the second and third, with the further suggestion that higher ribosomal activity was driving the higher rates (Lyndon and Brechin, 1999). These differences may reflect differences in physiological functions between holobranchs and differences in requirements for cellular turnover (Lyndon and Houlihan, 1998; Lyndon and Brechin, 1999).

Limited research suggests that protein synthesis is considerably higher in the head kidney than in the kidney (Haschemeyer, 1983). Differences in rates of protein synthesis in the head kidney (3.4%/day) and the kidney (1.0%/day) in icefish may be related to the synthesis of antifreeze glycoproteins (AFGP) in the head kidney (Haschemeyer, 1983). However, a twofold difference between head kidney and the kidney was measured for another Antarctic fish (*Notothenia corriceps*) that did not appear to synthesize AFGP (Haschemeyer, 1983). The head kidney has important functions in the immune response of fish and the immunological status of the fish could have an impact on rates of protein synthesis in this tissue.

Protein metabolism in the gastrointestinal tract of fish has rarely been studied but is likely to be of considerable importance in terms of both its contribution to whole-body protein synthesis as well as to its requirement for absorbed nutrients, particularly amino acids (Nieto and Lobley, 1999; Reeds *et al.*, 1999). Fractional rates of protein synthesis in Atlantic cod at maintenance were 1.6 and 4.3%/day in the stomach and intestine, respectively (Fig. 2). Similarly, in Atlantic salmon that had been unfed for several days, the rate of protein synthesis in the stomach was less than 2%/day (Martin *et al.*, 1993). In feeding fish, rates of protein synthesis were higher in the stomach than in the intestine and higher in both tissues than in unfed fish, ranging between 5 and 21%/day (Table III). In refed Atlantic cod rates changed following feeding and only exceeded 4%/day at 12 h when they peaked at 7.8%/day (Lyndon *et al.*, 1992). Fed and refed rainbow trout had similar rates in both the stomach and intestine and these were significantly higher than in unfed fish (McMillan and Houlihan, 1989). Recent work on mammals suggests that the fate of amino acids in the intestine may depend on their origin, with newly absorbed amino acids being used preferentially for protein synthesis (Reeds *et al.*, 1999). Similar investigations that require complex experimental procedures would be difficult if not impossible to conduct in fish.

Protein synthesis in muscle involves mainly the synthesis of myofibrillar protein with a greater proportion of the total due to myofibrillar protein in the white and red muscle (60–64%) than in the heart (40–51%) (Fauconneau *et al.*, 1995). Comparison of muscle protein synthesis rates ranks cardiac (atrium > ventricle) > red muscle > white muscle rates; differences in their oxidative capacity offer a partial explanation for this order (Fauconneau *et al.*, 1995). This is supported, to some extent, by the fact that mitochondrial protein fractions make a relatively larger contribution to the total protein in the cardiac tissue than in the skeletal muscle and the reverse for the myofibrillar protein. It is likely that both biotic and abiotic factors will influence oxidative capacity and effect changes in protein synthesis in the mitochondrial and postmitochondrial fractions that significantly impact the overall rates of muscle protein synthesis (Fauconneau *et al.*, 1995). *In vitro* studies on isolated fish hearts showed higher rates of protein synthesis in the atrium and ventricle after doubling the cardiac output (Houlihan *et al.*, 1988a). Different regions of the heart, the atrium (4 increasing to 10%/day), ventricle (2 increasing to 6%/day) and bulbous arteriousus (no increase, 3.5–4.0%/day), were ranked, demonstrating that there are differences in protein synthesis in the

Table III
Factors That Modify Fractional Rates of Protein Synthesis (%/day) for Different Tissues Where More Than One Tissue Was Measured

								Fractional rates of protein synthesis (%/day)							
Fish	Weight (g)	Temp. (°C)	Feeding[a]	Experiment	Liver	GIT[b]	Gill	White muscle	Red muscle	Heart	Brain	Kidney, spleen	Gonad	Skin, Scales	Source[c]
Environmental															
Temperature															
Rainbow trout	108	10	F	10°C	39.7	33.1									(a)
	118	18	F	18°C	135.2	75.9									(b)
Common carp	25–90	8	F	8°C				0.2	0.9						
		8	F	28°C				2.0	12.8						
		28	F	8°C				0.2	0.4						
		28	F	28°C				1.2	7.7						
Toadfish	480	10	S	10°C	2.2		2.0	0.1		0.7		2.1, 0.8	0.2		(c)
		20	S	20°C	13.7		7.7	0.2		2.3		15.2, 5.0	1.4		
Atlantic cod	166	5	FA	5°C		5.5	8.0			2.8					(d)
	247	15	FA	15°C		6.5	9.0			3.2					
Triggerfish	200–	20	?	20°C			5.5	0.4	1.0						(e)
	300	26	?	26°C			15	0.7	3.2						
		30	?	30°C			14	1.1	2.8						
Anoxia															
Crucian carp	18	10	FA	Control	10.0			0.3	0.8	1.3	0.4				(f)
				48 h Anoxia	0.5			0.1	0.4	0.6	0.3				
				24 h Recovery	8.5					0.9	0.3				
				168 h Anoxia	0.4					0.7	0.3				

Category / Species	Weight	n	Cond.	Treatment									Ref
Pollutants													
Rainbow trout	6	15	UF	Control	10.5		7.2						(g)
				Acid (15d)	8.6*		6.3						
				Acid + aluminum(15d)									
Dab	270	7	UF	Control	10.9		6.8				3.0, 2.5		(h)
				Sewage sludge	2.5			0.16			3.4, 2.5		
					2.6			0.14					
Biotic													
Species													
Moonfish	205	26–28	?				16	0.6	2.2				(i)
Triggerfish	275						15	0.7	3.2				
Jack	180						14	1.3	3.5				
Life history													
Rainbow trout	2038	8	F	Female diploid	43.7	6.9		0.38		0.7		1.8	(j)
	1922			Male diploid	28.6	7.0		0.59		31.3		1.2	
	1786			"Female" triploid	16.3	9.2		0.23				1.1	
	1673			Male triploid	30.8	9.0		0.26				1.4	
Atlantic salmon	2571	15	S	Female, July	8	(2)	5	2	3	2	2.5		(k)
	1965	8	S	Female, October	13	(9)	6	2	2	2.5	12.5		
Body weight													
Rainbow trout	100	12	UF	Scaling (100g)			8.1	0.43	1.36	2.0			(l)
Atlantic Cod	300	10	UF	Scaling [300g at k_c(max)]	31.6	9.0, 7.0	15.0	3.5		8.0	5.0		(m)
Manipulation													
Rainbow trout	84	12	UF	Control: exercise			9.1	0.5	1.3	2.2			(n)
				Control: train			11.1	1.2	2.7	3.9			
				Train & exercise			9.5	0.6	1.3	1.6			

45

(*continued*)

Table III (Continued)

Fish	Weight (g)	Temp. (°C)	Feeding[a]	Experiment	Liver	GIT[b]	Gill	White muscle	Red muscle	Heart	Brain	Kidney, spleen	Gonad	Skin, Scales	Source[c]
Rainbow trout	50	11	UF	Control	14.4	14.4, 4.4	6.9			3.4					(o)
				Ovine GH	16.9	15.0, 7.2	14.0			4.7					
Nutrition															
Feed intake															
Rainbow trout	200–400	12	F	Fed daily	17.4		9.1	0.4							(p)
			S	Unfed 15 d	15.2		7.3	0.2							
Rainbow trout	150–259	14	F	Fed daily				0.7	1.7						(q)
			S	Unfed 7 d				0.4	1.6						
			S	Unfed 14 d				0.2	0.9						
			S	Unfed 56 d				0.2	0.3						
Rainbow trout	50	15	F	Fed daily	17.2	21.1, 7.8	14.0	0.54	2.28	ND					(r)
			S	Unfed 6 d	17.9	8.4, 4.6	8.5	0.15	1.10	3.0					
			RF	Refed	30.6	18.2, 10.9	11.7	0.27	1.47	4.0					
N.coriiceps	600–1100	2	F	Fed	10.4		1.6	0.4		1.4	0.4	3.5, 2.1	2.6		(s)
			UF	Unfed	11.5		2.9	0.2		1.1	0.5	4.5, 1.9	3.1		
Time after meal															
Atlantic salmon	33.8	17	F	2 h	8.71	3.99	5.92	0.19							(t)
				18 h	7.84	2.43	5.83	0.24							

Species			Regimen	Feed/Time			Ref
Atlantic cod	180	10	F	0 h	1.15	1.16	(u)
				3 h	8.53		
				6 h	18.32	7.82	
				12 h	6.96	3.82	
				18 h	2.86	1.38	
				24 h	1.65	1.24	
Feeds							
Rainbow trout	228	15	UF	High protein (40%)	8.0		(v)
	153			Low protein (20%)	9.0	1.12	
						0.27	
Grass carp	23	22	UF	Pellet	2.47	2.47	(w)
	27			Lettuce	1.07	1.07	
European eel	56	25	?	Fish meal	6.2	0.49	(x)
				Meat meal	10.2	0.14	
				Sunflower meal	9.9	0.30	
				Sunflower meal + amino acids	4.3	0.45	

[a] Feeding regimen: S, starving (unfed for greater than 3 days); FA, fasted (unfed for 2–3 days); UF, unfed (unfed on previous day); F, fed within 24 h; RF, refed.
[b] GIT or separate for intestine, stomach.
[c] (a), Fauconneau and Arnal (1985); (b), Watt *et al.* (1988); (c), Pornjic *et al.* (1983); (d), Foster *et al.* (1992); (e), Haschemeyer *et al.* (1979); (f), Smith *et al.* (1996); (g), Wilson *et al.* (1996); (h), Houlihan *et al.* (1994); (i), Smith *et al.* (1980); (j), Fauconneau (1990); (k), Martin *et al.* (1993); (l), Houlihan *et al.* (1986); (m), Houlihan *et al.* (1988b); (n), Houlihan and Laurent (1987); (o), Foster *et al.* (1991); (p), Smith (1981); (q), Loughna and Goldspink (1984); (r), McMillan and Houlihan (1989); (s), Haschemeyer (1983); (t), Fauconneau *et al.* (1989); (u), Lyndon *et al.* (1992); (v), Peragon *et al.* (1994); (w) Carter *et al.* (1993a); (x), De la Higuera *et al.* (1999).

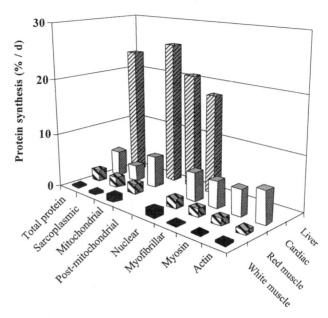

Fig. 3. Fractional rates of protein synthesis (%/day) of cellular subfractions and proteins in the liver (McMillan and Houlihan, 1989) and in cardiac, red and white muscle tissue of juvenile rainbow trout, *O. mykiss* (Fauconneau *et al.*, 1995).

same organ as well as differences in response to input pressure (Houlihan *et al.*, 1988a).

Other organs have been studied, but their ranking in terms of protein synthesis is difficult to assess because of differences in experimental procedures. The brain has low rates of protein synthesis, less than 1%/day (Haschemeyer, 1983; Sayegh and Lajtha, 1989; Smith *et al.*, 1996), which is approximately the same as the white muscle rates for fish under standard conditions. The ratio between fractional rates of protein synthesis in the brain and white muscle increased two- to threefold under conditions such as starvation or anoxia, where muscle rates decreased and brain rates were maintained (Haschemeyer, 1983; Smith *et al.*, 1996). In Antarctic species, the spleen had similar rates of protein synthesis to the kidney (1–2%/day) and about four times the white muscle rates (Haschemeyer, 1983).

The relative contribution of different cellular fractions and proteins to the whole-body protein pool was discussed above. There are also differences in rates of protein synthesis between the different cellular fractions and individual proteins and between the different muscle and tissues types (Fig. 3). Rates of protein synthesis in all of the components rank liver > cardiac muscle > red muscle > white muscle, showing that each component reflected the overall rate of protein synthesis in that organ. Rates in the liver mitochondrial fraction tended to be greater than total protein rates, due mainly to synthesis in polysomes (McMillan and Houlihan, 1989). In muscle, differences between fractions within the same muscle were less clear (Fauconneau *et al.*, 1995).

D. Whole-Body Protein Synthesis and Organ Integration

Clearly, the sum of the amount of protein synthesized by each organ would be expected to equal the amount of protein synthesized in the whole body over the same time period. There are few studies that provide rates of protein synthesis (and protein contents) for both the whole body and a sufficient number of the major organs for the calculation to be attempted (Fig. 4). In the juvenile rainbow trout (100 g) and the small cod, a large proportion of protein synthesis is attributed to the white muscle, GIT, and liver. The white muscle has the lowest fractional rates of protein synthesis, but is the largest protein mass in the whole body, whereas the liver is much smaller but has very high rates of synthesis. Mature salmonids in different stages of reproduction have large differences between the relative magnitude of protein synthesis in different tissues. The very high protein synthesis in the liver of the female rainbow trout is presumably the result of vitellogenin synthesis, whereas in the male rainbow trout and the female salmon resources are being partitioned into gonadal tissue.

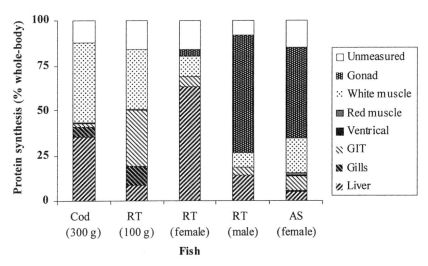

Fig. 4. The contribution of protein synthesis in tissues to the total whole-body protein synthesis per day in cod, *G. morhua* (Houlihan *et al.*, 1988b), juvenile rainbow trout, *O. mykiss* (Houlihan *et al.*, 1986), mature female and male rainbow trout (Fauconneau *et al.*, 1990), and mature female Atlantic salmon, *S. salar* (Martin *et al.*, 1993).

In comparison to fractional rates of protein synthesis, the protein accretion of organs is similar, although described by different allometric relationships, which means that the retention of synthesized protein is very different between tissues (Houlihan et al., 1986, 1988b). In cod, synthesis retention efficiency (k_g/k_s) was ranked white muscle > stomach > ventricle > intestine > gill > liver and shown to decrease with increasing growth rate (Houlihan et al., 1988b). At maximum growth rates, synthesis retention efficiency was approximately 60% for white muscle, compared with 20% for the intestine, 10% for gills, and less than 10% for the liver (Houlihan et al., 1988b).

There has also been interest in using organ protein synthesis as an indicator of whole-body synthesis and growth (Fauconneau et al., 1990; Houlihan et al., 1993a, 1995a). The analysis of muscle protein synthesis is more straightforward than that for the whole body because it involves a few hundred milligrams of tissue and avoids the complications of obtaining representative samples from the entire carcass. There is an expectation of a significant correlation between rates of white muscle synthesis and whole-body synthesis or growth because white muscle represents approximately 50% of the total whole-body protein and the retention of synthesized protein is the highest compared with all other tissues. There is a significant linear relationship between white muscle and whole-body rates of protein synthesis when data from three different species at different stages in their life cycle, but held under standard conditions, are analyzed (Fig. 5a). At very low levels of white muscle synthesis the relationship is not clear and may not be linear. It is highly unlikely that white muscle synthesis would be zero so there must be a minimum value with an associated and higher whole-body rate. Fig. 5a suggests that whole-body rates are approximately double the rate of white muscle protein synthesis; a value nearer to four times was reported for a single species, rainbow trout (McCarthy et al., unpublished).

Several studies have demonstrated a linear relationship between fractional rates of protein accretion and protein synthesis in individual fish (Section II.D). The relationship between protein synthesis and growth can be investigated using mean values (Fig. 5b). This approach is useful because white muscle protein synthesis can be used to calculate the growth rate of fish where this is not known (caught from the wild or taken from large populations). Additionally, the establishment of a generalized relationship will permit estimates of protein synthesis rates of the whole animal from white muscle data. Specific growth rate and white muscle synthesis rates have been measured more frequently than whole-body protein synthesis rates; hence, the analysis of this relationship was restricted to rainbow trout (Fig. 5b). There is a significant linear relationship, although the data show a relatively high degree of scatter. The intercept suggests that in rainbow trout the white muscle synthesis rate is 0.12%/day at maintenance. A similar relationship was constructed for rainbow trout and predicted a maintenance rate of

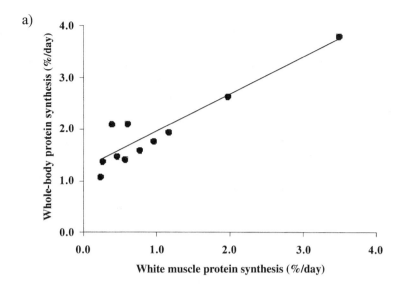
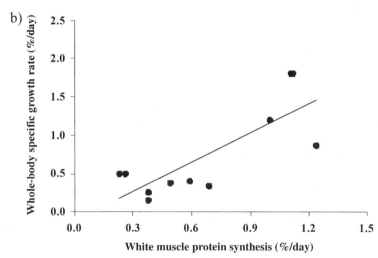

Fig. 5. (a) The relationship between white muscle protein synthesis (WMPS) and whole-body protein synthesis (WBPS) fractional rates of protein synthesis (%/day) described by WBPS = 0.722WMPS + 1.218 ($n = 11$; $R^2 = 0.86$; $P < 0.001$) [cod, *G. morhua* (Houlihan *et al.,* 1988b, 1989); rainbow trout, *O. mykiss* (Fauconneau *et al.,* 1990); wolffish, *Anarhichas lupus* (McCarthy *et al.,* 1999)]. (b) The relationship between white muscle fractional rates (%/day) of protein synthesis (WMPS) and whole-body growth rate (WBSGR) for rainbow trout under standard conditions and described by WBSGR = 1.272 WMPS − 0.120 ($n = 11$; $R^2 = 0.62$; $P < 0.01$) (Smith, 1981; Loughna and Goldspink, 1984; Houlihan and Laurent, 1987; Fauconneau *et al.,* 1990; Peragon *et al.,* 1994, 1999).

approximately 0.1%/day (Fauconneau et al., 1990). The slope of the relationship is 1.8 in the latter study and 1.3 from our data (Fig. 5b), both slopes indicating a high efficiency in comparison to farm animals (Fauconneau et al., 1990). This approach was expanded to describe white muscle protein synthesis (WMPS) in relation to temperature (T), fish weight (W) and specific growth rate (SGR) so that WMPS = $1.53 + 0.48 SGR - 0.10 T - 1.8610^{-4} W$ ($n = 14$; $R^2 = 0.86$; $P < 0.0001$) (Houlihan et al., 1995a).

Increasing the rRNA (C_s) and/or increasing ribosomal activity (k_{RNA}) provides a mechanism for increasing rates of protein synthesis (Fig. 6). Consideration of these variables in different tissues supports the case for both the amount and the activity of RNA influencing protein synthesis. Thus, tissues such as the liver have higher RNA capacities and efficiencies than tissues such as white muscle where rates of protein synthesis, C_s and k_{RNA} are typically less than 2%/day, 10 mg RNA/g protein, and 2 g protein synthesized /g RNA /day, respectively.

Although the study of whole-body rates of protein synthesis does not provide detailed information on specific pathways, it has a very important place in understanding many aspects of fish physiology in relation to different abiotic and biotic factors (Fauconneau, 1984; Houlihan et al., 1988b; McCarthy et al., 1994; Carter et al., 1998).

E. Amino Acid Requirements and Protein Accretion

Amino acid requirements are considered as a special case of nutrient requirements because of their direct relationship to the synthesis of protein. Ten amino acids (arginine, histidine, isoleucine, leucine, lysine, methionine, phenylalanine, threonine, tryptophan, valine) are essential for fish with an additional two being considered semiessential (cystine and tyrosine) (Wilson, 1989; also see Chapter 3). Quantitative requirements have been determined using standard dose–response experiments that show the optimum range of dietary amino acid content, either expressed as a proportion of the diet or in relation to the energy content of the diet. An important feature of amino acid requirements is that for each amino acid to be retained with maximum efficiency all the essential amino acids need to be in an optimum ratio with respect to each other. If one amino acid is in excess of this ratio, the assumption is that it will be in excess in the free amino acid pool and will be oxidized and excreted (Nose et al., 1978; Arzel et al., 1995). If one amino acid is deficient (limiting), then it will shift the remaining essential amino acids to being in excess and they will be oxidized and excreted. Essential amino acid requirements can be estimated by considering their ratio in proteins that most resemble the requirement profile. "Predictor proteins" for amino acid requirements have included the whole body, muscle, and fish eggs (Wilson, 1989). Comparison of known requirements with predictor proteins shows that in juvenile fish

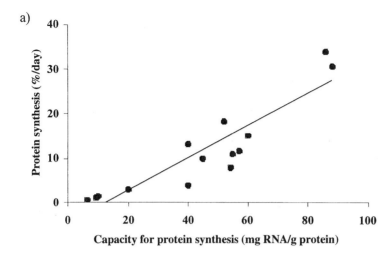
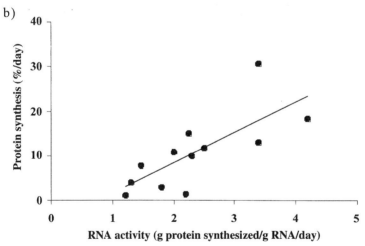

Fig. 6. The relationship between fractional rates of protein synthesis (%/day) and (a) capacity for protein synthesis described by PS = $0.36C_s - 4.57$ ($n = 14$; $R^2 = 0.82$; $P < 0.001$) and (b) RNA activity described by PS = $6.78k_{RNA} - 5.27$ ($n = 12$; $R^2 = 0.55$; $P < 0.01$) of different tissues from rainbow trout, *O. mykiss* (McMillan and Houlihan, 1988, 1989; Foster *et al.*, 1991; Peragon *et al.*, 1994).

the whole body provides the strongest correlation (Mambrini and Kaushik, 1993). Further points to consider are that essential amino acid requirements are divided into requirements for maintenance and for growth (Millward and Rivers, 1988; Millward, 1998) and that different species or life-history stages (see Chapter 5) may show small differences in requirements due to small differences in the whole-body balance of specific proteins and, therefore, amino acids.

The relationship between protein and energy intake is critical to the efficiency with which dietary protein can be partitioned into growth. It is well established that maximum protein accretion will occur over a narrow range of dietary protein to energy ratios. Research, particularly for salmonid species that can use high-fat diets, suggests that protein requirements are 16–18 g digestible protein per megajoule of digestible energy (g DP/MJ DE) (Hillestad and Johnsen, 1994; Einen and Roem, 1997). Outside the optimum range the efficiency of retaining amino acid nitrogen decreases and the ultimate fate of amino acids is deamination and catabolism. However, the effect of unbalanced protein intake on protein metabolism is extremely interesting because the influx of amino acids into the free pools may be very large but animals clearly regulate the free amino acid concentrations (Carter *et al.,* 1995a). It appears that protein synthesis may be used as a mechanism for regulation of free amino acid pool concentration. Protein synthesis and protein degradation both increase as digestible protein intake increases in common carp (*Cyprinus carpio*) (Meyer-Burgdorff and Rosenow, 1995b), which means that protein retention does not show a similar linear response. This is further illustrated by relationships between the dietary protein energy ratio and the anabolic stimulation efficiency and synthesis retention efficiency using data from a variety of studies on salmonid species (Fig. 7). It appears that as dietary protein increases above the optimum ratio relative to energy, there is an increase in the stimulation of protein synthesis but that this protein is not retained as growth and, therefore, the retention of synthesized protein decreases. The relationships are probably not linear, and both anabolic stimulation and synthesis retention efficiency may be higher at the optimum dietary ratio than predicted by linear extrapolation of the data (Fig. 7). Although the response of common carp to a graded range of protein energy ratios was not investigated, the data do support this hypothesis (Meyer-Burgdorff and Rosenow, 1995b). Feeding diets with protein energy ratios that were either lower or higher than the diet producing the highest protein retention resulted in higher anabolic stimulation efficiency (1.3 to 2.1 times higher) and lower synthesis retention efficiency (1.5 to 2.4 times lower).

Evidence for changes in protein turnover in relation to changes in essential amino acid intake is very limited for fish. Two identical diets, except for differing lysine content, fed to rainbow trout resulted in significantly higher protein accretion due to lower protein turnover (k_d) at similar rates of synthesis (Campbell, unpublished). Consequently, anabolic stimulation efficiency was similar (61 and

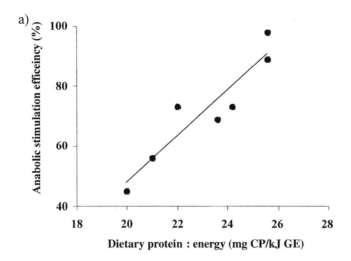
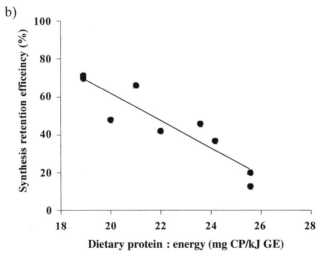

Fig. 7. The relationship between dietary protein energy (P:E, milligrams crude protein per kilojoule gross energy: mg CP/kJ GE) and (a) anabolic stimulation efficiency (%) described by ASE = 7.65P:E − 105.3 ($n = 7$; $R^2 = 0.87$; $P < 0.001$) and (b) synthesis retention efficiency (%) described by SRE = −7.16 P:E + 204.8 ($n = 9$; $R^2 = 0.84$; $P < 0.001$) for rainbow trout, *O. mykiss* [Fauconneau and Arnal, 1985; McCarthy *et al.*, 1994; Bolliet *et al.*, 2000 (SRE only)] and Atlantic salmon, *S. salar* (Carter *et al.*, 1993b; Carter *et al.*, unpublished).

57%), whereas synthesis retention efficiency was significantly higher (74 compared with 60%) on the better performing diet. The better performing diet had a higher lysine content that was closer to a 1:1 ratio with the white muscle amino acid concentration (% of total protein). The lysine content (0.8 g lysine/MJ GE) of the poorer performing diet was close to the dietary requirement, and it is likely that the diet became deficient in lysine following digestion and absorptive losses. Such studies highlight the need to combine measures of protein metabolism with dose–response amino acid requirement trials in order to understand protein metabolism over a spectrum of dietary amino acid intake. In a more extreme example, four diets in which fish meal was replaced with a bacterial single-cell protein (BSCP) were fed to rainbow trout. Feeding the diet containing no fish meal resulted in extremely low protein accretion and retention efficiency (Perera *et al.*, 1995b) and this was explained by absorption of some essential amino acids (lysine, methionine, phenylalanine, threonine) below requirements (Perera *et al.*, 1995a). The effect on protein metabolism was that white muscle protein synthesis rates were more than double those of the best performing diet; consequently, anabolic stimulation efficiency was higher (100 versus 51%) and synthesis retention efficiency lower (11 versus 60%) on the poorer diet (Perera, 1995).

The relationship between essential amino acid requirements, the optimum dietary protein energy ratio, and protein synthesis is of interest because it will highlight the mechanisms governing maximum protein accretion as well as having the potential to provide a rational assessment of requirements based on physiological understanding. Protein synthesis may show a variety of responses to differences in the intake and balance of essential amino acids. The dynamic nature of protein turnover in relation to amino acid nutrition has been considered for human adults under conditions where $k_g = 0$ and $k_t = k_s = k_d$ (Young *et al.*, 1991). If the maintenance requirement for essential amino acids is not met, an "adapted state" is reached where intake and oxidation are lower but protein turnover (synthesis and degradation) is maintained. In young mammals and fish, protein accretion is the more normal situation and oxidation will be less than intake so that $k_g > 0$ and $k_s > k_d$. Possible responses in protein synthesis can be described in growing fish (animals) under conditions where protein intake stays the same but the balance of essential amino acids changes and under conditions where protein intake is above the optimum level (Table IV). There is some evidence for an adapted state at suboptimum amino acid intake (Perera, 1995). Comparison of the ideal essential amino acid balance with that of a slightly imbalanced diet shows similar protein retention efficiency that is a consequence of increased anabolic stimulation (Table IV). A more imbalanced diet results in lower protein retention due to the larger decrease in the retention of synthesized protein because the available essential amino acids do not meet the requirements for tissue proteins. In the adapted state it is possible that higher protein turnover reduces metabolism of the more

Table IV
Hypothetical Relationships between Essential Amino Acid (EAA) Intake and Protein Intake, Turnover, and Retention (Arbitrary Units) in Growing Juvenile Fish Where Ideal Protein Intake Is 100 Units

	Protein intake (units)	Anabolic stimulation (%)	Synthesis retention (%)	Protein retention (%)	Relative values
Dietary EAA intake					
Large EAA deficiency	100	100	10	10	Perera (1995)
Mid EAA deficiency	100	60	60	36	Campbell (unpublished)
Slight EAA deficiency	100	60	75	45	"
Ideal EAA balance	100	50	90	45	Projected
Dietary protein intake					
Large DP excess	100	100	10	10	Fig. 7
Mid DP excess	100	70	40	28	"
Slight DP excess	100	50	60	30	"
Ideal DP:DE	100	50	90	45	Projected

Source: Adapted from Young *et al.* (1991).

limiting amino acids, making them available for synthesis of tissue proteins of the types that are retained.

IV. FACTORS THAT MODIFY PROTEIN SYNTHESIS

Numerous abiotic and biotic factors have the potential to modify rates of protein synthesis in the whole body or in specific organs. The majority of early work focused on the effect of temperature and the mechanisms that allow fish to acclimatize to different temperatures (Haschemeyer, 1973, 1983). More recently, issues such as global warming have stimulated interest in temperature effects (McCarthy and Houlihan, 1997). Feeding and nutrition are considered separately because the relationship between nutrition and protein synthesis has received little attention in fish; future work should focus on these interactions because of the practical implications in terms of aquaculture production. Much of the research to date has been on the effect of feed intake rather than of specific nutrients or dietary ingredients. Other environmental factors such as oxygen concentration, salinity, and potential toxicants have been investigated but data are few, and further description of the effects of these factors as well as other factors, such as photoperiod, on protein synthesis would, therefore, be of interest. A consequence of the high energy requirements associated with protein synthesis is that protein synthesis may change when environmental factors are related to changes in energy demands compared with normal conditions.

A. Environmental Factors

1. Temperature

The effects and consequences of temperature on protein synthesis in fish have been the subject of a sizable proportion of the literature on fish, and the area has been reviewed recently (McCarthy and Houlihan, 1997). As for many physiological processes, protein synthesis changes with temperature and is predicted to exhibit an asymmetrical response: rates increasing to a maximum near an upper temperature limit and, as the temperature increases beyond this value, declining rapidly (McCarthy and Houlihan, 1997). It has been suggested that maximum rates of protein synthesis occur at the optimum temperature for growth (Loughna and Goldspink, 1985; Pannevis and Houlihan, 1992; McCarthy and Houlihan, 1997). The evidence for this is the observation that the rates of synthesis in carp and rainbow trout muscle (Loughna and Goldspink, 1985) and in isolated trout hepatocytes (Pannevis and Houlihan, 1992) are maximized near to the optimum temperature for growth. Indices of growth and growth efficiency in Atlantic wolffish (*A. lupus*) show maximum values at 11°C, whereas whole-body and white muscle protein synthesis show an increasing linear relationship over the range 5–14°C (McCarthy *et al.*, 1999).

Thermal history influences protein synthesis when fish are exposed to different temperatures. Fish previously exposed to lower temperatures exhibit higher rates of protein synthesis than fish held at higher temperatures before protein synthesis measurements (Watt *et al.*, 1988). It has also been pointed out that because feed intake also increases with increasing temperature, protein synthesis may increase independently of temperature (Foster *et al.*, 1992; McCarthy and Houlihan, 1997). By ensuring equal food intake at 5 and 15°C it was demonstrated that Atlantic cod were able to maintain rates of protein synthesis and growth at the lower temperature through higher concentrations of RNA (Foster *et al.*, 1992). In contrast, rainbow trout acclimatized to a higher temperature and allowed to feed to appetite had higher rates of protein synthesis than those at a lower temperature despite having lower RNA concentrations. Under these circumstances differences in protein synthesis and growth were due to differences in k_{RNA} (McCarthy and Houlihan, 1997).

When a wide variety of animals are considered an exponential increase in rates of protein synthesis and k_{RNA} is observed with increasing temperature (Watt *et al.*, 1988; McCarthy and Houlihan, 1997; McCarthy *et al.*, 1999). This relationship appears to be consistent despite observations having been made on different species, different tissues, and under different thermal regimes. Of further interest, although there is an order of magnitude difference between liver and white muscle synthesis rates, there is no difference in k_{RNA} between the two tissues (McCarthy and Houlihan, 1997). This suggests that C_s (the amount of RNA) provides the mechanism for differences in protein synthesis between tissues and that, in this property, fish are similar to other animals.

2. OXYGEN

Low oxygen levels may lead to decreases in protein synthesis (Jackim and La Roche, 1973) but this is not necessarily the case, and fish may adopt a variable strategy, depending on the severity of the oxygen depletion, and show different responses in tissues (Fauconneau, 1985). The extreme case of anoxia (for 48 h) has been studied in the crucian carp (*Carassius carassius*), demonstrating that part of the ability of this fish to downregulate metabolism is expressed through large decreases in protein synthesis in the liver and muscle. Because these tissues constitute the majority of whole-body energy expenditure and protein synthesis is responsible for the majority of tissue energy expenditure, this strategy will result in a large reduction in overall metabolism (Smith *et al.*, 1996).

3. SALINITY

When this topic was last reviewed (Houlihan, 1991), few studies had considered the effects of salinity on protein synthesis and this still appears to be the case. Regulation in and adaptation to different salinities by euryhaline and stenohaline species remain of considerable interest. In coho salmon (*Oncorhynchus kisutch*) in spring the rate of incorporation of radio-labeled glutamate into muscle in parr in freshwater was found to be higher than in smolts in seawater (Guillaume *et al.*, 1984). However, this type of measurement does not clearly show that protein synthesis was lower, since amino acid-free pool concentration may have differed and, hence, the specific activity of the glutamate may have differed in the tissues. Furthermore, in smolts there is evidence to suggest white muscle protein synthesis increases in seawater (Jungham and Jurss, 1988).

4. POLLUTANTS

Pollutants may act as general environmental stressors and impose extra energy demands on organisms exposed to them. Sewage sludge added 5 days a week to tanks of dab (*Limanda limanda* L.) over 3 months did not affect whole-body or tissue protein synthesis (Table III). However, it had a significant effect on protein accretion, which resulted in significantly lower efficiencies of protein retention and retention of synthesized protein than the control fish (Houlihan *et al.*, 1994). It therefore appears that exposure to sewage sludge results in increased protein degradation and a probable higher energetic cost associated with this difference in protein metabolism. Exposure to acid and aluminum results in temporary changes in protein metabolism of rainbow trout that were evident during the first 7 days but had largely disappeared after 32 days (Wilson *et al.*, 1996). In this study whole-body protein synthesis decreased after 7 days and was mainly explained by the large decrease in feed intake that occurred immediately following exposure to the change in water quality. Gill protein synthesis and degradation increased after 7 days but were not different by 15 days, perhaps reflecting the period of greatest damage and repair of gill tissue (Wilson et al., 1996). In contrast, liver protein synthesis was not affected after 7 days and although it had decreased by 32 days, this did not impact the proportion of protein synthesized in the liver.

B. Biotic Factors

1. Species

It is difficult to make comparisons between species, first, because there have been very few studies aimed at comparing different species using the same techniques and, second, because approaches vary so much there tend to be several differences between key variables that might in themselves have a major impact on protein synthesis (Smith et al., 1980).

Haschemeyer (1983) compared two Antarctic fish with quite different physiology (Table III). Icefish blood does not contain hemoglobin and this species also synthesizes large amounts of AFGP, whereas the other fish, Notothenia, does have hemoglobin in its blood. Tissue protein synthesis rates were 43% lower in the icefish. This was explained by the low oxidative capacity of this fish and the need to reduce energy demanding processes. The protein synthesis rate in the icefish epaxial muscle was 23% of the rate in Notothenia, whereas the rate in the head kidney was reduced by less, to 76%. This indicated the relative importance of protein synthesis in the head kidney, possibly the site of AFGP synthesis. A selection of tropical fish species was used to measure protein synthesis under broadly similar conditions and produced similar relative values for gill and red and white muscle tissue (Smith et al., 1980).

Indications of species differences have been reported with active pelagic species having higher rates than less active benthic/reef dwelling species (Smith et al., 1980). However, despite measurements being conducted concurrently, interpretation remains difficult due to individual differences in size or growth rate. Several relationships have been constructed that use data from different species and these tend to suggest that species differences in protein synthesis are likely to be explained largely by weight, temperature, and feeding regime (Section III.D).

2. Life-History Stage

Protein synthesis has been measured over a range of life-history stages of fish (Table V), including yolk-sac larvae (Conceicao et al., 1997a), feeding larvae (Houlihan et al., 1992, 1995d), juveniles of several species and in mature fish at various stages in their reproductive cycle (Fauconneau et al., 1990; Martin et al., 1993). No studies have attempted to investigate changes in protein synthesis over the lifetime of one species, although the data from rainbow trout cover a wide size

Table V
Summary of Whole-Body Indices of Protein Turnover Calculated for Adult and Juvenile Fish Fed Formulated Feeds and for Larval Fish at Different Stages of Development and under Optimum Conditions

Fish	Weight (g)	Temp. (°C)	k_r	k_s (%/day)	k_g	k_s/k_r	$k_g/k_r r$ (%/day)	k_g/k_r	C_s	k_{RNA}	Source[a]
Atlantic cod[a]	300	10	7.3	3.8	2.0	52.0	53.0	22.4	8.8	4.3	(a)
Atlantic salmon[a]	180	13	5.3	3.8	1.7	71.7	44.7	32.1	5.0	7.6	(b)
Halibut	109	8–13		2.0	1.3		65.0				(c)
Rainbow trout	80	10	6.0	4.4	1.9	73.3	43.2	31.7	7.1	6.2	(d)
Atlantic wolffish[b]	65	11	2.9	1.8	1.0	62.1	55.6	34.5	10.6	1.7	(e)
Flounder[a]	60	7	6.0	4.1	1.4	68.3	34.1	23.3			(f)
Grass carp	23	22	6.0	2.9	1.7	48.3	58.6	28.3			(g)
Sea bass	3.5	18	11.4	5.1	2.4	44.7	47.0	21.1	6.9	7.4	(h)
Tilapia	1.0	25–27		0.9	0.5		55.6				(i)
Rainbow trout (fry)	0.31	15		3.1	2.3		74.0		27.3	2.3	(j)
Tilapia	0.01	25–27		10.3	4.9		47.6				(i)
Larval fish (feeding)[c]	Weight (mg)										
Nase (Artemia)	50	20		30.2	15.0		49.7				(k)
Herring (ZP)	0.40	8	11.1	13.5	7.0	122	51.9	63.1			(l)
African catfish (Y-S)	0.315	28		106	96		69.6		106.7	12.9	(m)
Turbot (ZP)	17 days	18	45.0	69.8	41.7	155	59.7	92.7	59.7	48.8	(n)
Turbot (ZP)	11 days	18	51.8	32.5	30.7	62.7	93.9	59.3			(n)

[a] Calculated from regression equations for maximum ration.
[b] Calculated from regression equations for treatment nearest to optimum temperature.
[c] Feeding: Y-S, yolk-sac; ZP, zooplankton.

[d] (a), Houlihan et al. (1988b, 1989); (b), Carter et al. (1993b); (c), Fraser et al. (1998); (d), McCarthy et al. (1994); (e), McCarthy et al. (1999); (f), Carter et al. (1998); (g), Carter et al. (1993c); (h), Langer et al. (1993); (i), Houlihan et al. (1993b); (j), Mathers et al. (1993); (k), Houlihan et al. (1992); (l), Houlihan et al. (1995d); (m), Conceicao et al. (1997a); (n), Conceicao et al. (1997b).

range (Table V). Protein synthesis in eggs, yolk-sac larvae, feeding larvae, and small juveniles have been investigated in whitefish (*Coregonus schinzii pallea*) (Fauconneau, 1985). The results of early studies have been reviewed (Fauconneau, 1985) but the very high rates of protein synthesis, on the order of 300%/day, with correspondingly low synthesis retention efficiencies, are not reflected in more recent studies (Table V). Nevertheless, larval fish appear to have rates of synthesis that are an order of magnitude higher than in juvenile fish, although low rates in herring larvae show that this is not always the case (Houlihan *et al.*, 1995d). Turbot larvae grow rapidly and fractional rates of protein accretion are greater than 30%/day over the first 17 days following hatching; protein synthesis is correspondingly high but synthesis retention is similar to juvenile fish after 17 days (Conceicao *et al.*, 1997b). This study showed that protein degradation was correspondingly higher in larval fish compared to juveniles. However, the 11-day turbot had a very much higher retention of synthesized protein, suggesting a rapid increase in protein turnover between 11 and 17 days after hatching. Higher rates of larval protein synthesis were due to both higher total RNA concentration and RNA activity (Table V).

Young Atlantic salmon in freshwater follow one of two distinct life-history strategies and either migrate to sea after one winter (early migrants) or after more than one winter in freshwater (late migrants). Salmon from the early migrant group had significantly higher protein accretion at similar rates of protein synthesis compared with the late migrant group, suggesting that growth was maximized by minimising protein turnover (Morgan *et al.*, 2000).

3. BODY WEIGHT

The observation that protein synthesis decreases with a progression of life-history stages has been expressed in terms of scaling relationships that have been established for the key indices of protein metabolism (Table VI). Scaling has been examined assuming an allometric relationship of the form $Y = aX^b$ and using data from rainbow trout with a wide range of weights (approximately 0.2–300 g) but held under similar conditions (Houlihan *et al.*, 1995c). Protein growth and synthesis were found to have similar weight exponents that indicated a parallel decrease as weight increased. Interestingly, the inference that synthesis retention efficiency does not change with weight is confirmed by the lack of trend when different species are considered (Table V). Weight exponents of -0.26 and -0.25 for synthesis and growth, respectively, are very similar to weight exponents for physiological processes such as oxygen consumption and ammonia excretion (Houlihan *et al.*, 1995c). This is also the case for total RNA concentration (-0.204) and supports the case for rRNA being the primary factor controlling rates of protein synthesis especially as over a wide range of weights RNA activity did not appear to change in relation to weight (Table V).

Table VI
Summary of Weight Scaling Coefficients for Indices of Protein Turnover for Whole-Body (WB) and White Muscle (WM) Measurements[a]

Parameter	Weight range (g)	Intercept (a)	Slope (b)	Source[b]
Protein synthesis (%/day)				
Rainbow trout (WB)	0.2–300	0.77	−0.259	(a)
Tilapia (WB)	0.016–8.09	0.90	−0.53	(b)
Rainbow trout (WM)	25–363	4.01	−0.49	(c)
Protein degradation (%/day)				
Tilapia (WB)	0.016–8.09	0.64	−0.55	(b)
Rainbow trout (WM)	25–363	0.83	−0.42	(c)
RNA: Protein (mg RNA/g protein)				
Various (WB)	0.0001–1498	1.31	−0.163	(a)
Rainbow trout (WB)	0.2–300	1.12	−0.204	(a)
Tilapia (WB)	0.016–8.09	1.27	−0.06	(b)
k_{RNA} *(g protein synthesized/gRNA/day)*				
Rainbow trout (WB)	0.2–300	0.45	−0.033	(a)
Tilapia (WB)	0.016–8.09	0.57	−0.41	(b)

[a] Where $\log_{10} Y = \log_{10} a + b \log_{10} W$.
[b] (a), Houlihan et al. (1995c); (b), Houlihan et al. (1993b); (c), Houlihan et al. (1986).

4. Social Hierarchy

Differences in feed intake between individuals explain much of the variation in growth found within groups of fish (McCarthy et al., 1992, 1993). However, differences in protein metabolism also influence the efficiency with which individuals use food (Carter et al., 1993b; McCarthy et al., 1994). Consequently, there may be subtle differences in whole-body or organ-specific protein synthesis that relate to social rank and reflect the stress imposed by a particular rank.

5. Exercise

Exercise training over several weeks does not result in any differences in rates of protein synthesis in the gills, ventricle, red muscle, or white muscle compared with untrained control animals when both groups are held in stationary water (Houlihan and Laurent, 1987). However, trained fish that had undergone 6 weeks of continuous swimming have significantly higher protein accretion in these tissues. Swimming stimulates protein synthesis and, to a lesser extent, degradation in all the muscle tissues, which explains the increased accretion (Houlihan and Laurent, 1987). Increased growth is not due to increased synthesis retention efficiency, implying that training would demand increases in feed intake to meet the increased demands for energy and protein synthesis.

6. Hormones

Intraperitoneal administration of ovine growth hormone (oGII) to juvenile rainbow trout fed the same rations as control fish results in increased rates of tissue protein synthesis in the heart, gill, liver, and stomach that are linked to increased C_s and k_{RNA} in the heart but only increased k_{RNA} in the gills (Foster et al., 1991). In this study rates of whole-body synthesis and degradation were 117 and 93%, respectively, compared to the controls, which explains the significant difference in protein accretion. Rates of white muscle protein synthesis in rainbow trout (human GH) and Atlantic salmon (oGH) increase with hormonal treatment whereas anti-salmon GH decreases whole-body growth and white muscle protein synthesis in trout (Fauconneau et al., 1996).

C. Feeding and Nutrition

1. Feed Intake

Information on feed intake is vital for a detailed understanding of how various other factors can affect protein synthesis. Consequently, there are major difficulties in the interpretation and comparison of many different effects on protein synthesis because the feeding regimen was not controlled. Methods for doing this have been reviewed elsewhere (McCarthy et al., 1993; Jobling et al., 1995), but it is important to point out that the different methods for measuring protein synthesis in fish can be adapted for use with different feeding regimens. For example, end-product methods in which the tracer is applied in the food can be used with single fish or groups of fish; if groups are used, an additional method of assessing individual intake can be used or group intake measured.

Several studies have constructed relationships between protein intake and indices of protein turnover (Section II.D) based on individual responses (Houlihan et al., 1988b, 1989; Carter et al., 1993a,b, 1998; McCarthy et al., 1994). Alternatively, groups can be considered and the mean response reported. There are no published examples where the effect of different group rations on mean rates of protein synthesis has been determined; the expectation would be for the relationships to be very similar to those based on individuals (Carter et al., 1995b).

2. Starvation and Refeeding

Many species of fish undergo periods of starvation or very low food intake during which time they exhibit comparatively small decreases in physical capability. Evidence suggests that changes in protein synthesis reflect the differing roles of tissues. For example, white muscle represents a large store of energy as well as being important in burst swimming, whereas red muscle is a small tissue in continuous use at all swimming speeds. The effects of starvation on rainbow trout white and red muscle studied over 56 days (Loughna and Goldspink, 1984) show differences over time and between the two muscle types (Table III). Common carp starved for 70 days show similar responses between red and white muscle, although the decrease is not as marked as in trout (Watt *et al.*, 1988). In this study white and red muscle rates of protein synthesis were maintained at rates of fed fish for 3 and 7 days, respectively. In the white muscle rates of protein synthesis then decreased to a stable level after 14 days that was approximately 20% of the fed rates, whereas rates in red muscle continued to decrease and were 52 and 19% of fed rates after 14 and 56 days, respectively (Loughna and Goldspink, 1984). The adaptive significance of the difference may have related to the balance between maintenance of function, a general decrease in activity, or the relative importance of tissues as energy stores. Similar arguments could be extended to other tissues, such as the liver and gills. In rainbow trout the rates of protein synthesis in both tissues are maintained after 15 days of starvation, indicating the need for synthesis of export proteins and for continuous turnover of proteins in both tissues (Smith, 1981). However, liver protein synthesis may decrease over a shorter timescale of 4 days prior to a return to fed rates, suggesting some complexity in the temporal response to starvation (McMillan and Houlihan, 1992). In addition the starvation response is sensitive to prior feed intake with fish fed at maintenance having significantly higher rates of liver synthesis than well-fed fish due to a greater reliance on muscle amino acids as an energy source (McMillan and Houlihan, 1992).

Refeeding stimulates protein synthesis in most tissues, although there are differences in the magnitude of the change. Rates of protein synthesis in rainbow trout are over 3 times higher in the stomach and intestine but approximately double in the liver, gill, and muscle (McMillan and Houlihan, 1988). Compared with continuously fed trout, protein synthesis (at 6 h after feeding) is similar in the gills, stomach, and intestine, but lower in the red and white muscle (McMillan and Houlihan, 1988). The complexities in the response to refeeding in these studies are probably due to the functions of the tissues, the normal rates of synthesis, and the effect of a 6-day fast. Thus, in muscle the protein synthesis rates are low, they decrease in response to fasting and take longer to return to levels measured in feeding fish. The stomach and intestine appear to respond rapidly to refeeding, indicating an ability to approach normal function following the first meal and, therefore, have an ability to adapt to temporally separated meals. Liver protein synthesis rates are high, they are maintained over a 6-day fast, and show hyperstimulation on refeeding. Rates were 50 and 25% higher at 3 and 6 h after refeeding, respectively, than in continuously fed fish and are driven through increased k_{RNA} not C_s (McMillan and Houlihan, 1988). This, again emphasizes the ability of rainbow trout to maintain function over a short period of time without food.

3. NUTRIENT REQUIREMENTS

Amino acid and protein requirements in relation to energy requirements were discussed in Section III.E. Few other data exist on nutrient requirements and protein synthesis. The absence of carbohydrate in a rainbow trout diet was compared to a diet containing 23% digestible carbohydrate (Peragon et al., 1999). Protein metabolism in white muscle was investigated and a difference in fractional rates of protein accretion was found, which was explained by increased degradation at similar rates of synthesis. It was proposed that the higher rate of degradation was related to significantly elevated rates of gluconeogenesis from amino acids in fish fed the diet lacking carbohydrate (Peragon et al., 1999). Because the carbohydrate provided extra energy compared to the noncarbohydrate diet, changes in protein metabolism cannot be unequivocally attributed to the lack of carbohydrate and could also have been related to differences in energy intake.

4. Feed Ingredients

Protein quality can have a major influence on protein accretion, and this may be due to differences in protein synthesis and degradation (Langer et al., 1993; De la Higuera et al., 1999). The replacement of fish meal with greaves meal (defatted collagen) and feather meal results in a reduction in dietary essential amino acid content causing reduced protein growth and protein retention efficiency (Langer et al., 1993). These decreases are accompanied by significant increases in both whole-body protein synthesis and degradation as well as in C_s, but not in k_{RNA} (Langer et al., 1993). These results offer further support for the importance of protein synthesis in removing excess dietary amino acids into a short-term protein pool that is not retained as long-term growth (Section III.E). Although the responses in protein synthesis, degradation and retention efficiency were similar in a subsequent experiment, which again used greaves meal to replace fish meal, neither protein growth or C_s was affected by diet (Langer and Guillaume, 1994).

The source and form of amino acids may also have a comparable effect, since the form of dietary amino acids in fish diets is important and can have a major influence on overall protein accretion. Some species do not use dietary free amino acids effectively. In the study of Langer et al. (1993) cited above, the use of a fish protein hydrolysate did not affect the dietary essential amino acid balance compared to a fish meal control diet, but did result in lower protein retention efficiency in the sea bass. Protein synthesis, C_s, and k_{RNA} were not significantly different, but protein degradation was significantly higher than in the sea bass fed the fish meal control diet, which is in contrast to protein metabolism when greaves and feather meals were used (see above). Common carp fed a diet supplemented with crystalline lysine grew more slowly and had lower protein retention efficiency than those fed a diet supplemented with a coated lysine; performance was generally better on the control diet that contained only protein-bound lysine (De la Higuera et al., 1998). At 18°C the feed intake for these three diets was found to be the same, which allowed an assessment of the effect of only diet composition (not feed intake) on protein synthesis in this study. Protein accretion was significantly higher on the control diet than on the crystalline-lysine diet, reflected by significantly higher rates of white muscle protein synthesis and degradation and higher

k_{RNA} at similar C_s values; fish fed the coated-lysine diet had intermediate values (De la Higuera et al., 1998). Supplementation of sunflower meal with essential amino acids resulted in increased feed intake, growth, and feed efficiency to the same level as for the control diet that contained only fish meal. Compared with the diet containing sunflower meal but no essential amino acid supplements, rates of protein synthesis and degradation were significantly lower in the livers of fish fed the supplemented diet (De la Higuera et al., 1999). This suggests that when the amino acid intake is imbalanced, liver protein synthesis may increase synthesis of temporary proteins that are not retained as growth (see Section III.E).

V. ENERGETIC COST OF PROTEIN SYNTHESIS IN FISH

Protein synthesis is clearly a central process in growth, but it also has an impact due to its high energetic cost. The minimum energetic cost of protein synthesis is generally assumed to be 50 mmol ATP per g of protein synthesized (40 mmol plus 10 mmol for transport) (Houlihan et al., 1988b). The cost of protein synthesis, expressed as part of the total energy expenditure has been estimated in several studies (Table VII): values range between 11 and 24% at maintenance to between 19 and 42% for feeding and growing fish. In juvenile fish, protein synthesis represents a higher proportion of metabolism as feed intake increases (Houlihan et al., 1988b; Carter et al., 1993a; Meyer-Burgdorff and Rosenow, 1995c). This may be due, in part, to protein synthesis having a variable cost that decreases as rates of protein synthesis increase, explained by the need to meet fixed costs associated with activation of tRNA and synthesis of rRNA (Houlihan et al., 1995a). Cycloheximide has been used to inhibit protein synthesis and measurements made with and without inhibition (Table VII). In larval fish between 31 and 79% of energy expenditure is associated with protein synthesis. Protein synthesis is, therefore, a major energy-demanding process in life-history stages that have the capacity for rapid somatic growth.

VI. SUMMARY

Protein synthesis is fundamental to all living organisms and it has been studied intensively and at varying levels of complexity. This chapter provides a comprehensive review of research on protein synthesis in fish, examines data to produce simple models describing protein synthesis in terms of key variables, and provides explanations for variations from expected or predicted rates of protein synthesis. The underlying theme is to integrate information at the organismal level. A variety of methods for measuring protein synthesis have been used and comparison sug-

Table VII
Energetic Cost of Protein Synthesis (% of Total Metabolism) in Fish

Fish	Weight (g)	Temp. (°C)	Feeding[a]	k_g (%/day)	Protein synthesis (mg protein/day)	Total metabolism (mg O_2/day)	Cost of protein synthesis (% total metabolism)	Source[b]
African catfish	0.0003	28	Y-S	38		0.06	43[c]	(a)
Herring	0.0005	8	F	<5		0.01	79[c]	(b)
Tilapia	0.016	27	F	31		1.32	31[c]	(c)
Nase	0.050	20	F	15	1.71	1.42	31[d]	(d)
Grass carp	23	22	R_{main}	0	127	153	11[d]	(e)
			F	1.7	180	217	22[d]	
Common carp	30–60	23	F	2.5	361	378	25[d]	(f)
			F	4.1	624	545	30[d]	
Atlantic cod	180	10	RF (18 h)		580	324	19[d]	(g)
	300	10	R_{main}	0	570	644	24[d]	(h)
			R_{max}	2.0	1920	1228	42[d]	

[a] Feeding: Y-S, yolk-sac stage; F, feeding; R_{main}, maintenance ration; R_{max}, maximum ration for experiment; RF, refeeding after 6 days starvation and measured at 18 h after feeding.

[b] (a), Conceicao et al. (1997a); (b), Houlihan et al. (1995d); (c); Houlihan et al. (1993b); (d), Houlihan et al. (1992); (e), Carter et al. (1993a); (f), Meyer-Burgdorff and Rosenow (1995c) ; (g), Lyndon et al. (1992); (h), Houlihan et al. (1988b).

[c] Direct estimate of reduction in oxygen consumption following cycloheximide administration.

[d] Indirect calculation based on separate measurements of protein synthesis and oxygen consumption and assuming 50 mmol ATP/g protein synthesized and 5.24 mg O_2/mmol ATP.

gests they give similar results for fish. Major influences on protein synthesis are species, life-history stage, temperature, feeding, and nutrition. The effects of other factors such as pollutants, anoxia, salinity, and hormones have also been investigated. In growing fish between 20 and 50% of energy expenditure is associated with protein synthesis. Protein synthesis is, therefore, a major energy-demanding process in fish that is influenced by many environmental and biotic factors.

ACKNOWLEDGEMENTS

We are grateful to S. A. M. Martin for his contribution to the section on protein degradation. This research was supported by NERC, BBSRC, and MAFF (United Kingdom) and by ARC and FRDC (Australia).

REFERENCES

Arzel, J., Metailler, R., Kerleguer, C., Delliou, H., and Guillaume, J. (1995). The protein requirement of brown trout (*Salmo trutta*) fry. *Aquaculture* **130,** 67–78.

Attaix, D., Combaret, L., and Taillandier, D. (1999). Mechanisms and regulation in protein degradation. *In* "Protein Metabolism and Nutrition" (G. E. Lobley, A. White, and J. C. MacRae, eds.), pp. 51–67. Wageningen Pers, Wageningen, The Netherlands.

Bolliet, V., Cheewasedtham, C., Houlihan, D. F., Gelineau, A., and Boujard, T. (2000). Effect of feeding time of digestibility, growth performance and protein metabolism in the rainbow trout *Oncorhynchus mykiss:* interactions with dietary fat level. *Aquat. Living Resour.* **13,** 107–111.

Carter, C. G., Houlihan, D. F., Brechin, J., and McCarthy, I. D. (1993a). The relationships between protein intake and protein accretion, synthesis and retention efficiency for individual grass carp, *Ctenopharyngodon idella* (Val.). *Can. J. Zool.* **71,** 393–400.

Carter, C. G., Houlihan, D. F., Buchanan, B., and Mitchell, A. I. (1993b). Protein-nitrogen flux and protein growth efficiency of individual Atlantic salmon (*Salmo salar* L.). *Fish Physiol. Biochem.* **12,** 305–315.

Carter, C. G., Owen, S. F., He, Z.-Y., Watt, P. W., Scrimgeour, C., Houlihan, D. F., and Rennie, M. J. (1994). Determination of protein synthesis in rainbow trout, *Oncorhynchus mykiss,* using a stable isotope. *J. Exp. Biol.* **189,** 279–284.

Carter, C. G., He, Z. Y., Houlihan, D. F., McCarthy, I. D., and Davidson, I. (1995a). Effect of feeding on tissue free amino acid concentrations in rainbow trout (*Oncorhynchus mykiss* Walbaum). *Fish Physiol. Biochem.* **14,** 153–164.

Carter, C. G., McCarthy, I. D., Houlihan, D. F., Fonseca, M., Perera, W. M. K., and Sillah, A. B. S. (1995b). The application of radiography to the study of fish nutrition. *J. Appl. Ichthy.* **11,** 231–239.

Carter, C. G., Houlihan, D. F., and Owen, S. F. (1998). Protein synthesis and nitrogen excretion and long-term growth of flounder *Pleuronectes flesus. J. Fish Biol.* **53,** 272–284.

Conceicao, L., Houlihan, D. F., and Verreth, J. (1997a). Fast growth, protein turnover and costs of protein metabolism in yolk-sac larvae of the African catfish (*Clarias gariepinus*). *Fish Physiol. Biochem.* **16,** 291–302.

Conceicao, L., van der Meen, T., Verreth, J., Evjen, M. S., Houlihan, D. F., and Fyhn, H. J. (1997b). Amino acid metabolism and protein turnover in larval turbot (*Scophthalmus maximus*) fed natural zooplankton or *Artemia*. *Mar. Biol.* **129,** 255–265.

Das, A. B., and Proser, C. L. (1967). Biochemical changes in tissues of goldfish acclimated to high and low temperatures. I. Protein synthesis. *Comp. Biochem. Physiol.* **21,** 449–467.

De la Higuera, M., Garzon, A., Hidalgo, M. C., Peragon, J., Cardenete, G., and Lupianez, J. A. (1998). Influence of temperature and dietary-protein supplementation either with free or coated lysine on the fractional protein-turnover rates in the white muscle of carp. *Fish Physiol. Biochem.* **18,** 85–95.

De la Higuera, M., Akharbach, H., Hidalgo, M. C., Peragon, J., Lupianez, J. A., and Garcia-Gallego, M. (1999). Liver and white muscle protein turnover rates in the European eel (*Anguilla anguilla*): effects of dietary protein quality. *Aquaculture* **179,** 203–216.

Einen, O., and Roem, A. J. (1997). Dietary protein/energy ratios for Atlantic salmon in relation to fish size: growth, feed utilization and slaughter quality. *Aquacult. Nutr.* **3,** 115–126.

Fauconneau, B. (1984). The measurement of whole body protein synthesis in larval and juvenile carp (*Cyprinus carpio*). *Comp. Biochem. Physiol.* **78B,** 845–850.

Fauconneau, B. (1985). Protein synthesis and protein deposition in fish. *In* "Nutrition and Feeding in Fish" (C. B. Cowey, A. M. Mackie, and J. G. Bell, eds.), pp. 17–46. Academic Press, London.

Fauconneau, B., and Arnal, M. (1985). *In vivo* protein synthesis in different tissues and the whole body rainbow trout (*Salmo gairdneri* R.). Influence of environmental temperature. *Comp. Biochem. Physiol.* **82A,** 179–187.

Fauconneau, B., and Tesseraund, S. (1990). Measurement of plasma leucine flux in rainbow trout (*Salmo gairdneri* R.) using miniosmotic pumps. Preliminary investigations on influence of diet. *Fish Physiol. Biochem.* **8,** 29–44.

Fauconneau, B., Breque, J., and Bielle, C. (1989). Influence of feeding on protein metabolism in Atlantic salmon (*Salmo salar*). *Aquaculture* **79,** 29–36.

Fauconneau, B., Aguirre, P., and Blanc, J. M. (1990). Protein synthesis in different tissues of mature rainbow trout (*Salmo gairdneri* R.). Influence of triploidy. *Comp. Biochem. Physiol.* **97C,** 345–352.

Fauconneau, B., Gray, C., and Houlihan, D. F. (1995). Assessment of individual protein turnover in three muscle types of rainbow trout. *Comp. Biochem. Physiol.* **111B,** 45–51.

Fauconneau, B., Mady, M. P., and LeBail, P. Y. (1996). Effect of growth hormone on muscle protein synthesis in rainbow trout (*Oncorhynchus mykiss*) and Atlantic salmon (*Salmo salar*). *Fish Physiol. Biochem.* **15,** 49–56.

Foster, A. R., Houlihan, D. F., Gray, C., Medale, F., Fauconneau, B., Kaushik, S. J., and Le Bail, P. Y. (1991). The effects of ovine growth hormone on protein turnover in rainbow trout. *Gen. Comp. End.* **82,** 111–120.

Foster, A. R., Houlihan, D. F., Hall, S. J., and Burren, L. J. (1992). The effects of temperature acclimation on protein synthesis rates and nucleic acid content of juvenile cod (*Gadus morhua* L.). *Can. J. Zool.* **70,** 2095–2102.

Fraser, K. P. P., Lyndon, A. R., and Houlihan, D. F. (1998). Protein synthesis and growth in juvenile Atlantic halibut, *Hippoglossus hippoglossus* (L.): application of 15N stable isotope tracer. *Aquacult. Res.* **29,** 289–298.

Fuller, M. F., and Garlick, P. J. (1994). Human amino acid requirements: can the controversy be resolved? *Annu. Rev. Nutr.* **14,** 217–241.

Garlick, P. J., McNurlan, M. A., and Preddy, V. R. (1980). A rapid and convenient technique for measuring the rate of protein synthesis in tissues by injection of ^{3}H phenylalanine. *Biochem. J.* **217,** 507–516.

Garlick, P. J., McNurlan, M. A., Essen, P., and Wernerman, J. (1994). Measurement of tissue protein synthesis rates *in vitro:* a critical analysis of contrasting methods. *Am. J. Physiol.* **266,** E287–E297.

Green, R., and Noller, H. F. (1997). Ribosomes and translation. *Annu. Rev. Biochem.* **66,** 679–716.
Guillaume, J., Stephen, G., Messager, J. L., and Garin, D. (1984). Preliminary studies of protein synthesis, gluconeogenesis and lipogenesis with labelled U ^{14}C glutamate in coho salmon: effect of sea transfer. *J. World Mar. Soc.* **14,** 203–209.
Haschemeyer, A. E. V. (1968). Compensation of liver protein synthesis in temperature acclimated toadfish, *Opsanus beta. Biol. Bull.* **135,** 130–140.
Haschemeyer, A. E. V. (1973). Control of protein synthesis in the acclimation of fish to environmental temperature changes. *In* "Responses of Fish to Environmental Changes" (W. Chavin, ed.), pp. 3–30. Charles C. Thomas, Springfield, IL.
Haschemeyer, A. E. V. (1983). A comparative study of protein synthesis in nototheniids and icefish at Palmer station, Antarctica. *Comp. Biochem. Physiol.* **76B,** 541–543.
Haschemeyer, A. E. V., and Smith, M. A. K. (1979). Protein synthesis in liver, muscle and gill of mullet (*Mugil cephalus* L.) *in vivo. Biol. Bull.* **156,** 93–102.
Haschemeyer, A. E. V., Pursell, P., and Smith, M. A. K. (1979). Effect of temperature on protein synthesis in fish of the Galapagos and Perlas Islands. *Comp. Biochem. Physiol.* **64B,** 91–95.
Hershko, A., and Ciechanover, A. (1998). The ubiquitin system. *Annu. Rev. Biochem.* **67,** 425–479.
Hillestad, M., and Johnsen, F. (1994). High-energy/ low-protein diets for Atlantic salmon: effects on growth, nutrient retention and slaughter quality. *Aquaculture* **124,** 109–116.
Houlihan, D. F. (1991). Protein turnover in ectotherms and its relationships to energetics. *Adv. Comp. Environ. Physiol.* **7,** 1–43.
Houlihan, D. F., and Laurent, P. (1987). Effects of exercise training on the performance, growth, and protein turnover of rainbow trout (*Salmo gairdneri*). *Can. J. Fish. Aquat. Sci.* **44,** 1614–1621.
Houlihan, D. F., McMillan, D. N., and Laurent, P. (1986). Growth rates, protein synthesis, and protein degradation rates in rainbow trout: effects of body size. *Physiol. Zool.* **59,** 482–493.
Houlihan, D. F., Agnisola, C., Lyndon, A. R., Gray, C., and Hamilton, N. M. (1988a). Protein synthesis in a fish heart: responses to increased power output. *J. Exp. Biol.* **137,** 565–587.
Houlihan, D. F., Hall, S. J., Gray, C., and Noble, B. S. (1988b). Growth rates and protein turnover in Atlantic cod, *Gadus morhua. Can. J. Fish. Aquat. Sci.* **45,** 951–964.
Houlihan, D. F., Hall, S. J., and Gray, C. (1989). Effects of ration on protein turnover in cod. *Aquaculture* **79,** 103–110.
Houlihan, D. F., Wieser, W., and Foster, A. R. (1992). *In vivo* protein synthesis rates in larval nase, *Chondrostoma nasus* L. *Can. J. Zool.* **70,** 2436–2440.
Houlihan, D. F., Mathers, E., and Foster, A. R. (1993a). Biochemical correlates of growth rate in fish. *In* "Fish Ecophysiology" (J. C. Jensen and F. B. Rankin, eds.), pp. 45–71. Chapman & Hall, London.
Houlihan, D. F., Pannevis, M. C., and Heba, H. M. A. (1993b). Protein synthesis in juvenile tilapia, *Oreochromis mossambicus. J. World Aquacult. Soc.* **24,** 145–151.
Houlihan, D. F., Costello, M. J., Secombes, C. J., Stagg, R., and Brechin, J. (1994). Effects of sewage sludge exposure on growth, feeding and protein synthesis of dab, *Limanda limanda. Mar. Environ. Res.* **37,** 331–353.
Houlihan, D. F., Carter, C. G., and McCarthy, I. D. (1995a). Protein synthesis in fish. *In* "Biochemistry and Molecular Biology of Fishes" (P. Hochachka and P. Mommsen, eds.), Vol. 4, pp. 191–219. Elsevier Science, Amsterdam.
Houlihan, D. F., Carter, C. G., and McCarthy, I. D. (1995b). Protein turnover in animals. *In* "Nitrogen and Excretion" (P. J. Wright and P. A. Walsh, eds.), pp. 1–29. CRC Press, Boca Raton, FL.
Houlihan, D. F., McCarthy, I. D., Carter, C. G., and Marttin, F. (1995c). Protein turnover and amino acid flux in fish larvae. *ICES Mar. Sci. Symp.* **201,** 87–99.
Houlihan, D. F., Pedersen, B. H., Steffensen, J. F., and Brechin, J. (1995d). Protein synthesis, growth, and energetics in larval herring (*Clupea harengus*) at different feeding regimes. *Fish Physiol. Biochem.* **14,** 195–208.

Jackim, E., and La Roche, G. (1973). Protein synthesis in *Fundulus heteroclitus* muscle. *Comp. Biochem. Physiol.* **44A,** 851–866.

Jobling, M., Arnesen, A. M., Baardvik, B. M., Christiansen, J. S., and Jørgensen, E. H. (1995). Monitoring feeding behaviour and food intake: methods and applications. *Aquacult. Nutr.* **1,** 131–143.

Jungham, I., and Jurss, K. (1988). Effect of feeding and salinity on protein synthesis in the white epaxial muscle of the rainbow trout, *Salmo gairdneri* Richardson. *Comp. Biochem. Physiol.* **89B,** 329–333.

Kimball, S. R., Vary, T. C., and Jefferson, L. S. (1994). Regulation of protein synthesis by insulin. *Annu. Rev. Physiol.* **56,** 321–348.

Langer, H., and Guillaume, J. (1994). Effect of feeding pattern and dietary protein source on protein synthesis in European sea bass (*Dicentrarchus labrax*). *Comp. Biochem. Physiol.* **108A,** 461–466.

Langer, H., Guillaume, J., Metailler, R., and Fauconneau, B. (1993). Augmentation of protein synthesis and degradation by poor dietary amino acid balance on European sea bass (*Dicentrarchus labrax*). *J. Nutr.* **123,** 1754–1761.

Lobley, G. E., White, A., and MacRae, J. C. (eds.). (1999). "Protein Metabolism and Nutrition." Wageningen Pers, Wageningen, The Netherlands.

Loughna, P. T., and Goldspink, G. (1984). The effects of starvation upon protein turnover in red and white myotomal muscle of rainbow trout. *J. Fish Biol.* **25,** 223–230.

Loughna, P. T., and Goldspink, G. (1985). Muscle protein synthesis rates during temperature acclimation in a eurythermal (*Cyprinus carpio*) and a stenohaline (*Salmo gairdneri*) species of teleost. *J. Exp. Biol.* **118,** 267–276.

Love, R. M. (1957). The biochemical composition of fish. *In* "The Physiology of Fishes" (M. E. Brown, ed.), pp. 401–418. Academic Press, New York.

Lyndon, A. R., and Brechin, J. G. (1999). Evidence for partitioning of physiological functions between holobranchs: protein synthesis rates in flounder gills. *J. Fish Biol.* **54,** 1326–1328.

Lyndon, A. R., and Houlihan, D. F. (1998). Gill protein turnover: costs of adaptation. *Comp. Biochem. Physiol.* **199A,** 27–34.

Lyndon, A. R., Houlihan, D. F., and Hall, S. J. (1992). The effect of short-term fasting and a single meal on protein synthesis and oxygen consumption in cod, *Gadus morhua*. *J. Comp. Physiol.* **B162,** 209–215.

Mambrini, M., and Kaushik, S. (1993). Indispensible amino acid requirements of fish: correspondence between quantitative data and amino acid profiles of tissue proteins. *J. Appl. Ichthy.* **11,** 240–247.

Martin, N. B., Houlihan, D. F., Talbot, C., and Palmer, R. M. (1993). Protein metabolism during sexual maturation in female Atlantic salmon (*Salmo salar* L.). *Fish Physiol. Biochem.* **12,** 131–141.

Mathers, E. M., Houlihan, D. F., McCarthy, I. D., and Burren, L. J. (1993). Rates of growth and protein synthesis correlated with nucleic acid content of fry of rainbow trout, *Oncorhynchus mykiss*: effects of age and temperature. *J. Fish Biol.* **43,** 245–263.

McCarthy, I. D., and Houlihan, D. F. (1997). The effect of water temperature on protein metabolism in fish: the possible consequences for wild Atlantic salmon (*Salmo salar* L.) stocks in Europe as a result of glocal warming. *In* "Global Warming: Implications for Freshwater and Marine Fish" (C. M. Wood and D. G. McDonald, eds.), pp. 51–77. Cambridge University Press, Cambridge.

McCarthy, I. D., Carter, C. G., and Houlihan, D. F. (1992). The effect of feeding hierarchy on individual variability in daily feeding of rainbow trout, *Oncorhynchus mykiss* (Walbaum). *J. Fish Biol.* **41,** 257–263.

McCarthy, I. D., Houlihan, D. F., Carter, C. G., and Moutou, K. (1993). Variation in individual food consumption rates of fish and its implications for the study of fish nutrition and physiology. *Proc. Nutr. Soc.* **52,** 411–420.

McCarthy, I. D., Houlihan, D. F., and Carter, C. G. (1994). Individual variation in protein turnover and growth efficiency in rainbow trout, *Oncorhynchus mykiss* (Walbaum). *Proc. Royal Soc. London B* **257,** 141–147.

McCarthy, I. D., Moksness, E., Pavlov, D., and Houlihan, D. F. (1999). Effects of water temparature on protein synthesis and protein growth in juvenile Atlantic wolffish (*Anarhichas lupus*). *Can. J. Fish. Aquat. Sci.* **56,** 231–241.

McMillan, D. N., and Houlihan, D. F. (1988). The effect of re-feeding on tissue protein synthesis in rainbow trout. *Physiol. Zool.* **61,** 429–441.

McMillan, D. N., and Houlihan, D. F. (1989). Short-term responses of protein synthesis to re-feeding in rainbow trout. *Aquaculture* **79,** 37–46.

McMillan, D. N., and Houlihan, D. F. (1992). Protein synthesis in trout liver is stimulated by both feeding and fasting. *Fish Physiol. Biochem.* **10,** 23–34.

Meyer-Burgdorff, K. H., and Rosenow, H. (1995a). Protein turnover and energy metabolism in growing carp. 1. Method of determining N-turnover using a 15N-labelled casein. *J. Anim. Physiol. Anim. Nutr.* **73,** 113–122.

Meyer-Burgdorff, K. H., and Rosenow, H. (1995b). Protein turnover and energy metabolism in growing carp. 2. Influence of feeding level and protein energy ratio. *J. Anim. Physiol. Anim. Nutr.* **73,** 123–133.

Meyer-Burgdorff, K. H., and Rosenow, H. (1995c). Protein turnover and energy metabolism in growing carp. 3. Energy cost of protein deposition. *J. Anim. Physiol. Anim. Nutr.* **73,** 134–139.

Millward, D. J. (1998). Metabolic demands for amino acids and the human dietary requirement: Millward and Rivers (1988) revisited. *J. Nutr.* **128,** 2563S-2576S.

Millward, D. J., and Rivers, J. (1988). The nutritional role of indispensible amino acids and the metabolic basis for their requirements. *Eur. J. Clin. Nutr.* **42,** 367–393.

Millward, D. J., Garlick, P. J., Stewart, R. J. C., Nnanyelugo, D. O., and Waterlow, J. C. (1975). Skeletal-muscle growth and protein turnover. *Biochem. J.* **150,** 235–243.

Millward, D. J., Bates, P. C., de Benoist, B., Brown, J. G., Cox, M., Halliday, D., Odedra, B., and Rennie, M. J. (1983). Protein turnover the nature of the phenomenon and its physiological regulation. *In* "IV Int. Symp. Protein metabolism and nutrition" Clermont-Ferrand (France), 5–9 Sept. INRA Pub.

Morgan, I. J., McCarthy, I. D. and Metcalfe, N. B. (2000). Life-history strategies and protein metabolism in overwintering Atlantic salmon: growth is enhanced in early migrants through lower protein turnover. *J. Fish Biol.* **56,** 637–647.

Negatu, Z., and Meier, A. H. (1993). Daily variation of protein synthesis in several tissues of the gulf killifish, *Fundulus grandis* Baird and Girard. *Comp. Biochem. Physiol.* **106A,** 251–255.

Nieto, R., and Lobley, G. E. (1999). Integration of protein metabolism within the whole body and between organs. *In* "Protein Metabolism and Nutrition" (G. E. Lobley, A. White, and J. C. MacRae, eds.), pp. 127–153. Wageningen Pers, Wageningen, The Netherlands.

Nose, T., Lee, D.-L., and Arai, S. (1978). The effects of the withdrawal of single free amino acid from an amino acid diet on the free amino acid composition of skeletal muscle in young carp. *Bull. Freshw. Fish. Lab. Tokyo* **28,** 255–263.

Owen, S. F., McCarthy, I. D., Watt, P. W., Ladero, V., Sanchez, J. A., Houlihan, D. F., and Rennie, M. J. (1999). *In vivo* rates of protein synthesis in Atlantic salmon (*Salmo salar* L.) smolts determined using a stable isotope flooding dose technique. *Fish Physiol. Biochem.* **20,** 87–94.

Pannevis, M. C., and Houlihan, D. F. (1992). The energetic cost of protein synthesis in isolated hepatocytes of rainbow trout (*Oncorhynchus mykiss*). *J. Comp. Physiol.* **162B,** 393–400.

Peragon, J., Barroso, J. B., Garcia-Salguero, L., de la Higuera, M., and Lupianez, J. A. (1994). Dietary protein effects on growth and fractional protein synthesis and degradation rates in liver and white muscle of rainbow trout (*Oncorhynchus mykiss*). *Aquaculture* **124,** 35–46.

Peragon, J., Barroso, J. B., Garcia-Salguero, L., de la Higuera, M., and Lupianez, J. A. (1999). Carbohydrates affect protein-turnover rates, growth, and nucleic acid content in the white muscle of rainbow trout (*Oncorhynchus mykiss*). *Aquaculture* **179,** 425–437.

Perera, W. M. K. (1995). Growth performance, nitrogen balance and protein turnover of rainbow trout,

Oncorhynchus mykiss (Walbaum), under different dietary regimens. Ph.D thesis, University of Aberdeen.

Perera, W. M. K., Carter, C. G., and Houlihan, D. F. (1995a). Apparent absorption efficiencies of amino acids in rainbow trout, *Oncorhynchus mykiss* (Walbaum), fed diets containing bacterial single cell protein. *Aquacult. Nutr.* **1,** 95–103.

Perera, W. M. K., Carter, C. G., and Houlihan, D. F. (1995b). Feed consumption, growth and growth efficiency of rainbow trout, *Oncorhynchus mykiss* (Walbaum), fed diets containing bacterial single cell protein. *Br. J. Nutr.* **73,** 591–603.

Pornjic, Z., Mathews, R. W., Rappaport, S., and Haschemeyer, A. E. V. (1983). Quantitative protein synthesis rates in various tissues of a temparate fish *in vivo* by the method of phenylalanine swamping. *Comp. Biochem. Physiol.* **74B,** 735–738.

Reeds, P. J., Fuller, M. F., and Nicholson, B. A. (1985). Metabolic basis of energy expenditure with particular reference to protein. *In* "Substrate and Energy Metabolism" (J. S. Garrow and D. Haliday, eds.), pp. 46–57. J. Libbey, London.

Reeds, P. J., Burrin, D. G., Stoll, B., and van Goudoever, J. B. (1999). Consequences and regulation of gut metabolism. *In* "Protein Metabolism and Nutrition" (G. E. Lobley, A. White, and J. C. MacRae, eds.), pp. 127–153. Wageningen Pers, Wageningen, The Netherlands.

Rennie, M. J., Smith, K. & Watt, P. W. (1994). Measurement of human tissue protein synthesis: an optimal approach. *Am. J. Physiol.* **266,** E298–E307.

Sayegh, J. F., and Lajtha, A. (1989). *In vivo* rates of protein synthesis in brain, muscle, and liver of five vertebrate species. *Neurochem. Res.* **14,** 1165–1168.

Sikorski, Z. E. (1994). The myofibrillar proteins in seafoods. *In* "Seafood Proteins" (Z. E. Sikorski, B. Sun Pan, and F. Shahidi, eds.), pp. 40–57. Chapman and Hall, New York.

Sikorski, Z. E., and Borderias, J. A. (1994). Collagen in the muscles and skin of marine animals. *In* "Seafood Proteins" (Z. E. Sikorski, B. Sun Pan, and F. Shahidi, eds.), pp. 58–70. Chapman and Hall, New York.

Smith, M. A. K. (1981). Estimation of the growth potential by measurement of protein synthetic rates in feeding and fasting rainbow trout, *Salmo gairdneri* Richardson. *J. Fish Biol.* **19,** 213–220.

Smith, M. A. K., Mathews, R. W., Hudson, A. P., and Haschemeyer, A. E. V. (1980). Protein metabolism of tropical reef and pelagic fish. *Comp. Biochem. Physiol.* **65B,** 415–418.

Smith, R. W., Houlihan, D. F., Nilsson, G. E., and Brechin, J. G. (1996). Tissue-specific changes in protein synthesis rates *in vivo* during anoxia in crucian carp. *Am. J. Physiol.* **271,** R897–R904.

Sugden, P. H., and Fuller, S. J. (1991). Regulation of protein turnover in skeletal and cardiac muscle. *Biochem. J.* **273,** 21–37.

Taylor, P. M., and Brameld, J. M. (1999). Mechanisms and regulation of transcription and translation. *In* "Protein Metabolism and Nutrition" (G. E. Lobley, A. White, and J. C. MacRae, eds.), pp. 25–50. Wageningen Pers, Wageningen, The Netherlands.

Tischler, M. E. (1992). Estimation of protein synthesis and proteolysis *in vitro*. *In* "Modern Methods in Protein Nutrition and Metabolism" (S. Nissen, ed.), pp. 225–248. Academic Press, San Diego.

Venugopal, V., Martin, A. M., and Patel, T. R. (1994). Extractability and stability of washed capelin (*Mallotus villus*) muscle in water. *Food Hydrocolloids* **8,** 135–145.

Von der Decken, A., Espe, M., and Lied, E. (1992). Growth and physiological properties in white muscle of two anadromous populations of Arctic charr (*Salvelinus alpinus*). *Fisk. Dir. Skr. Ser. Eraering.* **5,** 49–57.

Waterlow, J. C. (1995). Whole-body protein turnover in humans—past, present, and future. *Annu. Rev. Nutr.* **15,** 57–92.

Waterlow, J. C. (1999). The mysteries of nitrogen balance. *Nutr. Res. Rev.* **12,** 25–54.

Watt, P. W., Marshall, P. A., Heap, S. P., Loughna, P. T., and Goldspink, G. (1988). Protein synthesis in tissues of fed and starved carp, acclimated to different temperatures. *Fish Physiol. Biochem.* **4,** 165–173.

Webster, A. J. F. (1988). Comparative aspects of energy exchange. *In* "Comparative Nutrition" (K. Blaxter and I. MacDonald, eds.), pp. 37–54. J. Libbey, London.

Weisner, R. J., and Zak, R. (1991). Quantitative approaches for studying gene expression. *Am. J. Physiol.* **260,** L179-L188.

Wilson, R. P. (1989). Amino acids and proteins. *In* "Fish Nutrition," 2nd ed. (J. E. Halver, ed.), pp. 112–151. Academic Press, London.

Wilson, R. P., and Cowey, C. B. (1985). Amino acid composition of whole body tissue of rainbow trout and Atlantic salmon. *Aquaculture* **48,** 373–376.

Wilson, R. W., Wood, C. M., and Houlihan, D. F. (1996). Growth and protein turnover during acclimation to acid and aluminium in juvenile rainbow trout (*Oncorhynchus mykiss*). *Can. J. Fish. Aquat. Sci.* **53,** 802–811.

Wolfe, R. R. (1992). "Radioactive and Stable Isotope Tracers in Biomedicine." Wiley-Liss, New York.

Young, V. R., Yu, Y.-M., and Krempf, M. (1991). Protein and amino acid turnover using the stable isotopes 15N, 13C, and 2H as probes. *In* "New Techniques in Nutritional Research" (R. G. Whithead and A. Prentice, eds.), pp. 17–72. Academic Press, San Diego.

3

AMINO ACID METABOLISM

J. S. BALLANTYNE

I. Introduction
II. Digestion and Uptake of Amino Acids
III. Delivery Systems
IV. Tissues
   A. Liver
   B. Gill
   C. Kidney
   D. Blood
   E. Muscle
   F. Brain
   G. Intestine
V. Exercise
VI. Hormonal Regulation
   A. Cortisol
   B. Insulin and Glucagon
   C. Thyroid Hormones
VII. Diurnal Rhythms in Amino Acid Metabolism
VIII. Diet and Starvation
IX. Temperature
X. Salinity
XI. Anoxia
XII. Summary
    References

I. INTRODUCTION

Amino acids have numerous functions in fish. The most familiar role amino acids play is as the building blocks of proteins. Protein synthesis can account for a substantial portion (20–42%) of the energy expenditures of growing fish (Houlihan *et al.,* 1993, 1995). Most (85%) fish species are carnivorous (Love, 1980) and optimal growth requires very high dietary protein levels (30–55% crude protein) (Wilson, 1989). Fish display an efficiency of conversion of dietary protein to tissue protein that is up to 20-fold higher than that observed for chickens, pigs, and cattle (Tacon and Cowey, 1985). The efficiency of conversion of dietary protein into tissue protein is somewhat paradoxical given that fish rely extensively on amino acids as energy sources and thus burn the precursors for proteins. Several hypotheses have been put forward to explain this. Certainly, the direct excretion of ammonia as the end product of nitrogen metabolism is more energetically favorable than expending energy to convert it to less toxic forms such as urea or uric acid. The direct oxidation of dietary amino acids also avoids the energetic expense of synthesizing storage molecules such as glucose and lipids for subsequent use.

In energetic terms, a major function of amino acids is as catabolic substrates to provide ATP for biomechanical, synthetic, and transport processes. Amino acids provide 14–85% of the energy requirements of teleost fish (van Waarde, 1983) and 10–40% in nonteleost species such as sturgeon (Dabrowski et al., 1987). In rainbow trout, 35–40% of a representative amino acid such as leucine is oxidized while the remainder is converted to protein. This is a substantially higher rate of catabolism than in mammals (20%) (Fauconneau and Arnal, 1985).

The role of one amino acid in particular, glutamine, differs in fish compared to mammals. Glutamine does not serve as a nitrogen store for ammonium under normal conditions in fish as it does in mammals, and circulating levels of this amino acid are therefore lower in fish. Synthesis of glutamine in tissues, such as muscle, for export to other tissues also does not occur in fish. These differences in glutamine metabolism have a substantial impact on the metabolism of other amino acids in fish and are discussed in more detail in subsequent sections.

Despite many years of research on the metabolism of fishes, little is known of the factors regulating the flow of amino acids into catabolic or anabolic pathways. It has been suggested that some of the partitioning of amino acids toward growth rather than catabolism in fish may be due to the 10-fold lower Michaelis constant (K_m) for amino acids of aminoacyl tRNA synthetases compared to the K_m of the transaminases that funnel amino acids into oxidative pathways (Walton, 1985). This hypothesis has not been tested and the K_m for any tRNA synthetases has not been measured in fish. In spite of this, the pool size of free amino acids has been suggested to play an important role in regulating protein synthesis (Houlihan et al., 1993).

Most previous reviews of amino acid metabolism in fish have focused primarily on the applied nutritional aspects as they relate to aquaculture (Tacon and Cowey, 1985; Walton, 1985; Cowey and Walton, 1989; Wilson, 1989; Cowey, 1993). More recently Jurss and Bastrop (1995) reviewed other aspects of fish amino acids metabolism. This current review examines some aspects of the intermediary metabolism of amino acids in different tissues of fish with particular reference to the factors influencing and controlling the pathways involved.

II. DIGESTION AND UPTAKE OF AMINO ACIDS

Proteolysis is initiated in the stomach by pepsin, with further hydrolysis of peptides and aminopeptides taking place in the lumen of the intestine by the action of trypsin, chymotrypsin, and leucine aminopeptidase. In the rabbitfish all of these enzymes have alkaline pH optima, as in mammals (Sabapathy and Teo, 1995).

Amino acids are taken up from the gut as free amino acids or as dipeptides. Dipeptides are transported independent of cation gradients with hydrolysis to their component amino acids taking place immediately after transport (Reshkin and Ahearn, 1991). Plasma free amino acid levels peak about 12 h after feeding and then return to prefeeding values (Murai *et al.,* 1987; Navarro *et al.,* 1997).

Adaptation of the intestinal transport system to different dietary levels of amino acids is known to occur in fish. For example, the K_m for the transporters of amino acids in the guts of herbivorous and omnivorous fishes are lower than those for carnivores, presumably an adaptation to the lower gut levels of amino acids (Ferraris and Ahearn, 1984). Four sodium-dependent amino acid carrier systems have been identified in the intestinal brush-border membrane of the eel: (1) a carrier for the cationic amino acids, lysine and arginine; (2) an anionic carrier for glutamate and aspartate; (3) a carrier for proline and *N*-methylated amino acids; and (4) a neutral amino acid carrier for alanine, glycine, serine, cysteine, and others (Storelli *et al.,* 1989).

Transfer of absorbed amino acids to the blood side occurs at the basolateral surface of intestinal cells. The basolateral membrane has at least three amino acid carriers (Reshkin *et al.,* 1988; Collie and Ferraris, 1995). Alanine, lysine, and phenylalanine are transported by sodium-independent carriers (Reshkin *et al.,* 1988), proline and glutamate transport occur by the same sodium-dependent carrier, and glycine transport uses the outward potassium gradient and inward sodium gradient (Reshkin *et al.,* 1988). Due to the direction of the electrochemical gradients for sodium and potassium, these carriers likely function to bring amino acids into the cell rather than the reverse (Reshkin *et al.,* 1988). This may be necessary to provide intestinal cells with synthetic precursors and oxidative substrates between periods of feeding. Future research should be directed toward understanding the regulation and coupling of amino acid transport and metabolism in the intestine.

III. DELIVERY SYSTEMS

Whether they are newly arrived from the intestine or derived from proteolysis in other tissues, amino acids are delivered to most cells as free amino acids by the

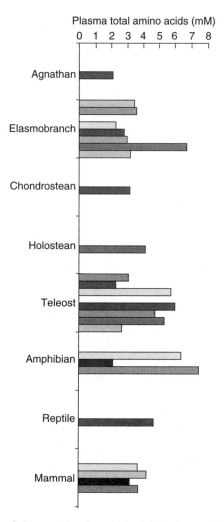

Fig. 1. Comparison of plasma total amino acid levels in various species of vertebrates. Values from top to bottom for each species are given by the following references. *Agnathan:* Sea lamprey, *Petromyzon marinus* (FW) (LeBlanc *et al.,* 1995). *Elasmobranch:* dogfish shark, *Squalus acanthias* (Chamberlin and Ballantyne, 1992); black tip reef shark, *Carcharhinus melanopterus* (Ballantyne, unpublished data); grey reef shark, *Carcharhinus amblyrhynchos* (Ballantyne, unpublished data); epaullete shark, *Centroscyllium fabricius* (Ballantyne, unpublished data); tawny nurse shark, *Nebrius ferrugineus* (Ballantyne, unpublished data); and blue spotted ray, *Taeniura lymma* (Ballantyne, unpublished data). *Chondrostean:* lake sturgeon, *Acipenser fulvescens* (Gillis and Ballantyne, 1996). *Holostean:* Florida gar, *Lepisosteur platyrhincus* (Frick, Bystriansky, and Ballantyne, unpublished data). *Teleost:* Arctic char, *Salvelinus alpinus* (Bystriansky and Ballantyne, unpublished data); coho salmon, *Oncorhynchus kisutch* (Ogata and Arai, 1985); rainbow trout, *Oncorhynchus mykiss* (Ogata and Arai, 1985); goldfish, *Carassius auratus* (Carrillo *et al.,* 1980, or van der Boon *et al.,* 1992); carp, *Cyprinus*

blood. The levels of circulating amino acids differ slightly in different parts of the circulation with levels being slightly higher in the hepatic portal vein compared to the more routine sampling site of the caudal vein/artery complex (Murai *et al.,* 1987). Levels of plasma total free amino acids in fish vary little from species to species and do not differ substantially from those of other vertebrates (Fig. 1). The turnover of amino acids in the plasma of fish, however, is relatively high compared to that of other vertebrates. In the kelp bass, 14% of the free amino acid pool is replaced every minute (Bever *et al.,* 1981). This is of similar magnitude to that of the rat, but considering the much lower metabolic rates of fish, the proportional metabolic utilization of amino acids is much higher (Bever *et al.,* 1981; Weber and Zwingelstein, 1995). This attests to the efficiency of the delivery system for amino acids, but also indicates the metabolic importance of amino acids in fish.

The pool of free amino acids in erythrocytes can represent a significant proportion (15–30%, depending on the species) of the amino acids in whole blood (Ogata and Arai, 1985). According to Ogata and Arai certain amino acids, such as arginine, lysine, and methionine, are transported mainly in the plasma. Other amino acids, such as aspartate, glutamate, and taurine, are transported primarily in the erythrocytes and others, such as tyrosine, phenylalanine, and tryptophan, are transported in both. The uptake of amino acids by erythrocytes is regulated by insulin in fish (Canals *et al.,* 1995).

Plasma glutamine levels in fish are lower than those of higher vertebrates, and of the fishes, elasmobranchs have the lowest levels (Fig. 2). It was previously thought that plasma glutamine levels of elasmobranchs were lower than those of any other vertebrate (Ballantyne, 1997). However, this was based on a small number of measurements of captive animals; more recent investigations of plasma glutamine in freshly caught elasmobranchs indicate that, although levels are low, they overlap those of many other teleost fishes (Fig. 2). In elasmobranchs, the substantial requirement of glutamine for urea synthesis likely contributes to low plasma levels, especially in captivity when dietary intake of amino acids may be low.

Some aspects of the transport of amino acids in the plasma of fish differ from those of higher vertebrates. For example, unlike birds and mammals that utilize albumin to transport part of the plasma pool of tryptophan, fish and other cold-

---

carpio (Ogata and Arai, 1985); northern pike, *Esox lucius* (Bystriansky, Rosenberger, and Ballantyne, unpublished data); and brown bullhead, *Ictalurus nebulosus* (Ogata and Arai, 1985). *Amphibian:* wood frog, *Rana sylvatica* (Storey *et al.,* 1992); African clawed frog, *Xenopus laevis* (Balinsky *et al.,* 1967); and African clawed frog, *Xenopus laevis* (Unsworth and Crook, 1967). *Reptiles:* American alligator, *Alligator mississippiensis* (Smith and Campbell, 1987); and American alligator, *Alligator mississippiensis* (Lemieux *et al.,* 1984). *Mammal:* rat, *Rattus norvegicus* (Carbo *et al.,* 1994); human, *Homo sapiens* (Mantagos *et al.,* 1989); horse, *Equus caballus* (Silver *et al.,* 1994); and black bear, *Ursus americanus* (Wright *et al.,* 1999).

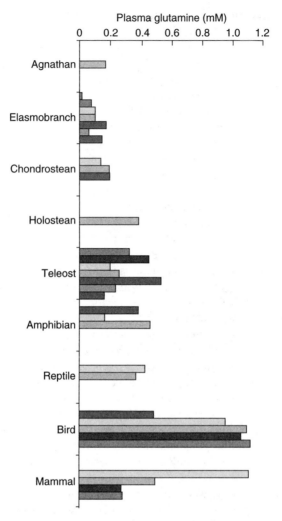

Fig. 2. Comparison of plasma glutamine levels in various species of vertebrates. Values from top to bottom for each species are given by the same references given in Fig. 1, except for birds and mammals, which follow. *Birds:* rock dove, *Columba livia* (Shihabi *et al.*, 1989); goose, *Anser* (Sitbon *et al.*, 1982); chicken, *Gallus domesticus* (Elkin *et al.*, 1988); mallard duck, *Anser platyrhynchos* (Elkin *et al.*, 1988); and turkey, *Meleagris gallopavo* (Elkin *et al.*, 1988). *Mammals:* rat, *Rattus norvegicus* (Henrickson, 1991); human, *Homo sapiens* (Graham *et al.*, 2000); horse, *Equus caballus* (Silver *et al.*, 1994); and black bear, *Ursus americanus* (Wright *et al.*, 1999).

blooded vertebrates transport this amino acid only in the free form (Fellows and Hird, 1982). It has been proposed that the development of a binding site for tryptophan on albumin arose to regulate the amount of free tryptophan accessible to the brain (Fellows and Hird, 1982). Tryptophan is a precursor for 5-hydroxytryptamine, the hormone involved in regulating sleep (Yamamoto et al., 1997). The development of a method for regulating plasma tryptophan levels with a separate binding site on albumin may have been a requirement for animals such as mammals with very high metabolic rates. Although some fish species appear to "sleep" (personal observation), the biochemical factors regulating this have not been examined.

IV. TISSUES

A. Liver

1. CATABOLISM

The liver is the main organ for amino acid catabolism and this may explain the high concentrations of free amino acids in liver tissue (Van den Thillart and van Raaij, 1995). Amino acids are more important oxidative substrates than lipids in the liver (French et al., 1981). The rate of oxidation of amino acids exceeds that of their role in other pathways, such as gluconeogenesis. For example, in Atlantic salmon the rate of oxidation of alanine exceeds its rate of incorporation into glucose by more than 10-fold (Mommsen et al., 1985) and in skipjack tuna hepatocytes, alanine is oxidized at rates 5 times the rate of lactate (Buck et al., 1992).

Amino acid catabolism results in the production of ammonia in fish (Chapter 4). It has been estimated that the liver is responsible for 50–70% (Van den Thillart and van Raaij, 1995) or as much as 99% (van Waarde, 1981) of ammonia production in goldfish. Several mechanisms exist for the production of ammonia from amino acids. The purine nucleotide cycle is thought to contribute little to ammonia production in liver, because hadicidin, an inhibitor of the purine nucleotide cycle, has no effect on ammonia production by isolated hepatocytes of goldfish (van Waarde and Kesbeke, 1981). The primary mechanism for catabolism of amino acids in fish liver is transdeamination in which the amino group of a variety of amino acids is transferred to α-ketoglutarate to form glutamate, which is then deaminated by glutamate dehydrogenase (GDH). GDH is a mitochondrial enzyme in fish and is allosterically activated by ADP, AMP, and leucine and inhibited by ATP and GTP (Jurss and Bastrop, 1995). Thus, during periods when ATP levels are low and ADP is high, oxidative deamination of amino acids will be higher. Leucine preferentially activates the aminating direction (Bittorf, 1985, as cited by Jurss and Bastrop, 1995), but the physiological significance of this remains to be determined. Direct deamination of certain specific amino acids involves enzymes such as histidase, asparginase, serine dehydratase, threonine dehydratase, and glutaminase, as well as GDH.

The subcellular distribution of two of the deaminating enzymes is known. Because glutaminase and GDH are both located in the mitochondria, ammonia is produced intramitochondrially. Ammonia may exit the mitochondrial matrix as NH_4^+ (Campbell and Vorhaben, 1983), rather than as NH_3, since alkalinization of the medium does not occur in mitochondria oxidizing glutamate. The nature of the carrier remains to be determined.

Comparisons of the rates of oxidation of various amino acids by isolated hepatocytes and hepatocyte mitochondria have been used to establish which steps limit the catabolism of amino acids in liver. In catfish hepatocytes, ammonia production from amino acids is greatest with asparagine and lower with other amino acids in the order glutamine > alanine, serine > aspartate, glutamate. Isolated mitochondria display a different order of preference, with glutamate oxidized at higher rates than alanine, serine, and asparagine (Campbell and Vorhaben, 1983). This indicates that transport of glutamate into the hepatocyte is rate limiting and, thus, very little exogenous glutamate is catabolized. Most *in vivo* catabolism of glutamate is likely due to transamination of other amino acids in the cytosol or mitochondrion. Asparagine is oxidized at higher rates in isolated hepatocytes than in mitochondria, indicating that cytosolic components of its catabolism or mitochondrial transport are limiting, although the subcellular distribution of asparginase is uncertain (Jurss and Bastrop, 1995).

Glutamate metabolism in fish differs from that of mammals. Glutamate is primarily deaminated in fish with production of ammonia, while in mammals most glutamate is transaminated to aspartate (Walton and Cowey, 1977; Campbell and Vorhaben, 1983). The rate of glutamate deamination by isolated catfish liver mitochondria can account for the total ammonia excreted (Campbell and Vorhaben, 1983). The metabolic fate of glutamate does vary with species. Glutamate is oxidized primarily via GDH in liver of goldfish (van Waarde and Henegouwen, 1982; van Waarde and de Wilde-Van Berghe Henegouwen, 1982), rainbow trout (Walton and Cowey, 1977) and lungfish (Janssens and Cohen, 1968), whereas in trout and catfish liver mitochondria 40% of the glutamate is transaminated to aspartate with the remainder deaminated (Campbell and Vorhaben, 1983).

The early studies of serine metabolism in fish established the liver as the main site of catabolism (Walton and Cowey, 1981), and this occurs primarily via reactions catalyzed by serine hydroxymethyltransferase and serine pyruvate transaminase (Walton and Cowey, 1981). Subsequent studies using transaminase inhibitors such as aminooxyacetate confirmed that serine is not transdeaminated in the mitochondria of catfish hepatocytes (Campbell and Vorhaben, 1983).

Due to the central importance of the liver in amino acid catabolism, a more detailed understanding of the pathways and regulation of the degradation of other amino acids, particularly glutamine, in this tissue is needed.

2. GLUCONEOGENESIS

Liver is an important site of gluconeogenesis in fish, with amino acids being the main source of carbon. As is the case in mammals (Newsholme and Leech, 1983), gluconeogenesis from alanine is quantitatively the most important pathway in fish (French *et al.*, 1981). The rates of glucose synthesis from amino acids by isolated trout hepatocytes occur in the order serine > aspatagine > alanine > glycine > proline > valine (French *et al.*, 1981). Although it has not been established how these relate to the *in vivo* rates, it suggests the possibility that serine is more important than alanine for gluconeogenesis in fish. The importance of serine has been downplayed due to low plasma levels of serine (Suarez and Mommsen, 1987). Fish differ from most mammals in the pathway for gluconeogenesis from serine. Serine transaminase is present at high levels in fish liver, but serine dehydratase activity is insignificant (Suarez and Mommsen, 1987). Thus, it is thought that in fish gluconeogenesis from serine occurs via serine pyruvate transaminase and glycerate dehydrogenase rather than by the serine dehydratase pathway that predominates in mammals (Walton and Cowey, 1981). The resulting production of D-glycerate in fish bypasses the reactions catalyzed by pyruvate carboxylase (PC) and phosphoenolpyruvate carboxykinase (PEPCK), which are the rate-limiting steps in the gluconeogenic pathway for serine in mammals (Suarez and Mommsen, 1987). This may explain the high rates of gluconeogenesis from serine in fish compared to mammals.

The relative importance of amino acids for gluconeogenesis may vary depending on the availability of lactate. Recovery from periods of anoxia or exercise requires resynthesis of glycogen from lactate (Hochachka and Somero, 1984). The rates of gluconeogenesis from amino acids in trout hepatocytes are slower compared to the rate with lactate as the carbon source (French *et al.*, 1981). On the other hand, gluconeogenesis from alanine is much greater than that from lactate in skipjack tuna hepatocytes (Buck *et al.*, 1992).

The incorporation of amino acid carbon into glucose is rapid, but not all of the carbon actually ends up in glucose in some fish species (Bever *et al.*, 1981). It has been suggested that a significant portion of the carbon from gluconeogenic amino acids is utilized for mucopolysaccharide production for mucus (Bever *et al.*, 1981; Weber and Zwingelstein, 1995). Fauconneau and Tesseraud (1990) estimated that 11% of ingested nitrogen is utilized for mucous protein production in rainbow trout. This is a significant drain on nitrogen resources. The importance of this in other fish species has not been studied. It would be interesting to know what the cost of mucous production is in very active fish such as tuna and in terrestrial fish such as mudskippers where mechanical loss of mucus may be high.

3. LIPOGENESIS

The liver is the primary site of lipid synthesis in fish. The few existing studies of the role of amino acids in lipogenesis are consistent in showing that amino acids are the preferred carbon source for lipogenesis in liver (Henderson and Tocher, 1987). Both alanine (Nagai and Ikeda, 1973; Henderson and Sargent, 1981) and leucine (van Raaij et al., 1994) are incorporated into lipids at higher rates than glucose in rainbow trout, carp, and goldfish. Considering the importance of both lipids and amino acids to the energetics of fish, further research into the coupling of these two metabolic systems is needed.

B. Gill

The gill is the primary site of ammonia excretion in fish (Chapter 4). The origin of the ammonia excreted by this tissue is not entirely derived from other tissues. The gill is a metabolically active tissue requiring 7% of a fish's total oxygen consumption (Mommsen, 1984). The quantitative importance of ammonia production by the gill is controversial, as is the identity of the source of amino acids for deamination. Alanine is oxidized to CO_2 in the rainbow trout gill at significant rates (Mommsen, 1984). Mommsen reports that glutamate, glutamine, alanine, aspartate, serine, and histidine are not taken up by the gill. Walton and Cowey (1977), however, provided evidence for glutamate uptake from the blood by the gill of rainbow trout. Current estimates for the role of the gill in ammonia production are low (Pequin and Serfaty, 1963; Goldstein et al., 1964; Payan and Matty, 1975; Cameron and Heisler, 1983) or none at all (Payan and Pic, 1977).

Enzyme activities such as that of phosphate-dependent glutaminase (PDG) (Goldstein and Forster, 1961; Chew and Ip, 1987; Ip et al., 1990; Chamberlin et al., 1991), transaminase (at high levels; Mommsen, 1984), and GDH (Chamberlin et al., 1991; Chamberlin and Ballantyne, 1992) are present in the gill. The high K_m for ammonia of gill GDH led Fields et al. (1978) to suggest that the enzyme is only involved in deamination. GDH activity in gill is moderate, but the enzyme is regulated in a similar way to that of other fish GDHs (Mommsen, 1984).

The gill can also synthesize glutamine. Glutamine synthetase (GS) is present in gills of a freshwater holostean and a teleost (Chamberlin et al., 1991). The presence of both GS and glutaminase provides a potential futile cycle that may have value in providing sensitive control of flux of glutamine and hence release or uptake of NH_4^+ (Mommsen, 1984). Certainly, the organization and regulation of glutamine metabolism in the fish gill need to be examined in greater detail.

C. Kidney

Amino acids are important substrates for the fish kidney for both oxidation and gluconeogenesis. Kidney is the second most important tissue, after the liver, for gluconeogenesis from amino acids (Jurss and Bastrop, 1995). Little is known of the pathways and regulation of this process in fish. The rate of oxidation of alanine is more than 10-fold higher than the rate of incorporation into glucose in

Atlantic salmon kidney (Mommsen et al., 1985). The levels of both PDG and GS are high in the kidneys of fish (King and Goldstein, 1983a; Chamberlin et al., 1991). King and Goldstein (1983a) have proposed a substrate cycle involving glutaminase and GS that regulates ammonia production in the kidney of the dogfish shark. This pathway would involve uptake of glutamine by the kidney from the plasma and degradation to glutamate and α-ketoglutarate catalyzed by PDG and GDH. Resynthesis of glutamate and glutamine would occur via reactions catalyzed by GDH and GS. The extent to which one direction predominates would be a function of the up- or downregulation of the two enzymes. Based on kinetic analyses of GDH from the kidneys of Amazonian fishes, Storey et al. (1978) concluded that the enzyme functions in a glutamate-oxidizing direction. In agreement with this, the kidney form of GDH differs from the liver form in that the reductive amination direction is not as highly activated by leucine as is the oxidative deaminating direction (Jurss et al., 1985). High levels of leucine, perhaps indicative of high circulating amino acid levels, may enhance ammonia production in the kidney. In acidotic mammals, circulating glutamine arriving at the kidney is fully oxidized to NH_4^+ and HCO_3^-; NH_4^+ is excreted in the urine, whereas HCO_3^- is reabsorbed (Knepper et al., 1989). The role of glutamine in acid–base regulation in fish kidney has not been established; glutamine levels in the kidney of goldfish exposed to ammonia do not increase (Levi et al., 1974).

D. Blood

Walsh et al. (1990) concluded that amino acids are not important oxidative fuels in the erythrocytes of most fish. GDH and glutamate pyruvate transaminase (GPT) activities are undetectable in erythrocytes of rainbow trout and the levels of glutamate oxaloacetate transaminase (GOT) activity are low (2.2 units/ml) (Gaudet et al., 1975). The transport of amino acids into erythrocytes has been reviewed by Walsh et al. (1998). Carp erythrocytes oxidize exogenous glutamine and aspartate but at lower rates than some carbohydrate fuels such as lactate, pyruvate, and glucose (Tihonen and Nikinmaa, 1992). Lungfish erythrocytes, on the other hand, oxidize glutamate at higher rates than that of glucose (Mauro and Isaacks, 1989). The role of amino acids, especially glutamine, in the metabolism of fish leukocytes differs in several ways from that of mammals. Thrombocytes from rainbow trout use glucose and glutamine for less than 3% of their ATP production, which contrasts with the situation in mammalian platelets where glutamine is a major energy source (Guppy et al., 1999). In fish thrombocytes, 80% of the energy source was unaccounted for and could have been other amino acids. Oxidation of glucose by peripheral lymphocytes is not affected by glutamine in fish and this also differs from the situation in mammals (Albi et al., 1993). In addition, fish leukocytes stimulated by invading bacterial lectins do not require glutamine for proliferation (Ganassin et al., 1998). In spite of the apparent lack of need for glutamine by cells of the immune system in fish, plasma glutamine levels fall when fish are ill, as occurs in mammals (Walker et al., 1996). This may be more related to the reduced protein intake during illness than to an immune response.

E. Muscle

Due to its mass (~50% of body weight), white muscle of fish contains not only the largest pool of protein but also the largest pool of free amino acids. The pool size of free amino acids in white muscle may vary depending on growth rate. Free amino acid levels in white muscle of rapidly growing juvenile fish are lower than those of adults due to increased utilization for protein synthesis (Siddiqui et al., 1973).

Similar to the situation in mammals, fish muscle is the main site of oxidation of branched-chain amino acids (van der Boon et al., 1992). Amino acid metabolism in muscle of fish, however, differs from that of mammals in several respects. Mammalian muscle synthesizes glutamine that is then taken up by other tissues, such as the intestine and kidney. This is not the case in fish. The levels of GS are much lower in fish muscle than in mammalian muscle (Chamberlin et al., 1991), which in turn would affect the production of glutamine. Alanine, not glutamine, is released from muscle of elasmobranchs (Leech et al., 1979). Furthermore, mammalian muscle does not catabolize glutamine to a significant extent, whereas in fish, glutamine has been suggested to be an important oxidative substrate in red muscle (Chamberlin et al., 1991; Chamberlin and Ballantyne, 1992).

The pathways for catabolism of glutamine and other amino acids in fish are designed for efficiency. Malic enzyme may play a role in the catabolism of glutamine in fish red muscle by channeling some of the flux from α-ketoglutarate toward providing pyruvate for acetyl CoA production (Chamberlin et al., 1991) (Fig. 3). By having both 2- (acetyl) and 4-carbon (oxaloacetate) compounds provided by glutamine, Krebs cycle activity can be maintained without carbon input from other sources. A similar pathway could occur for any amino acids being transdeaminated via GDH. This strategy is more efficient in that amino acids can be catabolized alone with no need to use stored carbohydrate or lipid fuels to provide 2- or 4-carbon intermediates to maintain Krebs cycle flux. The cost of synthesis of these storage fuels would thus be saved. If such "autocatalytic" amino acid catabolism occurred in other tissues the overall energy savings may be significant.

The relative importance of amino acids such as glutamine as fuels in red muscle varies in different fish groups. Elasmobranchs have much higher levels of PDG in red muscle than other vertebrates (Ballantyne, 1997). This may be due to the inability of elasmobranchs to use lipid in red muscle (Ballantyne, 1997). On the other hand, teleost fish catabolize lipid in red muscle and this may reduce the

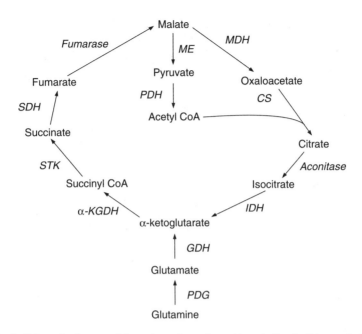

Fig. 3. Schematic diagram of the pathway for amino acid catabolism in fish muscle. The abbreviations refer to the following enzymes: ME, malic enzyme; PDH, pyruvate dehydrogenase; CS, citrate synthase; IDH, isocitrate dehydrogenase; GDH, glutamate dehydrogenase; PDG, phosphate-dependent glutaminase; α-KGDH, α-ketoglutarate dehydrogenase; STK, succinyl CoA thiokinase; SDH, succinate dehydrogenase.

reliance on amino acids in this tissue. Thus, in lake char (*Salvelinus namaycush*), red muscle mitochondria catabolize glutamine at lower rates than that of lipid substrates (Chamberlin *et al.*, 1991). In contrast, in isolated red muscle mitochondria of the Holostean bowfin (*Amia calva*), glutamine catabolism exceeds that of lipid substrates and is second only to that of pyruvate in oxidation rate (Chamberlin *et al.*, 1991).

Studies using isolated mitochondria indicate that exogenous glutamate does not have the same metabolic fate as exogenous glutamine. Glutamine oxidation is inhibited by only 20% in bowfin and lake char while glutamate oxidation is inhibited 86% by the transaminase inhibitor aminooxyacetate (Chamberlin *et al.*, 1991). Similarly, it has been reported that exogenous glutamate is oxidized mainly via transaminases in mitochondria isolated from red muscle of goldfish, a conclusion based on studies using the transaminase inhibitor aminooxyacetate (van Waarde and Henegouwen, 1982). It has been estimated that GDH accounts for about one half of the ammonia production of goldfish muscle with the purine nucleotide cycle accounting for the remainder (van Waarde, 1981). Glutamine oxidation by isolated muscle mitochondria from the mudskipper *Periophthalmus* is inhibited by aminooxyacetate and bromofuroate, an inhibitor of GDH, indicating catabolism by a transaminase and GDH, respectively (Ip *et al.*, 1990).

Gluconeogenesis from amino acids does not take place in the muscle itself. Although white muscle is the main source of alanine released into the blood during exercise, alanine is taken up by the liver for gluconeogenesis (Milligan, 1997). Free alanine levels in white muscle are usually higher than those of other amino acids, except taurine.

Proteolysis in muscle provides the free amino acids required for metabolic functions in other tissues. The enzymes of muscle proteolysis have been examined in a few species of fish. Calpain (E.C. 3.4.22.17), a calcium-dependent cysteine protease, is found in two forms (I and II) in carp muscle. Toyohara and Makinodan (1989) report that the concentrations of calcium required for activation of calpain in carp muscle are much higher than those required in mammals. They also report that the pH optima for both forms of calpain are in the physiologic range and the endogenous inhibitor of calpain, calpastatin, is also present in skeletal muscle tissue.

The levels of proteolysis in muscle vary with metabolic state and are higher during periods of reproduction (von der Decken, 1992; Martin *et al.*, 1993). This is due to elevated muscle protease activity (Ando *et al.*, 1986), which correlates with reduced levels of plasma trypsin inhibitor (Ando *et al.*, 1985). The muscle proteases, cathepsin, cathepsin D, and carboxypeptidase A, also increase in muscle of coho salmon during the spawning migration (Mommsen *et al.*, 1980). In spite of the excellent studies outlined above, the coupling of proteolysis and amino acid metabolism in fish muscle is not well understood. Further research is needed to establish how the balance between protein synthesis and degradation and the utilization of amino acids as energy sources is regulated.

F. Brain

Some amino acids are neurotransmitters (e.g., glutamate) and, therefore, their metabolism in the brain differs from that of other tissues. Brain integrity is defended against ammonia toxicity by high levels of GS. This enzyme reduces brain ammonia levels by aminating glutamate to glutamine. The regions of the brain of *Squalus acanthias* differ in the activity of GS (cerebellum > medulla oblongata > cerebrum > optic lobe > spinal cord) (Webb and Brown, 1980). Glutamine levels in brain of goldfish increase 10-fold with ammonia exposure, while other tissues showed much smaller increases (Levi *et al.*, 1974).

G. Intestine

The intestine is an important site of amino acid absorption and influences the availability of amino acids for other tissues. The intestine not only transports

3. AMINO ACID METABOLISM

amino acids from the gut lumen, but also uses them as endogenous energy sources. Although GDH activities in intestinal mucosa are about a fourth of those of liver in rainbow trout (Auerswald *et al.*, 1997), this still represents a significant catabolic capacity. The metabolism of fish intestine resembles that of mammals with respect to its reliance on glutamine as an oxidative substrate. Catabolism of exogenous amino acids, such as alanine and glutamine, is more important than exogenous glucose in providing the energy for transport in this tissue (Ando, 1988). Both alanine and glutamine stimulate Na^+,K^+-ATPase activity and respiration in isolated gut and elevate the potential difference across the tissue, whereas glucose does not (Ando, 1988). Exogenous alanine is catabolized via GPT because aminooxyacetate inhibits the stimulatory effects on the potential difference across the gut (Ando, 1988). The pathway for glutamine degradation in the intestine has not been established.

V. EXERCISE

Based on oxidation rates of radiolabeled substrates, Van den Thillart (1986) concluded that oxidation of amino acids accounts for 80% of the energy requirements at rest and 90% during sustained swimming with lipid catabolism accounting for the remainder. The ammonia quotient (moles ammonia produced/moles oxygen used) increases fourfold during exercise in *Tilapia mossambica*, which is indicative of elevated amino acid catabolism (Kutty, 1972). Recent studies, however, dispute the importance of amino acids as oxidative substrates during exercise. Utilization of amino acids as an energy source during exercise accounts for only 30% of the fuel utilization in juvenile rainbow trout and this does not change over a range of swimming speeds (Lauff and Wood, 1996a; Kieffer *et al.*, 1998). The discrepancies between these studies may relate to the degree of stress as mediated by cortisol effects on protein mobilization (Mommsen *et al.*, 1999).

Catabolism of amino acids postexercise may be important for recovery of ATP levels required for gluconeogenesis (Milligan, 1997). After exhaustive exercise, plasma amino acid levels rise in response to release of cortisol and provide substrates for gluconeogenesis in the liver (Milligan, 1997).

VI. HORMONAL REGULATION

A. Cortisol

The metabolic effects of cortisol have been recently reviewed by Mommsen *et al.* (1999). A variety of aspects of amino acid metabolism are affected by this hormone, including effects on protein synthesis, plasma amino acid levels and tissue enzyme and transporter activity. The action of cortisol in elevating plasma amino acids is rapid (minutes to hours) and likely acts through elevated proteolysis (Milligan, 1997). The liver may be the primary target for cortisol-mediated release of amino acids with branched-chain amino acids being the main amino acids released (Milligan, 1997). In fed fish treated with cortisol, elevated plasma amino acid levels may also be due to higher rates of transport of amino acids in the gut (Schreck, 1993). In liver, the metabolic consequences of these changes in enzyme activity and transport include enhanced gluconeogenesis from amino acids in rainbow trout (Vijayan *et al.*, 1994) and elevated amino acid oxidation rates in sea ravens (*Hemitripterus americanus*) (Vijayan *et al.*, 1993).

It has been suggested that some of the effects of cortisol on protein synthesis are mediated via changes in plasma triiodothyronine (T_3) levels (Mommsen *et al.*, 1999). These effects may be due to altered rates of enzyme synthesis. The effects of cortisol on the activities of specific enzymes of amino acid metabolism are species and tissue specific. Cortisol enhances liver arginase activity in sea raven (Vijayan *et al.*, 1996) and conditions that elevate cortisol levels induce cytosolic GS activity in liver (Walsh and Milligan, 1995) and brain (Mommsen *et al.*, 1999). Elevated GOT, GDH, and GS (but not GPT), tyrosine aminotransferase, ornithine carbamoyltransferase, and arginase activities were observed with dexamethasone (a cortisol analog) treatment in isolated hepatocytes of toadfish (*Opsanus beta*) (Mommsen *et al.*, 1992). Cortisol treatment, however, had no effect on the activities of GDH, GOT, and GPT in trout liver (Anderson *et al.*, 1991). Cortisol also induces tyrosine aminotransferase in some fish species (Whiting and Wiggs, 1977; Davis *et al.*, 1985), but not in others (Mommsen *et al.*, 1992). Some discrepancies have been rationalized as differences in the method of administration of the hormone (Mommsen *et al.*, 1999), but interspecies differences may also exist.

B. Insulin and Glucagon

Insulin plays an important role in mediating the dynamics of amino acid metabolism in fish (Matty and Lone, 1985; Christiansen and Klungsoyr, 1987). As is the case with cortisol, some of the effects involve changes in amino acid transport. Insulin stimulates the uptake of amino acids in liver, as demonstrated in isolated hepatocytes (Plisetskaya *et al.*, 1984; Canals *et al.*, 1995), but there is a decrease in the uptake of amino acids by erythrocytes (Canals *et al.*, 1995). Insulin treatment increases protein synthesis in muscle (Ince and Thorpe, 1976) and reduces liver gluconeogenesis from alanine in rainbow trout (Cowey *et al.*, 1977). The net effect of these processes would be to reduce plasma amino acid levels (Inui *et al.*, 1975; Ince and Thorpe, 1978; Matty and Lone, 1985).

Amino acids themselves may mediate insulin release. Some amino acids such as arginine, lysine, leucine, and phenylalanine stimulate insulin release (Ince and Thorpe, 1977; Matty and Lone, 1985; Plisetskaya *et al.*, 1991), whereas others, such as histidine, inhibit insulin release (Ince and Thorpe, 1977; Plisetskaya

3. AMINO ACID METABOLISM 93

et al., 1991). The rationale for the effects of these specific amino acids has not been established, although the fact that all are essential but one (arginine) may be significant.

Glucagon reduces plasma amino acid levels and stimulates protein synthesis in liver (Matty and Lone, 1985). Glucagon injection elevates liver GOT and GS activities in rainbow trout, but has no effect on GDH, GPT, arginase, tyrosine aminotransferase, and ornithine transcarbamylase activities (Mommsen *et al.*, 1992).

C. Thyroid Hormones

The effects of thyroid hormones may be more related to the overall effects of these hormones in mediating metabolic rates via altered protein synthesis than in specifically targeting amino acid metabolism. T_3, but not thyroxine (T_4), stimulates protein synthesis in liver and gill of brook char (*Salvelinus fontinalis*) and rainbow trout (Narayansingh and Eales, 1975). T_4-treated goldfish display elevated ammonia excretion rates, muscle and liver free amino acid levels, and leucine uptake by muscle but not by liver (Thornburn and Matty, 1963). T_4 injection also increases plasma and tissue amino acids in several other fish species (Plisetskaya *et al.*, 1983). The effects of T_3 can be quite rapid. Injected T_3 elevates GDH activity in liver of dogfish after 3 h (Battersby *et al.*, 1996).

VII. DIURNAL RHYTHMS IN AMINO ACID METABOLISM

Plasma amino acids undergo diurnal changes (Carrillo *et al.*, 1980), as would be expected due to their importance in the overall metabolism of fish. In most cases, amino acid levels increase as the photoperiod progresses, peaking at the onset of the dark period and then declining through the dark period (Carillo *et al.*, 1980). The changes in amino acid concentration can be substantial. In goldfish, plasma glutamine levels increase more than fourfold from the onset of the photoperiod to the onset of the dark period (Carrillo *et al.*, 1980). Catabolism of glutamine also displays a circadian rhythm in liver (Vellas *et al.*, 1982). Activities of several enzymes of amino acid catabolism (GDH, glutaminase, arginase, uricase) display variations in activity related to circadian rhythms (Vellas *et al.*, 1982). Changes in the activities of these enzymes correlate with nitrogen excretion rates in rainbow trout. Peaks in plasma ammonia coincide with peaks in liver glutaminase activity and peaks in plasma urea coincide with peaks in liver arginase activity (Vellas *et al.*, 1982). Although the circadian rhythms may be entrained to feeding times rather than photoperiod, these studies illustrate the dynamic nature of the regulation of amino acid metabolism.

VIII. DIET AND STARVATION

The dietary protein requirements of fish are higher than those of mammals because of the substantial reliance on amino acids as energy sources (Cho and Kaushik, 1985). A high protein content in the diet not only allows high rates of protein deposition but also allows amino acid carbon to be channeled into other metabolic pathways. Lipogenesis from amino acids increases in carp fed high-protein diets (Nagai and Ikeda, 1973). Conversely, a high lipid content in the diet has a sparing effect on amino acid catabolism with a resultant increase in the amount of amino acid available for growth (Atherton and Aitken, 1970).

The specific amino acid composition of the diet can also influence the levels of free amino acids in tissues. The tissue levels of histidine (Knapp and Wieser, 1981) and taurine (Sakaguchi *et al.*, 1988) correlate with the amounts of these amino acids found in the diet of teleost fish. The activities of some amino acid metabolizing enzymes are responsive to the dietary content of the specific amino acid they metabolize. For example, liver and kidney arginase activities increase with dietary arginine levels in Atlantic salmon, *Salmo salar* (Lall *et al.*, 1994). The plasma levels of serine and alanine and of the essential amino acids lysine, histidine, arginine, and methionine are sensitive to dietary protein content in sturgeon (Kaushik *et al.*, 1994). The protein content in the diet also affects the activity of some enzymes related to amino acid metabolism such as tyrosine aminotransferase in brook trout (Whiting and Wiggs, 1977).

Many fish naturally undergo food deprivation, sometimes for prolonged periods. A reduction in overall metabolic rate may occur and the catabolism of amino acids may be reduced as a consequence. The use of protein as a metabolic fuel during starvation accounts for only 24% of the energy needs of rainbow trout even though total body protein content declines by 66.5% (Lauff and Wood, 1996b). This apparent discrepancy has been attributed to redistribution of protein carbon into lipid and carbohydrate (Lauff and Wood, 1996b). Enhanced gluconeogenesis from amino acids is also observed in hepatocytes isolated from food-deprived rainbow trout (French *et al.*, 1981; Suarez and Mommsen, 1987). Alanine is the main gluconeogenic precursor in nonfeeding migrating Pacific sockeye salmon (French *et al.*, 1983). Increased gluconeogenesis from amino acids correlates with a twofold elevation of liver GOT levels in starved eels (Larsson and Lewander, 1973) and of GDH, GOT, and GPT in starved rainbow trout (French *et al.*, 1981). Other species respond differently. GDH (Jurss *et al.*, 1984), GS, and arginase (Wright, 1993) increase in liver during starvation in tilapia (*Oreochromis mossambicus*), but GOT and GPT do not (Jurss *et al.*, 1984). Some of these differences may be related to the herbivorous nature of tilapia and further research is needed into the organization and regulation of amino acid metabolism in these and other herbivorous fish.

3. AMINO ACID METABOLISM

The effects of starvation and increased proteolysis may be reflected in changes in plasma total free amino acid levels in some species. Plasma essential amino acids decline in rainbow trout and carp during the early stages of starvation (Navarro and Gutierrez, 1995). Other species do not display such changes until much later. Total plasma amino acids of starved lake sturgeon (*Acipenser fulvescens*) were found to be unchanged during the first 45 days of starvation, but glutamine and alanine levels increase (Gillis and Ballantyne, 1996). Intracellular levels of amino acids also decline during starvation. The concentrations of total amino acids in muscle of rainbow trout decrease by more than 50% after 140 days of starvation (Timoshina and Shabalina, 1970). This may indicate that the requirement for amino acids exceeds the ability of the muscle to provide these compounds and may also reflect a reduced level of protein synthesis.

IX. TEMPERATURE

Temperature has a substantial effect on the metabolism of fishes. In response to decreasing environmental temperature, one common strategy is to increase the tissue activity of enzymes by inducing synthesis of more copies of the enzyme to compensate for the effects of low thermal energy on enzyme rates (Hochachka and Somero, 1984). This occurs in some of the enzymes of amino acid metabolism. The two main transaminases, GOT and GPT, are responsive to temperature change in some teleost fishes (Jurss, 1979). GPT does not increase in trout muscle acclimated to low temperature (Jurss, 1981), but is elevated in the cold-adapted muscle of the pond loach, *Misgurnus fossilis* (Mester *et al.*, 1973). Total liver GOT increases with cold acclimation (Jurss, 1981), although this is primarily due to an increase in liver size and hepatosomatic index. Liver GDH activity is not affected by thermal acclimation in rainbow trout (Vellas *et al.*, 1982). Arginase displays complete thermal compensation in rainbow trout liver, but GDH, uricase, and PDG do not (Vellas *et al.*, 1982).

Enhanced activity of amino acid-related enzymes at low temperatures may be due to factors other than a need to maintain metabolic fluxes as temperature declines. Metabolic changes associated with low temperature (e.g., elevated lipogenesis, reproduction) may also influence amino acid metabolism. In many fish species, enhanced lipogenesis is observed at colder temperatures. Fatty acid synthesis from amino acids is elevated with cold acclimation in carp liver (Shikata *et al.*, 1995).

Another strategy for temperature acclimation is to produce different enzyme isoforms that function better at the new temperature. Some evidence for thermal isoforms of GPT was provided in a study of the pond loach (Mester *et al.*, 1973). There is some evidence that thermal isoforms of the aminoacyl-tRNA synthetases exist in eurythermal fish (Haschemeyer, 1985), presumably for the purpose of maintaining rates of protein synthesis at different temperatures.

The dietary protein requirement of fish increases at higher temperatures and it has been suggested that this is due to increased oxidation of amino acids (DeLong et al., 1958), but this is not the case for all amino acids. While oxidation of alanine increases in cold-acclimated carp liver (Shikata et al., 1995), temperature acclimation does not change the proportion of leucine oxidized by rainbow trout (63% at 10°C and 61% at 18°C) (Fauconneau and Arnal, 1985). On the other hand, at 18°C, 21% of injected leucine is oxidized to CO_2, but this decreases to 16% at 10°C in rainbow trout (Fauconneau and Arnal, 1985). Cold acclimation has little effect on the utilization of protein as an energy source in resting juvenile rainbow trout, but reduces the utilization of protein as an energy source in exercise (Kieffer et al., 1998). The studies described above indicate that temperature may differentially affect the metabolism of specific amino acids as well as the overall importance of amino acids as energy sources. Little is known of the factors responsible for regulating these responses to temperature in any fish.

X. SALINITY

Euryhaline fish species travel between freshwater and seawater. The huge shifts in external salinity cause osmotic stresses that are counteracted in several ways. Enhanced Na^+,K^+-ATPase activity in the gill plays an important role in whole-body osmoregulation (McCormick and Saunders, 1987), but the individual cells in various tissues also make metabolic and osmotic adjustments. Amino acids play two key roles during metabolic adjustments to these different environments. First, as important intracellular solutes, their levels must be adjusted to maintain cell volume (King and Goldstein, 1983b). Second, as metabolic energy sources, amino acids can be oxidized to provide ATP for osmoregulation (Ballantyne and Chamberlin, 1988).

Tissue amino acid levels rise in response to elevated salinity in some teleost fishes (Assem and Hanke, 1983). Free amino acids in muscle of rainbow trout in seawater are elevated above those of freshwater controls (Kaushik and Luquet, 1979), *T. mossambica* (Venkatachari, 1974), *Anguilla anguilla* (Huggins and Colley, 1971), flounder *Pleuronectes flesus* (Lange and Fugelli, 1965), and skate *Raja erinacea* (Forster and Goldstein, 1979; Forster and Hannafin, 1980). Other tissues such as gill, liver, heart, and kidney also display elevated free amino acid levels in seawater-acclimated fish, such as *T. mossambica* (Venkatachari, 1974).

Even stenohaline species such as carp (*Cyprinus carpio*) display elevated liver alanine, glutamate and taurine and muscle taurine, glycine, alanine, and histidine upon transfer to 1.5% seawater (Hegab and Hanke, 1983). In skate, plasma levels of amino acids do not change even though intracellular levels do (Forster and Goldstein, 1979; Forster and Hannafin, 1980).

Some species do not display significant changes in plasma or tissue free amino acids. These include coho salmon, *Oncorhynchus kisutch* (Sweeting *et al.*, 1985), and Atlantic salmon, *S. salar*, migrating between freshwater and seawater (Cowey and Daisley, 1962). Because these species are excellent osmoregulators, the need for tissue-specific responses to changing salinity may be lower than in other species.

Taurine plays a particularly important role as an intracellular osmolyte, and levels of this amino acid have been shown to rise with environmental salinity in flounder erythrocytes (Fugelli and Zachariassen, 1976) and heart (Vislie and Fugelli, 1975) and in rainbow trout intestinal mucosa (Auerswald *et al.*, 1997). Interestingly, there was no difference in tissue taurine levels in another excellent osmoregulator, the chum salmon (*Oncorhynchus keta*), in freshwater or seawater (Sakaguchi *et al.*, 1988).

The changes in the metabolism of amino acids with changing salinity are selective. In acclimation to low salinities, certain osmotically important amino acids need to be reduced in concentration. This can be accomplished by increased oxidation. Sarcosine oxidation by isolated hepatocytes and isolated mitochondria from little skates (*R. erinacea*) is osmotically sensitive with the highest oxidation rates at about 500 mOsm (Ballantyne *et al.*, 1986). Moyes and Moon (1987) suggested that the sensitivity of the glycine cleavage system of elasmobranchs and teleosts to solutes may be an important regulatory mechanism to reduce intracellular amino acid concentrations during environmental dilution.

Elevated catabolism of amino acids at higher salinities may be important to offset the additional costs of osmoregulation. Nonessential amino acids may be preferentially catabolized, since the proportion of essential amino acids increases in whole blood in seawater-exposed rainbow trout (Kaushik and Luquet, 1979). The activities of some enzymes involved in the catabolism of amino acids are elevated in seawater-acclimated fish. Arginase activities increase in liver, but not in kidney or white muscle of rainbow trout in 20% seawater (Jurss *et al.*, 1987). Jurss *et al.* (1986) showed elevated GDH in liver of rainbow trout exposed to increasing salinity. Jurss *et al.* (1985) speculated that metabolic adaptation to elevated salinity may involve regulation by metabolite activation (e.g., activation of GDH by leucine). Thus, as amino acid levels (including leucine) rise in tissues as part of the osmotic adaptation, the activity of GDH would be enhanced. This would allow increased catabolism of amino acids to provide the energy needed for osmoregulation.

XI. ANOXIA

Many fish species, including *A. anguilla, Carassius auratus, C. carpio, O. mossambicus, Gillichthys mirabilis, Myxine glutinosa,* and *Typhlogobius californiensis,* are able to withstand the absence of oxygen for periods ranging from a few hours to days (Van den Thillart and van Waarde, 1985). Extensive studies of amino acid metabolism in anoxia-tolerant fish, such as goldfish and carp, have provided an understanding of this important survival strategy. Goldfish can survive complete anoxia for 15 h at 20°C and 120 days at 2°C (Van den Thillart and van Waarde, 1985). During anoxic periods, ammonia continues to be produced, indicating catabolism of amino acids as an energy source (van Waarde, 1983).

The importance of the liver in amino acid deamination declines under anoxic conditions (van Waarde, 1983). This is due to an inhibition of GDH via a reduced $NAD^+/NADH$ ratio in liver mitochondria (Van den Thillart et al., 1982). The production of ammonia by muscle increases under anoxic conditions coincident with elevated $NAD^+/NADH$ ratios in eel mitochondria (van Waarde, 1983). Alanine accumulation has been reported in red and white muscle and in liver of goldfish (van der Boon et al., 1992), whole body of crucian carp (*Carassius carassius*) (Johnston and Bernard, 1983), and brain of both goldfish and tilapia (van Ginneken et al., 1996). The quantitative importance of this pathway during anoxia is thought to be small (Van den Thillart and van Waarde, 1985). Aspartate (Van den Thillart and van Waarde, 1985) and alanine (van Waarde et al., 1982) have been suggested to be involved in ethanol production in anoxic goldfish, but the pathways involved have not been established.

The role of amino acids as neurotransmitters comes into play during periods of anoxia in fish. The responses observed serve to reduce brain activity to conserve energy. In general, the levels of inhibitory amino acids (e.g., γ-aminobutyric acid, glycine) increase in the brain during anoxia, while levels of the excitatory amino acids (e.g., aspartate, glutamate, and glutamine) decrease (Nilsson et al., 1991; van Ginneken et al., 1996). Although the link between the amino acids involved in neurotransmission and amino acid metabolism in general are only poorly understood, the regulation of nerve excitability by amino acids during anoxia may be a useful model for examining aspects of the regulation of amino acid metabolism.

XII. SUMMARY

The amino acids play an important role in the metabolism of fishes. Fish display a high efficiency of conversion of dietary protein into fish protein. This requires regulation of the flux of amino acids into other metabolic fates of amino acids such as oxidation, gluconeogenesis, and lipogenesis. Amino acids have a substantial role in energy metabolism as oxidative substrates in many tissues. The metabolism of glutamine in fish muscle, in particular, differs from that of the higher vertebrates. Rather than exporting glutamine to other tissues as occurs in mammals, fish muscle has a high capacity for the catabolism of this amino acid. This is especially pronounced in the elasmobranchs, which rely on glutamine as an energy source to offset the inability to catabolize lipids in muscle. The autocatalytic oxidative pathway for amino acids may contribute to the efficiency of their metabolic use of dietary protein. Little is known of the regulation of these functions although some aspects of the roles of hormones such as cortisol have been established. The relative importance of amino acids as energy sources in tissues is affected by growth rate and environmental factors such as temperature, salinity, and anoxia.

REFERENCES

Albi, J. L., Planas, J., and Sanchez, J. (1993). Glucose metabolism by brown trout peripheral blood lymphocytes. *J. Comp. Physiol.* **163B**, 118–122.

Anderson, D. E., Reid, S. D., Moon, T. W., and Perry, S. F. (1991). Metabolic effects associated with chronically elevated cortisol in rainbow trout (*Oncorhynchus mykiss*). *Can. J. Fish. Aquat. Sci.* **48**, 1811–1817.

Ando, M. (1988). Amino acid metabolism and water transport across the seawater eel intestine. *J. Exp. Biol.* **138**, 93–106.

Ando, S., Hatano, M., and Zama, K. (1985). Deterioration of chum salmon muscle during spawning migration—VI. Changes in serum protease inhibitory activity during spawning migration of chum salmon (*Oncorhynchus keta*). *Comp. Biochem. Physiol.* **82B**, 111–115.

Ando, S., Hatano, M., and Zama, K. (1986). Protein degradation and protease activity of chum salmon (*Oncorhynchus keta*) muscle during spawning migration. *Fish Physiol. Biochem.* **1**, 17–26.

Assem, H., and Hanke, W. (1983). The significance of the amino acids during osmotic adjustment in teleost fish—I. Changes in the euryhaline *Sarotherodon mossambicus*. *Comp. Biochem. Physiol.* **74A**, 531–536.

Atherton, W. D., and Aitken, A. (1970). Growth, nitrogen metabolism and fat metabolism in *Salmo gairdneri*, Rich. *Comp. Biochem. Physiol.* **36**, 719–747.

Auerswald, L., Jurss, K., Schiedek, D., and Bastrop, R. (1997). The influence of salinity acclimation on free amino acids and enzyme activities in the intestinal mucosa of rainbow trout, *Oncorhynchus mykiss* (Walbaum). *Comp. Biochem. Physiol.* **116A**, 149–155.

Balinsky, J. B., Choritz, E. L., Coe, C.G.L., and van der Schans, G. S. (1967). Amino acid metabolism and urea synthesis in naturally estivating *Xenopus laevis*. *Comp. Biochem. Physiol.* **22**, 59–68.

Ballantyne, J. S. (1997). Jaws: the inside story. The metabolism of elasmobranch fishes. *Comp. Biochem. Physiol.* **118B**, 703–742.

Ballantyne, J. S., and Chamberlin, M. E. (1988). Adaptation and evolution of mitochondria: osmotic and ionic considerations. *Can. J. Zool.* **66**, 1028–1035.

Ballantyne, J. S., Moyes, C. D., and Moon, T. W. (1986). Osmolarity affects oxidation of sarcosine by isolated hepatocytes and mitochondria from a euryhaline elasmobranch. *J. Exp. Zool.* **238**, 267–271.

Battersby, B. J., McFarlane, W. J., and Ballantyne, J. S. (1996). Short term effects of 3,5,3'-triiodothyronine on the intermediary metabolism of the dogfish shark *Squalus acanthias*: evidence from enzyme activities. *J. Exp. Zool.* **274**, 157–162.

Bever, K., Chenoweth, M., and Dunn, A. (1981). Amino acid gluconeogenesis and glucose turnover in kelp bass (*Paralabrax* sp.). *Am. J. Physiol.* **240**, R246-R252.

Buck, L. T., Brill, R. W., and Hochachka, P. W. (1992). Gluconeogenesis in hepatocytes isolated from the skipjack tuna (*Katsuwonus pelamis*). *Can. J. Zool.* **70**, 1254–1257.

Cameron, J. N., and Heisler, N. (1983). Studies of ammonia in the rainbow trout: physico-chemical parameters, acid–base behavior and respiratory clearance. *J. Exp. Biol.* **105,** 107–125.

Campbell, J. W., and Vorhaben, J. E. (1983). Mitochondrial ammoniagenesis in liver of the channel catfish *Ictalurus punctatus. Am. J. Physiol.* **244,** R709-R717.

Canals, P., Gallardo, M. A., and Sanchez, J. (1995). Effects of insulin on the uptake of amino acids by hepatocytes and red blood cells from trout (*Salmo trutta*) are opposite. *Comp. Biochem. Physiol.* **112C,** 221–228.

Carbo, D., Lopez-Soriano, F. J., and Argiles, J. M. (1994). The effects of tumour necrosis factor-α on circulating amino acids in the pregnant rat. *Cancer Lett.* **79,** 27–32.

Carrillo, M., Zanuy, S., and Herrera, E. (1980). Daily rhythms of amino acid levels in the plasma of goldfish (*Carassius auratus*). *Comp. Biochem. Physiol.* **67A,** 581–586.

Chamberlin, M. E., and Ballantyne, J. S. (1992). Glutamine metabolism in elasmobranch and agnathan muscle. *J. Exp. Zool.* **264,** 269–272.

Chamberlin, M. E., Glemet, H. C., and Ballantyne, J. S. (1991). Glutamine metabolism in a Holostean fish (*Amia calva*) and a teleost (*Salvelinus namaycush*). *Am. J. Physiol.* **260,** R159-R166.

Chew, S. F., and Ip, Y. K. (1987). Ammoniagenesis in mudskippers *Boleophthalmus boddaerti* and *Periophthalmodon schlosseri. Comp. Biochem. Physiol.* **87B,** 941–948.

Christiansen, D. C., and Klungsoyr, L. (1987). Metabolic utilization of nutrients and the effects of insulin in fish. *Comp. Biochem. Physiol.* **88B,**701–711.

Cho, C. Y., and Kaushik, S. J. (1985). Effects of protein intake on metabolizable and net energy values of fish diets. *In* "Nutrition and Feeding in Fish" (C. B. Cowey, A. M. Mackie, and J. G. Bell, eds.), pp. 95–117. Academic Press, New York.

Collie, N. L., and Ferraris, R. P. (1995). Nutrient fluxes and regulation in fish intestine. *In* "Biochemistry and Molecular Biology of Fishes, Volume 4, Metabolic Biochemistry" (P. W. Hochachka, and T. P. Mommsen, eds.), pp. 221–239. Elsevier Science, Amsterdam.

Cowey, C. B. (1993). Protein metabolism in fish. *In* "Aquaculture: Fundamental and Applied Research" (B. Lahlou and P. Vitiello, eds.), pp. 125–137. American Geophysical Union, Washington, DC.

Cowey, C. B., and Daisley, K. W. (1962). Study of amino acids, free or as components of protein, and of some B vitamins in the tissues of the Atlantic salmon, *Salmo salar,* during spawning migration. *Comp. Biochem. Physiol.* **7,** 29–38.

Cowey, C. B., and Walton, M. J. (1989). Intermediary metabolism. *In* "Fish Nutrition" (J. E. Halver, ed.), pp. 259–329. Academic Press, New York.

Cowey, C. B., De La Higuera, M., and Adron, J. W. (1977). The effect of dietary composition and of insulin on gluconeogenesis in rainbow trout (*Salmo gairdneri*). *Br. J. Nutr.* **38,** 385–395.

Dabrowski, K., Kaushik, S. J., and Fauconneau, B. (1987). Rearing of sturgeon (*Acipenser baeri* Brandt) larvae III. Nitrogen and energy metabolism and amino acid absorption. *Aquaculture* **65,** 31–41.

Davis, K. B., Torrance, P., Parker, N. C., and Suttle, M. A. (1985). Growth, body composition and hepatic tyrosine aminotransferase activity in cortisol-fed channel catfish, *Ictalurus punctatus* Rafinesque. *J. Fish Biol.* **27,** 177–184.

DeLong, D. C., Halver, J. E., and Mertz, E. T. (1958). Nutrition of salmonid fishes. VI. Protein requirements of chinook salmon at two water temperatures. *J. Nutr.* **65,** 589–599.

Elkin, R. G., Lyons, M. L., and Rogler, J. C. (1988). Comparative utilization of D- and L-methionine by the white pekin duckling (*Anas platyrhynchos*). *Comp. Biochem. Physiol.* **91B,** 325–329.

Fauconneau, B., and Arnal, M. (1985). Leucine metabolism in trout (*Salmo gairdneri* R.). Influence of temperature. *Comp. Biochem. Physiol.* **82A,** 435–445.

Fauconneau, B., and Tesseraud, S. S. (1990). Measurement of plasma leucine flux in rainbow trout (*Salmo gairdneri* R.) using osmotic pump. Preliminary investigations of diet. *Fish Physiol. Biochem.* **8,** 29–44.

Fellows, F. C. I., and Hird, F. J. (1982). A comparative study of the binding of L-tryptophan and bilirubin by plasma proteins. *Arch. Biochem. Biophys.* **216,** 93–100.

Ferraris, R. P., and Ahearn, G. A. (1984). Sugar and amino acid transport in fish intestine. *Comp. Biochem. Physiol.* **77A,** 397–413.

Fields, J.H.A., Driedzic, W. R., French, C. J., and Hochachka, P. W. (1978). Kinetic properties of glutamate dehydrogenase from the gills of *Arapaima gigas* and *Osteoglossum bicirrhosum*. *Can. J. Zool.* **56,** 809–813.

Forster, R. P., and Goldstein, L. (1979). Amino acids and cell volume regulation. *Yale J. Biol. Med.* **52,** 497–515.

Forster, R. P., and Hannafin, J. A. (1980). Osmotic and cell volume regulation in atrium and ventricle of the elasmobranch skate, *Raja erinacea*. *Comp. Biochem. Physiol.* **65A,** 445–451.

French, C. J., Mommsen, T. P., and Hochachka, P. W. (1981). Amino acid utilisation in isolated hepatocytes from rainbow trout. *Eur. J. Biochem.* **113,** 311–317.

French, C. J., Hochachka, P. W., and Mommsen, T. P. (1983). Metabolic organization of liver during spawning migration of sockeye salmon. *Am. J. Physiol.* **245,** R827–R830.

Fugelli, K., and Zachariassen, K. E. (1976). The distribution of taurine, gamma-aminobutyric acid and inorganic ions between plasma and erythrocytes in flounder (*Platichthys flesus*) at different plasma osmolarities. *Comp. Biochem. Physiol.* **55A,** 173–177.

Ganassin, R. G., Barlow, J., and Bols, N. C. (1998). Influence of glutamine on phytohemaglutinin stimulated mitochogenesis of leucocytes from rainbow trout head kidney. *Fish Shellfish Immunol.* **8,** 561–564.

Gaudet, M., Racicot, J. G., and Leray, C. (1975). Enzyme activities of plasma and selected tissues in rainbow trout *Salmo gairdneri* Richardson. *J. Fish Biol.* **7,** 505–512.

Gillis, T. E., and Ballantyne, J. S. (1996). The effects of starvation on plasma free amino acid and glucose concentrations in lake sturgeon, *Acipenser fulvescens*. *J. Fish Biol.* **49,** 1306–1316.

Goldstein, L., and Forster, R. P. (1961). Source of ammonia excreted by the gills of the marine teleost, *Myoxocephalus scorpius*. *Am. J. Physiol.* **200,** 1116–1118.

Goldstein, L., Forster, R. P., and Fanelli, G. M. (1964). Gill blood flow and ammonia excretion in the marine teleost, *Myoxocephalus scorpius*. *Comp. Biochem. Physiol.* **12,** 489–499.

Graham, T. E., Sgro, V., Friars, D., and Gibala, M. J. (2000). Glutamate ingestion: the plasma and muscle free amino acid pools of resting humans. *Am. J. Physiol.* **278,** E83–E89.

Guppy, M., Hill, D. J., Arthur, P., and Rowley, A. F. (1999). Differences in fuel utilization between trout and human thrombocytes in physiological media. *J. Comp. Physiol.* **169B,** 515–520.

Haschemeyer, A.E.V. (1985). Multiple amino-tRNA synthetases (translases) in temperature acclimation of eurythermal fish. *J. Exp. Mar. Biol. Ecol.* **87,** 191–198.

Hegab, S. A., and Hanke, W. (1983). The significance of the amino acids during osmotic adjustment in teleost fish—II. Changes in the stenohaline *Cyprinus carpio*. *Comp. Biochem. Physiol.* **74A,** 537–543.

Henderson, R. J., and Sargent, J. R. (1981). Lipid biosynthesis in rainbow trout, *Salmo gairdneri,* fed diets of differing lipid content. *Comp. Biochem. Physiol.* **69C,** 31–37.

Henderson, R. J., and Tocher, D. R. (1987). The lipid composition and biochemistry of freshwater fish. *Prog. Lipid Res.* **26,** 281–347.

Henrickson, J. (1991). Effect of exercise on amino acid concentrations in skeletal muscle and plasma. *J. Exp. Biol.* **160,** 149–165.

Hochachka, P. W., and Somero, G. N. (1984). *In* "Biochemical Adaptation," p. 525. Princeton University Press, Princeton, NJ.

Houlihan, D. F., Mathers, E. M., and Foster, E. M. (1993). Biochemical correlates of growth rate in fish. *In* "Fish Ecophysiology" (J. C. Rankin, and F. B. Jensen, eds.), pp. 45–71. Chapman and Hall, London.

Houlihan, D. F., Carter, C. G., and McCarthy, I. D. (1995). Protein turnover in animals. *In* "Nitrogen

Metabolism and Excretion" (P. J. Walsh, and P. Wright, eds.), pp. 1–32. CRC Press, Boca Raton, FL.

Huggins, A. K., and Colley, L. (1971). The changes in the nonprotein nitrogenous constituents of muscle during the adaptation of the eel *Anguilla anguilla* L. from fresh water to sea water. *Comp. Biochem. Physiol.* **38B,** 537–541.

Ince, B. W., and Thorpe, A. (1976). The *in vivo* metabolism of ^{14}C-glucose and ^{14}C-glycine in insulin-treated northern pike (*Esox lucius* L.). *Gen. Comp. Endocrinol.* **28,** 481–486.

Ince, B. W., and Thorpe, A. (1977). Glucose and amino acid stimulated insulin release *in vivo* in the European silver eel (*Anguilla anguilla* L.). *Gen. Comp. Endocrinol.* **31,** 249–256.

Ince, B. W., and Thorpe, A. (1978). The effects of insulin on plasma amino acid levels in the Northern pike, *Esox lucius* L. *J. Fish Biol.* **12,** 503–506.

Inui, Y., Arai, S., and Yokote, M. (1975). Gluconeogenesis in the eel—VI. Effects of hepatectomy, alloxan, and mammalian insulin on the behavior of plasma amino acids. *Bull. Japan Soc. Sci. Fish.* **41,** 1105–1111.

Ip, Y. K., Chew, S. F., and Lim, R.W.L. (1990). Ammoniagenesis in the mudskipper, *Periophthalmus chrysospilos*. *Zool. Sci.* **7,** 187–194.

Janssens, P. A., and Cohen, P. P. (1968). Nitrogen metabolism in the African lungfish. *Comp. Biochem. Physiol.* **24,** 879–886.

Johnston, I. A., and Bernard, L. M. (1983). Utilization of the ethanol pathway in carp following exposure to anoxia. *J. Exp. Biol.* **104,** 73–78.

Jurss, K. (1979). Effects of temperature, salinity, and feeding on aminotransferase activity in the liver and white muscle of rainbow trout (*Salmo gairdneri* Richardson). *Comp. Biochem. Physiol.* **64B,** 213–218.

Jurss, K. (1981). Influence of temperature and ratio of lipid to protein in diets on aminotransferase activity in the liver and white muscle of rainbow trout (*Salmo gairdneri* Richardson). *Comp. Biochem. Physiol.* **68B,** 527–533.

Jurss, K., and Bastrop, R. (1995). Amino acid metabolism in fish. *In* "Biochemistry and Molecular Biology of Fishes, Volume 4, Metabolic Biochemistry" (P. W. Hochachka, and T. P. Mommsen, eds.), pp. 159–189. Elsevier Science, Amsterdam.

Jurss, K., Bittorf, T., Vokler, T., and Wacke, R. (1984). Biochemical investigation into the influences of environmental salinity on starvation of the tilapia, *Oreochromis mossambicus*. *Aquaculture* **40,** 171–182.

Jurss, K., Bittorf, T., and Vokler, T. (1985). Influence of salinity and ratio of lipid to protein in diets on certain enzyme activities in rainbow trout (*Salmo gairdneri* Richardson). *Comp. Biochem. Physiol.* **81B,** 73–79.

Jurss, K., Bittorf, T., and Vokler, T. (1986). Influence of salinity and food deprivation on growth, RNA/DNA ratio and certain enzyme activities in rainbow trout (*Salmo gairdneri* Richardson). *Comp. Biochem. Physiol.* **83B,** 425–433.

Jurss, K., Bittorf, T., Vokler, T., and Wacke, R. (1987). Effects of temperature, food deprivation and salinity on growth, RNA/DNA ratio and certain enzyme activities in rainbow trout (*Salmo gairdneri* Richardson). *Comp. Biochem. Physiol.* **87B,** 241–253.

Kaushik, S. J., and Luquet, P. (1979). Influence of dietary amino acid patterns on the free amino acid contents of blood and muscle of rainbow trout (*Salmo gairdneri* R.). *Comp. Biochem. Physiol.* **64B,** 175–180.

Kaushik, S. J., Breque, J., and Blanc, D. (1994). Apparent amino acid availability and plasma free amino acid levels in Siberian sturgeon (*Acipenser baeri*). *Comp. Biochem. Physiol.* **107A,** 433–438.

Kieffer, J. D., Alsop, D., and Wood, C. M. (1998). A respirometric analysis of fuel use during aerobic swimming at different temperatures in rainbow trout (*Oncorhynchus mykiss*). *J. Exp. Biol.* **201,** 3123–3133.

King, P. A., and Goldstein, L. (1983a). Renal ammoniagenesis and acid excretion in the dogfish, *Squalus acanthias. Am. J. Physiol.* **245,** R581–R589.
King, P. A., and Goldstein, L. (1983b). Organic osmolytes and cell volume regulation in fish. *Mol. Physiol.* **4,** 53–66.
Knapp, E., and Wieser, W. (1981). Effects of temperature and food on the free amino acids in tissues of roach (*Rutilus rutilus* L.) and rudd (*Scardinius erythrophthalmus* L.). *Comp. Biochem. Physiol.* **68A,** 187–198.
Knepper, M. A., Packer, R., and Good, D. W. (1989). Ammonia transport in the kidney. *Physiol. Rev.* **69,** 179–249.
Kutty, M. N. (1972). Respiratory quotient and ammonia excretion in *Tilapia mossambica. Mar. Biol.* **16,** 126–133.
Lall, S. P., Kaushik, S. J., LeBail, P. Y., Anderson, J. S., and Plisetskaya, E. (1994). Quantitative arginine requirement of Atlantic salmon (*Salmo salar*) reared in sea water. *Aquaculture* **124,** 13–25.
Lange, R., and Fugelli, K. (1965). The osmotic adjustment in the euryhaline teleosts, the flounder, *Pleuronectes flesus* L. and the three spined stickleback, *Gasterosteus aculeatus* L. *Comp. Biochem. Physiol.* **15,** 283–292.
Larsson, A., and Lewander, K. (1973). Metabolic effects of starvation in the eel, *Anguilla anguilla* L. *Comp. Biochem. Physiol.* **44A,** 367–374.
Lauff, R. F., and Wood, C. M. (1996a). Respiratory gas exchange, nitrogenous waste excretion, and fuel usage during aerobic swimming in juvenile rainbow trout. *J. Comp. Physiol.* **166B,** 501–509.
Lauff, R. F., and Wood, C. M. (1996b). Respiratory gas exchange, nitrogenous waste excretion, and fuel usage during starvation in juvenile trout, *Oncorhynchus mykiss. J. Comp. Physiol.* **165B,** 542–551.
LeBlanc, P. J., Gillis, T. E., Gerrits, M. F., and Ballantyne, J. S. (1995). Metabolic organization of liver and somatic muscle of landlocked sea lamprey, *Petromyzon marinus,* during the spawning migration. *Can. J. Zool.* **73,** 916–923.
Leech, A. R., Goldstein, L., Cha, C., and Goldstein, J. M. (1979). Alanine biosynthesis during starvation in skeletal muscle of the spiny dogfish, *Squalus acanthias. J. Exp. Zool.* **207,** 73–80.
Lemieux, G., Craan, A. G., Quenneville, A., Lemieux, C., Berkofsky, J., and Lewis, V. S. (1984). Metabolic machinery of the alligator kidney. *Am. J. Physiol.* **247,** F686–F693.
Levi, G., Morisi, G., Coletti, A., and Catanzaro, R. (1974). Free amino acids in fish brain: normal levels and changes upon exposure to high ammonia concentrations *in vivo,* and upon incubation of brain slices. *Comp. Biochem. Physiol.* **49A,** 623–636.
Love, R. M. (1980). "The Chemical Biology of Fishes," p. 943. Academic Press, London.
Mantagos, S., Moustogianni, A., Varvarigou, A., and Frimas, C. (1989). Effect of light on diurnal variation of blood amino acids in neonates. *Biol. Neonate* **55,** 97–103.
Martin, N. B., Houlihan, D. F., Talbot, C., and Palmer, R. M. (1993). Protein metabolism during sexual maturation in female Atlantic salmon (*Salmo salar* L.). *Fish Physiol. Biochem.* **12,** 131–141.
Matty, A. J., and Lone, K. P. (1985). The hormonal control of metabolism and feeding. *In* "Fish Energetics. New Perspectives" (P. Tytler and P. Calow, eds.), pp. 185–209. Johns Hopkins University Press, Baltimore, MD.
Mauro, N. A., and Isaacks, R. E. (1989). Relative oxidation of glutamate and glucose by vertebrate erythrocytes. *Comp. Biochem. Physiol.* **94A,** 95–97.
McCormick, S. D., and Saunders, R. L. (1987). Preparatory physiological adaptations for marine life of salmonids: osmoregulation, growth and metabolism. *Am. Fish. Soc. Symp.* **1,** 1–229.
Mester, R., Iordachescu, D., and Niculescu, S. (1973). The influence of adaptation temperature on the behavior of L-alanine:2- oxoglutarate amino transferase of the skeletal muscle of pond loach (*Misgurnus fossilis* L.). *Comp. Biochem. Physiol.* **45B,** 923–931.

Milligan, C. L. (1997). The role of cortisol in amino acid mobilization and metabolism following exhaustive exercise in rainbow trout (*Oncorhynchus mykiss* Walbaum). *Fish Physiol. Biochem.* **16,** 119–128.
Mommsen, T. P. (1984). Metabolism of the fish gill. *Fish Physiol.* **XB,** 203–238.
Mommsen, T. P., French, C. J., and Hochachka, P. W. (1980). Sites and patterns of protein and amino acid utilization during the spawning migration of salmon. *Can. J. Zool.* **58,** 1785–1799.
Mommsen, T. P., Walsh, P. J., and Moon, T. W. (1985). Gluconeogenesis in hepatocytes and kidney of Atlantic salmon. *Mol. Physiol.* **8,** 89–100.
Mommsen, T. P., Danulat, E., and Walsh, P. J. (1992). Metabolic actions of glucagon and dexamethasone in liver of the ureogenic teleost *Opsanus beta. Gen. Comp. Endocrinol.* **85,** 316–326.
Mommsen, T. P., Vijayan, M. M., and Moon, T. W. (1999). Cortisol in teleosts: dynamics, mechanisms of action, and metabolic regulation. *Rev. Fish Biol. Fish.* **9,** 211–268.
Moyes, C. D., and Moon, T. W. (1987). Solute effects on the glycine cleavage system of two osmoconformers (*Raja erinacea* and *Mya arenaria*) and an osmoregulator (*Pseudopleuronectes americanus*). *J. Exp. Zool.* **242,** 1–8.
Murai, T., Ogata, H., Hirashima, Y., Akiyama, T., and Nose, T. (1987). Portal absorption and hepatic uptake of amino acids in rainbow trout force-fed complete diets containing casein or crystalline amino acids. *Nippon Suisan Gakkaishi* **53,** 1847–1859.
Nagai, M., and Ikeda, S. (1973). Carbohydrate metabolism in fish—IV. Effect of dietary composition on metabolism of acetate-U-^{14}C and L-alanine-U-^{14}C in carp. *Bull. Japan Soc. Sci. Fish.* **39,** 633–643.
Narayansingh, T., and Eales, J. G. (1975). Effects of thyroid hormones on *in vivo* 1–^{14}C L-leucine incorporation into plasma and tissue protein of brook trout (*Salvelinus fontinalis*) and rainbow trout (*Salmo gairdneri*). *Comp. Biochem. Physiol.* **52B,** 399–405.
Navarro, I., and Gutierrez, J. (1995). Fasting and starvation. *In* "Biochemistry and Molecular Biology of Fishes, Volume 4, Metabolic Biochemistry" (P. W. Hochachka, and T. P. Mommsen, eds.), pp. 393–434. Elsevier Science, Amsterdam.
Navarro, I., Blasco, J., Banos, N., and Gutierrez, J. (1997). Effects of fasting and feeding on plasma amino acid levels in brown trout. *Fish Physiol. Biochem.* **16,** 303–309.
Newsholme, E. A., and Leech, A. R. (1983). "Biochemistry for the Medical Sciences," p. 952. Wiley, Toronto.
Nilsson, G. E., Lutz, P. L., and Jackson, T. L. (1991). Neurotransmitters and anoxia survival of the brain: a comparison of anoxia-tolerant and anoxia-intolerant vertebrates. *Physiol. Zool.* **64,** 638–652.
Ogata, H., and Arai, S. (1985). Comparison of free amino acid contents in plasma, whole blood and erythrocytes of carp, coho salmon, rainbow trout, and channel catfish. *Bull. Japan Soc. Sci. Fish.* **51,** 1181–1186.
Payan, P., and Matty, A. J. (1975). The characteristics of ammonia excretion by a perfused isolated head of trout (*Salmo gairdneri*): effect of temperature and CO_2-free ringer. *J. Comp. Physiol.* **96,** 167–184.
Payan, P., and Pic, P. (1977). Origine de l'ammonium excrete par les branchies chez la truite (*Salmo gairdneri*). *C.R. Acad. Sci. Paris* **284,** 2519–2522.
Pequin, L., and Serfaty, A. (1963). L'excretion ammoniacale chez un teleosteen dulcicole: *Cyprinus carpio* L. *Comp. Biochem. Physiol.* **10,** 315–324.
Plisetskaya, E., Woo, N. Y. S., and Murat, J. (1983). Thyroid hormones in cyclostomes and fish and their role in regulation of intermediary metabolism. *Comp. Biochem. Physiol.* **74A,** 179–187.
Plisetskaya, E., Bhattacharya, S., Dickhoff, W. W., and Gorbman, A. (1984). The effect of insulin on amino acid metabolism and glycogen content in isolated liver cells of juvenile coho salmon, *Oncorhynchus kisutch. Comp. Biochem. Physiol.* **78A,** 773–778.
Plisetskaya, E. M., Buchelli-Narvaez, L. I., Hardy, R. W., and Dickhoff, W. W. (1991). Effects of

injected and dietary arginine on plasma insulin levels and growth of Pacific salmon and rainbow trout. *Comp. Biochem. Physiol.* **98A,** 165–170.
Reshkin, S. J., and Ahearn, G. A. (1991). Intestinal glycyl-L-phenylalanine and L-phenylalanine transport in a euryhaline teleost. *Am. J. Physiol.* **260,** R563–R569.
Reshkin, S. J., Vilella, S., Cassano, G., Ahearn, G. A., and Storelli, C. (1988). Basolateral amino acid transport and glucose transport by the intestine of the teleost *Anguilla anguilla. Comp. Biochem. Physiol.* **91A,** 779–788.
Sabapathy, U., and Teo, L. H. (1995). Some properties of the intestinal proteases of the rabbitfish, *Siganus canaliculatus* (Park). *Fish Physiol. Biochem.* **14,** 215–221.
Sakaguchi, M., Murata, M., Daikoku, T., and Arai, S. (1988). Effects of dietary taurine on tissue taurine and free amino acid levels of the chum salmon, *Oncorhynchus keta,* reared in freshwater and seawater environments. *Comp. Biochem. Physiol.* **89A,** 437–442.
Schreck, C. B. (1993). Glucocorticoids: metabolism, growth and development. *In* "The Endocrinology of Growth, Development, and Metabolism in Vertebrates" (M. P. Schreibman, C. G. Scanes, and P. K. T. Pang, eds.), pp. 367–392. Academic Press, San Diego.
Shihabi, Z. K., Goodman, H. O., Holmes, R. P., and O'Connor, M. L. (1989). The taurine content of avian erythrocytes and its role in osmoregulation. *Comp. Biochem. Physiol.* **92A,** 545–549.
Shikata, T., Iwanaga, S., and Shimeno, S. (1995). Metabolic response to acclimation temperature in carp. *Fish. Sci.* **61,** 512–516.
Siddiqui, A. Q., Siddiqui, A. H., and Ahmad, K. (1973). Free amino acid contents of the skeletal muscle of carp at juvenile and adult stages. *Comp. Biochem. Physiol.* **44B,** 725–728.
Silver, M., Fowden, A. L., Taylor, P. M., Knox, J., and Hill, C. M. (1994). Blood amino acids in the pregnant mare and fetus: the effects of maternal fasting and intrafetal insulin. *Exp. Physiol.* **79,** 423–433.
Sitbon, G., Khemiss, F., and Boulanger, Y. (1982). Effects of total pancreatectomy and amino-acid treatment on plasma amino- acids and glucose in the goose. *J. Physiol. (Paris)* **78,** 258–265.
Smith, D. D., and Campbell, J. W. (1987). Glutamine synthetase in liver of the American alligator, *Alligator mississippiensis. Comp. Biochem. Physiol.* **86B,** 755–762.
Storelli, C., Vilella, S., Romano, M. P., Maffia, M., and Cassano, G. (1989). Brush-border amino acid transport mechanisms in carnivorous eel intestine. *Am. J. Physiol.* **257,** R506–R510.
Storey, K. B., Guderley, H. E., Guppy, M., and Hochachka, P. W. (1978). Control of ammoniagenesis in the kidney of water- and air-breathing osteoglossids: characterization of glutamate dehydrogenase. *Can. J. Zool.* **56,** 845–851.
Storey, K. B., McDonald, D. G., Perry, S. F., Harris, V. L., and Storey, J. M. (1992). Effect of freezing on the blood chemistry of the wood frog. *Cryo-Lett.* **13,** 363–370.
Suarez, R. K., and Mommsen, T. P. (1987). Gluconeogenesis in teleost fishes. *Can. J. Zool.* **65,** 1869–1882.
Sweeting, R. M., Wagner, G. F., and McKeown, B. A. (1985). Changes in plasma glucose, amino acid nitrogen and growth hormone during smoltification and seawater adaptation in Coho salmon, *Oncorhynchus kisutch. Aquaculture* **45,** 185–197.
Tacon, A. G. J., and Cowey, C. B. (1985). Protein and amino acid requirements. *In* "Fish Energetics. New Perspectives" (P. Tytler and P. Calow, eds.) pp. 155–183. Johns Hopkins University Press, Baltimore, MD.
Thornburn, C. C., and Matty, A. J. (1963). The effect of thyroxine on some aspects of nitrogen metabolism in the goldfish (*Carassius auratus*) and the trout (*Salmo trutta*). *Comp. Biochem. Physiol.* **8,** 1–12.
Tihonen, K., and Nikinmaa, M. (1992). Substrate utilization by carp (*Cyprinus carpio*) erythrocytes. *J. Exp. Biol.* **161,** 509–514.
Timoshina, L. A., and Shabalina, A. A. (1970). Effect of starvation on the dynamics of concentration of amino acid and free fatty acid in rainbow trout. *Hydrobiol. J.* **8,** 36–41.

Toyohara, H., and Makinodan, Y. (1989). Comparison of calpain I and calpain II from carp muscle. *Comp. Biochem. Physiol.* **92B,** 577–581.

Unsworth, B. R., and Crook, E. M. (1967). The effect of water shortage on the nitrogen metabolism of *Xenopus laevis*. *Comp. Biochem. Physiol.* **23,** 831–845.

Van den Thillart, G. (1986). Energy metabolism of swimming trout (*Salmo gairdneri*). Oxidation rates of palmitate, glucose, lactate, alanine, leucine and glutamate. *J. Comp. Physiol.* **156B,** 511–520.

Van den Thillart, G., and van Waarde, A. (1985). Teleosts in hypoxia: aspects of anaerobic metabolism. *Mol. Physiol.* **8,** 393–409.

Van den Thillart, G., and van Raaij, M. (1995). Endogenous fuels; noninvasive versus invasive. In "Biochemistry and Molecular Biology of Fishes, Volume 4, Metabolic Biochemistry" (P. W. Hochachka, and T. P. Mommsen, eds.), pp. 33–63. Elsevier Science, Amsterdam.

Van den Thillart, G., van Waarde, A., Dobbe, F., and Kesbeke, F. (1982). Anaerobic energy metabolism of goldfish, *Carassius auratus* (L.). Effects of anoxia on measured and calculated $NAD^+/NADH$ ratios in muscle and liver. *J. Comp. Physiol.* **146,** 41–49.

van der Boon, J., Eelkema, F. A., van den Thillart, G. E. E. J. M., and Addink, A. D. F. (1992). Influence of anoxia on free amino acid levels in blood, liver and skeletal muscles of the goldfish, *Carassius auratus*. *Comp. Biochem. Physiol.* **101B,** 193–198.

van Ginneken, V., Nieveen, M., Van Eersel, R., Van den Thillart, G., and Addink, A. (1996). Neurotransmitter levels and energy status in brain of fish species with and without the survival strategy of metabolic depression. *Comp. Biochem. Physiol.* **114A,** 189–196.

van Raaij, M. T. M., Van den Thillart, G., and Addink, A. (1994). Metabolism of $1-{}^{14}C$-acetate and $1-{}^{14}C$-leucine by anoxic goldfish (*Carassius auratus,* L.): evidence for anaerobic lipid synthesis. *Physiol. Zool.* **67,** 673–692.

van Waarde, A. (1981). Nitrogen metabolism in goldfish *Carassius auratus* (L.). Activities of transamination reactions, purine nucleotide cycle and glutamate dehydrogenase in goldfish tissues. *Comp. Biochem. Physiol.* **68B,** 407–413.

van Waarde, A. (1983). Aerobic and anaerobic ammonia production by fish. *Comp. Biochem. Physiol.* **74B,** 675–684.

van Waarde, A., and de Wilde-Van Berghe Henegouwen, M. (1982). Nitrogen metabolism in goldfish, *Carassius auratus* (L.). Pathway of aerobic and anaerobic glutamate oxidation in goldfish liver and muscle mitochondria. *Am. J. Physiol.* **244,** R709–R717.

van Waarde, A., and Henegouwen, M.D.W. (1982). Nitrogen metabolism in goldfish, *Carassius auratus* (L.) pathway of aerobic and anaerobic glutamate oxidation in goldfish liver and muscle mitochondria. *Comp. Biochem. Physiol.* **72B,** 133–136.

van Waarde, A., and Kesbeke, F. (1981). Nitrogen metabolism in goldfish *Carassius auratus* (L.). Influence of added substrates and enzyme inhibitors on ammonia production of isolated hepatocytes. *Comp. Biochem. Physiol.* **70B,** 499–507.

van Waarde, A., Van den Thillart, G., and Dobbe, F. (1982). Anaerobic metabolism of goldfish, *Carassius auratus* (L.). Influence of anoxia on mass-action ratios of transaminase reactions and levels of ammonia and succinate. *J. Comp. Physiol.* **147B,** 53–59.

Vellas, F., Parent, J., Bahamondes, I., and Charpenteau, M. (1982). Influence d'une augmentation de la temperature sur certains aspects du catabolism azote chez la truite arc-en-ciel (*Salmo gairdneri* Rich). *Rep. Nutr. Dev.* **22,** 851–864.

Venkatachari, S.A.T. (1974). Effect of salinity adaptation on nitrogen metabolism in the freshwater fish *Tilapia mossambica* I. Tissue protein and amino acid levels. *Mar. Biol.* **24,** 57–63.

Vijayan, M. M., Foster, G. D., and Moon, T. W. (1993). Effect of cortisol on hepatic carbohydrate metabolism and responsiveness to hormones in the sea raven, *Hemitripterus americanus*. *Fish Physiol. Biochem.* **12,** 327–335.

Vijayan, M. M., Reddy, P. K., Leatherland, J. F., and Moon, T. W. (1994). The effects of cortisol on hepatocyte metabolism in rainbow trout: a study using the steroid analogue RU486. *Gen. Comp. Endocrinol.* **96,** 75–84.

Vijayan, M. M., Morgan, J. D., Sakamoto, T., Grau, E. G., and Iwama, G. K. (1996). Food-deprivation affects seawater acclimation in tilapia: hormonal and metabolic changes. *J. Exp. Biol.* **199,** 2467–2475.

Vislie, T., and Fugelli, K. (1975). Cell volume regulation in flounder (*Platichthys flesus*) heart muscle accompanying an alteration in plasma osmolality. *Comp. Biochem. Physiol.* **52A,** 415–418.

von der Decken, A. (1992). Physiological changes in skeletal muscle by maturation—spawning of nonmigrating female Atlantic salmon, *Salmo salar. Comp. Biochem. Physiol.* **101B,** 299–301.

Walker, S. P., Keast, D., and McBride, S. (1996). Distribution of glutamine synthetase in the snapper (*Pagrus auratus*) and implications for the immune system. *Fish Physiol. Biochem.* **15,** 187–194.

Walsh, P. J., and Milligan, C. L. (1995). Effects of feeding and confinement on nitrogen metabolism and excretion in the gulf toadfish *Opsanus beta. J. Exp. Biol.* **198,** 1559–1566.

Walsh, P. J., Wood, C. M., Thomas, S., and Perry, S. F. (1990). Characterization of red cell metabolism in rainbow trout. *J. Exp. Biol.* **154,** 475–489.

Walsh, P. J., Wood, C. M., and Moon, T. W. (1998). Red blood cell metabolism. *In* "Fish Physiology, Volume 17, Fish Respiration" (S. F. Perry, and R. A. Tuft, eds.), pp. 41–73. Academic Press, San Diego.

Walton, M. J. (1985). Aspects of amino acid metabolism in teleost fish. *In* "Nutrition and Feeding in Fish." (C. B. Cowey, A. M. Mackie, and J. G. Bell, eds.), pp. 47–67. Academic Press, New York.

Walton, M. J., and Cowey, C. B. (1977). Aspects of ammoniogenesis in rainbow trout, *Salmo gairdneri. Comp. Biochem. Physiol.* **57B,** 143–149.

Walton, M. J., and Cowey, C. B. (1981). Distribution and some kinetic properties of serine catabolizing enzymes in rainbow trout *Salmo gairdneri. Comp. Biochem. Physiol.* **68B,** 147–150.

Webb, J. T., and Brown, G. W. (1980). Glutamine synthetase activity in subdivisions of brain of the shark, *Squalus acanthias. Experientia* **36,** 903–904.

Weber, J., and Zwingelstein, G. (1995). Circulatory substrate fluxes and their regulation. *In* "Biochemistry and Molecular Biology of Fishes, Volume 4, Metabolic Biochemistry" (P. W. Hochachka, and T. P. Mommsen, eds.), pp. 15–32. Elsevier Science, Amsterdam.

Whiting, S. J., and Wiggs, A. J. (1977). Effect of nutritional factors and cortisol on tyrosine aminotransferase activity in liver of brook trout, *Salveliinus fontinalis* Mitchill. *Comp. Biochem. Physiol.* **58B,** 189–193.

Wilson, R. P. (1989). Amino acids and proteins. *In* "Fish nutrition" (J. E. Halver, ed.), pp. 111–151. Academic Press, New York.

Wright, P. A. (1993). Nitrogen excretion and enzyme pathways for ureogenesis in freshwater tilapia (*Oreochromis niloticus*). *Physiol. Zool.* **66,** 881–901.

Wright, P. A., Obbard, M. E., Battersby, B. J., Felskie, A. K., LeBlanc, P. J., and Ballantyne, J. S. (1999). Lactation during hibernation in wild black bears: effects on plasma amino acids and nitrogen-end products. *Physiol. Biochem. Zool.* **72,** 597–604.

Yamamoto, T., Castell, L. M., Botella, J., Powell, H., Hall, G. M., Young, A., and Newsholme, E. A. (1997). Changes in the albumin binding of tryptophan during postoperative recovery: a possible link with central fatigue? *Brain Res. Bull.* **43,** 43–46.

4

AMMONIA TOXICITY, TOLERANCE, AND EXCRETION

Y. K. IP, S. F. CHEW, AND D. J. RANDALL

I. Introduction
II. Ammonia Toxicity to Fish
   A. Ammonia in Aqueous Solution
   B. Fish Ammonia Production and Excretion
   C. Why Is Ammonia Toxic?
   D. Toxicity Studies
   E. Problems with Present Ammonia Toxicity Criteria
III. Strategies to Reduce, Tolerate, or Avoid Ammonia Toxicity
   A. Reduction in Ammonia Production
   B. Partial Amino Acid Catabolism Leading to the Formation and Storage of Alanine
   C. Ammonia Detoxification and Storage as Glutamine and/or Urea
   D. Ammonia Excretion
IV. Summary
   References

I. INTRODUCTION

Ammonia is an unusual toxicant in that it is produced by, as well as being poisonous to, animals. Ammonia is excreted by many aquatic animals and is continually produced as a result of the decomposition of excreted wastes of living organisms and/or the decomposition of dead organisms, usually ending up in the environment as ammonium bicarbonate. A less important but natural source of gaseous ammonia is produced by volcanic activity. Ammonium salts are released into the environment through agricultural fertilization and industrial emissions; for example, burning coal can result in ammonia release. Ammonia enters the aquatic environment from both point and nonpoint sources. In the United States more than 95% of the aquatic point source emissions are from sewage treatment plants. Most of the remainder are from the steel, fertilizer, petroleum, and meat processing industries (American Petroleum Institute, 1981). The major nonpoint sources of ammonia in the aquatic environment in the United States include fertilizer and urban runoff, animal feedlots, animal wastes spread on the soil, and precipitation. Fish culture can be an important local site of ammonia production. Handy and Poxton (1993) report that in aquaculture systems between 52 and 95% of the nitrogen added to the system (usually in the diet) will ultimately pollute the environment.

Elevated ammonia levels in the aquatic environment are toxic. The threshold concentration of total ammonia ($[NH_3] + [NH_4^+]$) resulting in unacceptable biological effects, promulgated by the U.S. Environmental Protection Agency (EPA, 1998) was 3.48 mg N/liter at pH 6.5 and 0.25 mg N/liter at pH 9.0. These values are based on an analysis of data collected in freshwater rather than seawater experiments. Most of the data on ammonia toxicity in the aquatic environment refer to fresh water animals, usually fish. There is only a relatively small saltwater data set (Handy and Poxton, 1993).

The U.S. National Research Council Committee on Medical and Biologic Effects of Environmental Pollutants (1979) has reviewed ammonia toxicity and the World Health Organization (WHO, 1986) has compiled a similar review. The 1979 review is medically oriented, whereas the WHO review has more information on aquatic systems, including saltwater. The EPA (1984) reviewed ammonia toxicity in both freshwater and saltwater systems, producing an addendum to its 1984 document in 1998 (EPA, 1998). Another more specific review of seawater ammonia toxicity was produced subsequently (EPA, 1989). Stephan *et al.* (1985) discussed guidelines for developing water quality criteria. Handy and Poxton (1993) have reviewed ammonia toxicity in mariculture systems.

II. AMMONIA TOXICITY TO FISH

A. Ammonia in Aqueous Solution

In aqueous solution, ammonia exists as two species, NH_3 and NH_4^+. The equilibrium reaction can be written:

$$NH_3 + H_3O^+ \rightleftharpoons NH_4^+ + H_2O.$$

Total ammonia (T_{amm}) is the sum of $[NH_3] + [NH_4^+]$ and the pK of this ammonia/ammonium ion reaction is around 9.5. The amounts of each of the two species can be calculated from the Henderson-Hasselbalch equation if the pH and appropriate pK are known:

$$NH_4^+ = T_{amm} / \{1 + \text{antilog}(\text{pH} - \text{p}K)\} = T_{amm} - NH_3.$$

Emerson *et al.* (1975) reported pK values for ammonia in an aqueous solution of zero salinity using the following equation:

$$\text{p}K_{amm} = 0.09018 + 2729.92/T_k,$$

4. AMMONIA TOXICITY, TOLERANCE, AND EXCRETION

where T_k is temperature in degrees Kelvin (C + 273.15). Khoo *et al.* (1977) have investigated the dissociation of ammonium ions in seawater and report pK_{amm} values at various salinities and temperatures between 5 and 40°C. Boutilier *et al.* (1984) suggested that the pK of saline solutions could be calculated using the Davies approximation formula. There is little difference between measured and calculated values of pK_{amm} for trout plasma and 500 mM NaCl solutions (Cameron and Heisler, 1983) using this approach. Thus, reasonably accurate estimates of the pK of the ammonia/ammonium ion reaction in both freshwater and saltwater are available. Values of pK are affected by temperature, pressure, and ionic strength.

Pressure. There have been no studies of the effect of pressure on ammonia toxicity, nor have there been any studies of ammonia production and excretion by deep-sea animals living at high pressure. Pressure increases by about 1 atm every 10 m of depth. Therefore, marine animals are subject to a range of pressures. All ammonia toxicity studies have been carried out at pressures close to 1 atm. The effect of pressure on pK_{amm}, however, is minimal. The effect of pressure can be determined by the following equation from Whitfield (1974):

$$pK_{amm} \text{ (at pressure } P_2) = pK_{amm} \text{ (at 1 atm)} + 0.0415 \, P_2 / T_k,$$

where P_2 is in atmospheres and T_k is in degrees Kelvin. Thus, at 25C an increase in pressure of 100 atm (approximately 1000 m of depth) will increase the pK_{amm} by 0.014, resulting in a very small decrease in % [NH_3].

Temperature. The change in temperature from 30 to 10°C that occurs over the same change in depth of 1000 m in the ocean considered above will increase the pK_{amm} by 0.642, that is, from 9.206 to 9.848, causing a nearly 50-fold larger decrease in the % [NH_3] in the water compared with that caused by the pressure change alone.

Ionic strength. An increase in ionic strength increases pK_{amm}, such that if all other conditions remain constant the addition of salt will reduce the % [NH_3] in the water. A change from freshwater to seawater at any temperature between 5 and 35°C increases the pK value by about 0.114, only about one-third of the effect of a 10°C temperature change.

pH. Water pH, on the other hand, has a very marked effect on both the equilibrium of the ammonia/ammonium ion reaction and on ammonia toxicity. The effect of pH on the fraction (f) of NH_3 can be calculated from the following equation:

$$f \, NH_3 = 1/(10^{pK - pH} + 1).$$

Thus, water at pH 8 and 25°C will have slightly more than 4% of total ammonia in the un-ionized form. An increase in pH above this value has been shown to result in a large increase in ammonia toxicity expressed as total ammonia, largely because of the increase in the % [NH_3] in the water.

Ammonia is removed from the aquatic environment by conversion to less toxic nitrite and nitrate. These compounds are taken up by organisms, stimulate growth, and result in eutrophication of aquatic systems, which in turn causes hypoxia. Thus, elevated ammonia levels are often associated with hypoxia in aquatic systems, an increasingly common problem in coastal waters close to large human populations.

B. Fish Ammonia Production and Excretion

The two main waste products of metabolism are CO_2 and NH_3, with ammonia production being about a tenth to a third of the rate of carbon dioxide production. The main internal source of ammonia in fish is through catabolism of proteins. Ammonia is produced in the fish liver (Pequin and Serfaty, 1963), but other tissues are also capable of ammonia production (Walton and Cowey, 1977). Ammonia production occurs as the result of transamination of amino acids followed by the deamination of glutamate and/or by the deamination of adenylates in fish muscle during severe exercise (Driedzic and Hochachka, 1976). Most aquatic animals keep body ammonia levels low by simply excreting excess ammonia directly. Increased production results in increased accumulation of ammonia in the body, which leads to increased excretion. As a result there is a small delay between increased production and the subsequent rise in excretion. Ammonia excretion increases in fish following a meal (Handy and Poxton, 1993; Leung *et al.*, 1999), reflecting increased production associated with the breakdown of ingested protein, but remains low during starvation (Brett and Zala, 1975).

It is usually assumed that most nitrogenous excretion ($\approx 90\%$) takes place across the gills, predominantly in the form of un-ionized ammonia. Although this is true of many freshwater teleosts, this is not the case for marine and/or amphibious fish. Marine fish excrete more nitrogen via the skin, gut, and/or kidneys than freshwater fish, and only 50–70% of nitrogen excreted is across the gills (Sayer and Davenport, 1987a,b). In general, however, ammonia formed in liver, as well as other tissues, is cleared from the blood at the gills (Heisler, 1990; Wilkie, 1997). This overall mechanism implies that ammonia formed in the cytosolic and/or mitochondrial compartments of liver and other tissues simply diffuses down a NH_3 gradient across the boundary membranes of various body compartments. The rate of NH_3 excretion is determined by the magnitude of the NH_3 gradient between blood and water (Wilson *et al.*, 1994). Ammonia excretion is augmented by acidic conditions in the water because any NH_3 excreted into the water is rapidly converted and trapped as NH_4^+, maintaining the NH_3 gradient across the gills and augmenting ammonia excretion. Many freshwater fish actively excrete protons, forming an acid boundary layer next to the gill surface (Wright *et al.*, 1986; Lin and Randall, 1991), which augments ammonia excretion (Wright *et al.*, 1989).

Above water pH 9.0 ammonia excretion is reduced because of the absence of ammonium ion trapping (Wright et al., 1989), resulting in elevated plasma ammonia levels (Yesaki and Iwama, 1992). Thus, many animals have difficulty excreting ammonia when exposed to alkaline conditions. There is some evidence that ammonia production may be reduced if excretion is impaired (Wilson et al., 1998), but the mechanisms involved are not known.

Wright and Wood (1985) showed that exposure of freshwater trout (*Oncorhynchus mykiss*) to waters of pH 9.6 resulted in an inhibition of branchial excretion of NH_3 due to a reduction of the blood-to-water pNH_3 gradient. Urea excretion increases transiently and plasma ammonia levels remain greatly elevated (Wilkie and Wood, 1991). On the other hand, the Lahontan cutthroat trout (*Oncorhynchus clarki henshawi*) thrives in Pyramid Lake, Nevada, at pH 9.4 (Wilkie et al., 1993). Wright et al. (1993) reported that the majority of nitrogenous wastes were excreted as ammonia (56% through the gills, 10% through the kidney), while urea excretion accounted for 34% (32% through the gills, 2% through the kidneys). They concluded that *O. clarki henshawi* is able to survive in the alkaline lake due to lower rates of ammonia excretion, higher rates of urea excretion, a higher rate of renal ammonia excretion, greater plasma pH, and greater total ammonia levels, which facilitate the diffusive excretion of NH_3 across the gills. When the cutthroat trout was challenged with water at pH 10, survival was relatively poor, with more than 50% mortality after 72 h of exposure. The plasma pNH_3 level was greatly elevated and the plasma Na^+ and Cl^- levels were depressed in nonsurviving trout, suggesting that the cause of death was related to a combination of ammonia toxicity and ionoregulatory failure (Wilkie et al., 1994; Wilkie and Wood, 1996).

NH_3 solubility in water is approximately 1000 times that of CO_2, while its aqueous diffusion coefficient is similar to that of CO_2. Therefore, the diffusivity of NH_3 through water is about 1000 times higher than that of CO_2. This statement is often misinterpreted to mean that the diffusivity of NH_3 through fish tissues is equally high. Indeed, it is commonly assumed that NH_3 is highly lipophilic and therefore diffuses even more easily through cell membranes. However, the available data, while limited, indicate that the lipid versus water partition coefficient for NH_3 is less than 0.1 (Evans and Cameron, 1986). Diffusion through water-filled channels is likely to be much faster than diffusion through lipoprotein membranes. The number of water-filled channels in epithelia is often low, and under these conditions the lipid permeability of NH_3 is more relevant. Lande et al. (1995), using artificial vesicles, showed that NH_3 permeability decreased with increased lipid fluidity. Several epithelia in the mammalian kidney have low NH_3 permeability and this may be due to reduced fluidity of the membranes. The ammonium ion, NH_4^+, being charged and larger than NH_3, especially in the hydrated form, is normally thought to be much less diffusive and to move almost entirely through water-filled channels, particularly paracellular channels (McDonald et al., 1989). The NH_4^+ ion is considerably larger than Na^+ and slightly larger than K^+; nevertheless, it appears to be more permeant through epithelia than either of these cations (Wood, 1993).

More important than absolute diffusivity is the relative permeability (pNH_3/pNH_4^+) of biological membranes. While there are many estimates of this ratio in the literature (e.g., Cameron and Heisler, 1985; Evans and More, 1988; Evans et al., 1989), the outstanding characteristic is their variability, with pNH_3/pNH_4^+ ratios ranging from $<10:1$ to $>1000:1$. To a certain extent, these results may be tissue specific; for example, seawater gills appear to have greater NH_4^+ permeability than freshwater gills (Goldstein et al., 1982; Evans et al., 1989) due to the presence of more leaky paracellular channels and/or NH_4^+ transport mechanisms.

On a *net* basis, ammonia is produced as NH_3 by metabolism (in acid-base terms, this is equivalent to equimolar NH_4^+ and HCO_3^- production by the oxidation of proteins); therefore, by convention, the excretion of NH_3 is neutral with respect to acid–base balance, whereas the excretion of NH_4^+ represents the excretion of acidic equivalents. Many ammonoteles, especially freshwater teleosts, are carnivorous and sustain high rates of hepatic amino acid gluconeogenesis (Cowey et al., 1977; Bever et al., 1981). The resulting high rate of mitochondrial amino acid deamination should result in a high rate of efflux of ammonia across the inner mitochondrial membrane. There is a fundamental issue concerning which species of ammonia crosses the mitochondrial membrane. When the ionophore valinomycin is added to beef heart mitochondria respiring in NH_4Cl, there is an immediate uncoupling of electron transport-dependent phosphorylation as indicated by depletion of the medium O_2 (Brierly and Stoner, 1970). The postulated mechanism for this is that in the presence of valinomycin, the inner membrane, which is normally impermeable to cations, becomes permeable to NH_4^+, allowing it to penetrate into the matrix. Thus, acid equivalents are moved into the mitochondrial matrix and the proton gradient is reduced, thereby uncoupling phosphorylation (Brierly and Stoner, 1970; Campbell, 1991). It would appear, therefore, that the mitochondrial membrane is not very permeable to NH_4^+. If ammonia crosses the mitochondrial membrane as NH_3 (Flessner et al., 1991), it has the potential to raise pH in the cytosol and also reduce the proton gradient across the mitochondrial membrane. This effect will be mitigated normally by the parallel movement of CO_2. It is possible, however, that a transport protein may be present in ammonotelic liver mitochondria that transports NH_4^+ out of the matrix, augmenting the proton gradient established by the electrogenic H^+ pump. Thus, NH_4^+ production and transport, by enhancing a mitochondrial proton gradient, could increase ATP formation by the synthase complex (Campbell, 1997). There is considerable evidence for NH_4^+ translocation in several membrane systems involving substitution of NH_4^+ for K^+ in Na^+,K^+-ATPase and $Na^+/K^+/2\ Cl^-$ cotransport, and NH_4^+ for H^+ on apical membrane Na^+/H^+ exchangers (see Section III.D).

C. Why Is Ammonia Toxic?

Although NH_3 in water is considered the major toxic form in the environment, NH_4^+ is probably the major toxic element within the body. Body pH is around 7.0–7.8, varying with the tissue. Thus, more than 95% of total ammonia in the body exists as NH_4^+. As a result, most ammonia entering the body as NH_3 will be rapidly converted to NH_4^+. Elevated body ammonia levels are toxic and have both acute and chronic impacts that vary with the species.

Elevated water ammonia levels inhibit feeding and growth in Dover sole (Hampson, 1976; Alderson, 1979). Chronic ammonia toxicity results in gill hyperplasia (Burrows, 1964; Reichenbach-Klinke, 1967; Smart, 1976; Thurston *et al.*, 1978) and changes in mucous production, growth, and stamina (Lang *et al.*, 1987). Elevated water ammonia levels reduce swimming ability in coho salmon (Randall and Wicks, 2000), probably because increased ammonium ion levels in the fish cause muscle depolarization (Taylor, 2000). Ammonia can interfere with energy metabolism through impairment of the tricarboxylic acid (TCA) cycle (Arillo *et al.*, 1981). The decrease is due to inhibition of some key enzymes, including isocitrate dehydrogenase, α-ketoglutarate dehydrogenase, and pyruvate dehydrogenase (see review by Cooper and Plum, 1987). Ammonia stimulates glycolysis in fish by activation of phosphofructokinase in the muscle (Kloppick *et al.*, 1967). Fromm and Gilette (1968) observed increased pyruvate and lactate levels in the plasma of rainbow trout exposed to elevated water ammonia levels. It has also been suggested that ammonia affects the ionic balance in fish, reducing Na^+ influx and K^+ loss through substitution of NH_4^+ for K^+ in Na^+,K^+-ATPase and/or $Na^+/K^+/2\,Cl^-$ cotransport (see Wilkie, 1997, for review; Person-Le Ruyet *et al.*, 1998) and/or the substitution for H^+ in Na^+/H^+ exchanger [probably Na^+/H^+ exchangers type NHE2 and/or NHE3 (Randall *et al.*, 1999)].

Ammonia acts on the central nervous system of vertebrates, including fish, causing hyperventilation (Hillaby and Randall, 1979; McKenzie *et al.*, 1993), hyperexcitability, coma, convulsions, and finally death. Ammonia (NH_3) can cross the blood–brain barrier in mammals (Sears *et al.*, 1985) and high ammonia levels modify many aspects of the blood brain barrier (Cooper and Plum, 1987). In addition, NH_4^+ can substitute for K^+, affecting membrane potential in the squid giant axon (Binstock and Lecar, 1969). Elevated ammonia levels have been shown to impair glutamate and aspartate metabolism in mammals (Hindfelt *et al.*, 1977) and interfere with amino acid transport (Mans *et al.*, 1983). Smart (1978) suggested that the mechanism of ammonia toxicity in fish may be similar to the action of ammonia in mammals during hepatic encephalopathy. This action consists of depletion of phosphocreatine, glucose, glycogen, and ATP in the basilar region of the brain (Schenker *et al.*, 1967). In the past it was suspected to be a consequence of the suppression of the TCA cycle through depletion of α-ketoglutarate as it is converted to glutamate to remove ammonia. In some circumstances ammonia exposure may lead to the depletion of cerebral glutamate and thus remove an important neurotransmitter. Furthermore, the increased ATP demand for the glutamine synthetase reaction may account for the decrease in cerebral ATP noted in ammonia-exposed rainbow trout (Arillo et al., 1981). Because in all cases the cerebral pool of glutamine plus glutamate increases substantially, it is clear that anaplerotic reactions must be activated to refill the pool of C5 metabolites, including the reductive synthesis of glutamate, thus levying additional stress on the oxidative demands of the brain. To date, at least in mammals, hepatic encephalopathy has been shown to be related to the high level of extracellular glutamate (Felipo et al., 1994). High ammonia levels in the brain would induce an increase in the extracelluar level of glutamate due to increased release or decreased reuptake or both (Hilgier et al., 1991; Rao et al., 1992; Bosman et al., 1992, Schmidt et al., 1993). Increased extracellular glutamate results in increased activation of the NMDA receptor, leading to the entry of Ca^{2+} and Na^+ into the neuron (Marcaida et al., 1992). Increased intracellular Ca^{2+} would activate Ca^{2+}-dependent enzymes including protein kinases, protein phosphatases, and proteases. This would lead to decreased protein kinase C-mediated phosphorylation and concomitant activation of Na^+,K^+-ATPase. Increased ATPase activity would lead to consumption of larger amounts of ATP and could explain the depletion of brain ATP induced by the injection of large doses of ammonia.

D. Toxicity Studies

Acute toxicity has been determined for a number of species (EPA, 1984, 1989, 1998). There are fewer studies of chronic toxicity in aquatic species (Handy and Poxton, 1993) and this remains an important weakness in the available information. Experiments on chronic toxicity of total ammonia to fish in freshwater based on 30-day exposures measuring weight loss, change in total biomass, or hatchability indicate that chronic effects can be observed at one-half to one-twentieth of the acute toxicity level, the ratio of chronic to acute levels varying with the species. Chronic toxicity, expressed as total ammonia, is very pH dependent, increasing with pH in much the same way as acute toxicity.

Effects of temperature and pH. In many studies and reviews, ammonia toxicity is expressed as the concentration of NH_3 in water because the un-ionized NH_3 gradient is an important determinant of the rate of ammonia uptake, especially in freshwater animals. Water pH, and to a lesser extent, temperature, pressure, and ionic strength, have an effect on the NH_3/NH_4^+ equilibrium and, as a result, the [NH_3] in water (see Section II.A). In general, epithelial membranes are more permeable to ions in seawater than freshwater and so it might be expected that the contribution of ammonium ions in seawater to ammonia toxicity will be greater than in freshwater. Thus, the effects of pH on ammonia toxicity in seawater might be less than that observed in freshwater. Data from saltwater systems, how-

4. AMMONIA TOXICITY, TOLERANCE, AND EXCRETION 117

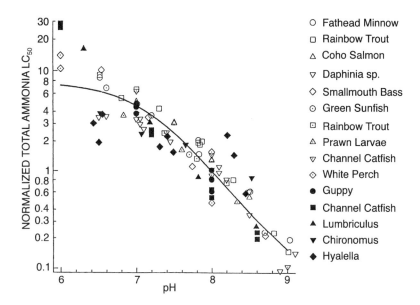

Fig. 1. Effect of pH on nomalized acute total ammonia toxicity.

ever, are insufficient to decide this matter. The acute and chronic toxicity of ammonia, when expressed as total ammonia in the water, increases with pH, and very low levels of ammonia are toxic if water pH exceeds 9.5 (Fig. 1).

In the EPA (1998) addendum, ammonia toxicity is expressed as milligrams total ammonia N/liter, whereas earlier criteria were expressed as mg NH_3, which varies with temperature. Data analysis indicates that toxicity expressed as total ammonia does not vary with temperature, eliminating the need for this factor in the new criteria (Fig. 2). In addition, most ammonia measurements are made as total ammonia and so expressing toxicity in terms of total ammonia rather than NH_3 eliminates the need to calculate un-ionized ammonia concentrations. Measurement of water pH is still required because it is an important determinant of ammonia toxicity.

Effects of water hardness. The hardness of ambient water can influence ammonia toxicity (Tomasso *et al.,* 1980; Soderberg and Meade, 1992). Addition of calcium to water with pH ranging from 6.5 to 9.0 decreases acute toxicity of ammonia to rainbow trout (Randall and Wicks, 2000). Elevated calcium levels activate apical membrane proton ATPase in the gills (Lin and Randall, 1995) and reduce gill permeability to ions. The former could increase the acid boundary layer around the gills and reduce ammonia entry into the fish. Exposure of Lahontan cutthroat trout to soft water leads to a reduction in ammonia excretion. Addi-

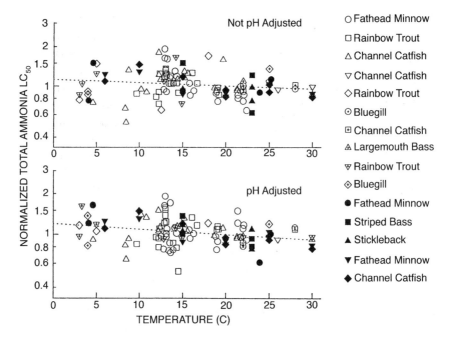

Fig. 2. Effect of temperature on normalized acute total ammonia toxicity.

tion of either Ca^{2+} or Mg^{2+} brings the ammonia excretion rate back to normal (Iwama et al., 1997). In addition, elevated calcium levels appear to reduce the cortisol response to stressful situations (Yesaki and Iwama, 1992). Cortisol stimulates protein catabolism; thus, reduced cortisol levels will reduce ammonia production by the fish and this may be the major way in which elevated calcium ameliorates ammonia toxicity (Wilson et al., 1998).

Effects of hypoxia. Reduced oxygen levels in the water increase ammonia toxicity in both freshwater (Merkens and Downing, 1957) and seawater (Wajsbrot et al., 1991). Thurston et al. (1981), however, could find no correlation between oxygen level and ammonia toxicity to fathead minnows. It is difficult to separate the toxic effects of hypoxia and ammonia, and there is some evidence for an additive toxic effect (Wajsbrot et al., 1991).

Effects of seawater. The body surface of marine animals is generally more permeable to ions than that of freshwater animals (Evans, 1984a). Thus, the passive flux of ammonium ions is likely to be greater in marine animals. Much more is known about ammonia toxicity in freshwater as compared with that for marine fish. Ammonia toxicity decreases with increasing salinity up to levels of 30–40% seawater, but then increases at higher salinity thereafter (Herbert and Shurben,

1965; Harader and Allen, 1983). Comparisons between freshwater and marine species are, to some extent, like comparing apples and oranges. Despite the difficulties of making such comparisons, the average of the species mean acute toxicity values for 32 freshwater species is 2.79 mg NH_3/liter compared with 1.86 mg NH_3/liter for 17 seawater species. The data to calculate these averages are taken from the 1984 and 1989 EPA reports, respectively. This comparison indicates that seawater species are more sensitive to ammonia toxicity than freshwater species. Average values for species mean acute toxicity for the five most sensitive species in the two data sets are 0.68 mg NH_3/liter for seawater species and 0.79 mg NH_3/liter for freshwater species, supporting the contention that seawater species are slightly more sensitive to ammonia toxicity. This is not the view of Haywood (1983), who suggested that the maximal permissible (harmless) level of un-ionized ammonia for marine fish was 40 μg/liter NH_3 in seawater, twice that suggested for nonsalmonid freshwater fish. In making these recommendations, he recognized the paucity of data with respect to marine species.

Criteria and standards. The criterion continuous concentration (CCC) is defined as the threshold value resulting in unacceptable effects, that is, more than a 20% reduction in survival, growth, and/or reproduction. The criterion maximum concentrations (CMC) is defined as half of the final mean acute value. Both the CCC and the CMC are pH dependent in freshwater, approaching zero environmental ammonia levels above pH 9.5 in freshwater animals. Except for a very sensitive species, the EPA (1998) considers it adequate to protect freshwater aquatic life if the following conditions are satisfied:

1. The 1-h average concentration of total ammonia nitrogen (in mg N/liter) does not exceed, more than once every 3 years on average, the CMC calculated as a function of site pH from:

   $$CMC = 0.385 \ (1 + 10^{7.15-pH})^{-1} + 63.1 \ (1 + 10^{pH-7.15})^{-1}.$$

2. The 30-day average concentration of total ammonia nitrogen (in mg N/liter) does not exceed, more than once every 3 years on the average, the CCC calculated as a function of site pH from:

   $$CCC = 0.0902 \ (1 + 10^{7.69-pH})^{-1} + 3.92 \ (1 + 10^{pH-7.69})^{-1}.$$

3. And, the highest 4-day average within the 30-day period does not exceed twice the CCC.

At pH 6.5 the revised CCC by the EPA (1998) is 3.48 mg N/liter total ammonia, whereas the new CMC is 32.5 (salmonids present) and 48.8 (salmonids absent) mg N/liter total ammonia (Fig. 3). The recommended CCC and CMC values are presented in Table I. For pH 8.0, the values are 1.27 and 8.40 mg N/liter total ammonia, respectively. Tests conducted in an experimental stream (Zischke and Arthur, 1987) indicated little change in biomass for numerous test

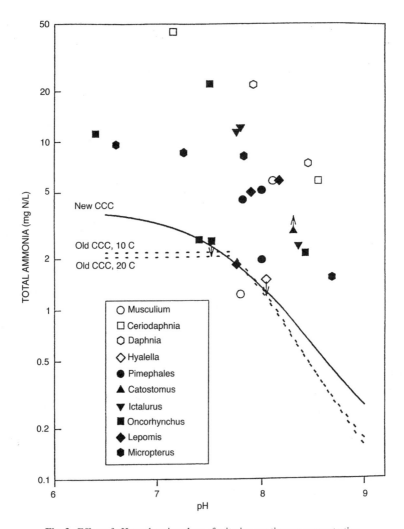

Fig. 3. Effect of pH on chronic values of criterion continuous concentration.

species unless the 4-day mean ammonia level exceeded the CCC. Therefore, based on the available data, the EPA standards for ammonia levels in freshwater appear to be appropriate. The national criteria promulgated by the EPA (1989) for saltwater are a CCC of 0.035 mg/liter of unionized ammonia and a CMC of 0.233 mg/liter un-ionized ammonia. This converts to a CCC of 0.99 mg N/liter total ammonia and a CMC of 6.58 mg N/liter total ammonia, somewhat less than the equivalent freshwater values of 1.27 and 8.4 (for nonsalmonids) mg N/liter total

Table I
U.S. EPA 1998 Ammonia Criteria Expressed as Criterion Continuous Concentration or Criterion Maxium Concentration[a]

Water pH	CCC	CMC for salmonids	CMC for non-salmonids
6.5	3.48	32.5	48.8
8.0	1.27	5.72	8.40
8.0 (seawater)	0.99	—	—
9.0	0.25	—	—

[a] CCC, the threshold value resulting in more than 20% reduction in survival, growth, and/or reproduction; CMC, the half mean acute value derived from 96-h LC_{50} tests; in mg N/liter.

ammonia, respectively, at pH 8.0. This is consistent with marine species being somewhat more sensitive to ammonia than freshwater species.

A number of other jurisdictions have derived ammonia criteria, several based on the recommendations of the EPA, coupled to site-specific criteria. In site-specific cases it might be required that monthly measurements of ammonia in the effluent be undertaken, associated with 96-h LC_{50} tests on the effluent. The discharge permit might require that specific ammonia levels (derived from the EPA-recommended chronic criteria multiplied by a calculated dilution factor) not be exceeded in the effluent at point of discharge and that 50% of the test fish survive in the effluent.

E. Problems with Present Ammonia Toxicity Criteria

Most criteria are based on the analysis of data carried out by the EPA (1984, 1989, 1998). The EPA criteria are based on North American species with an emphasis on Atlantic species. Only a few Pacific species are included in their data banks. They also state that the EPA criteria do not protect local sensitive species. Nevertheless, their criteria are the most exhaustive and, because the data were collected over a large geographical area, probably have some global applicability. The freshwater acute values are based on more data than those in saltwater and these in turn are better established than chronic values. Few data are available on chronic toxicity of ammonia to animals in saltwater.

Methods designated by the EPA for toxicity studies follow standard guidelines that require toxicity tests to be conducted under static conditions on unfed fish. Ammonia levels inside animals vary depending on the state of the animal; for example, ammonia levels are high following both feeding and stress. Elevated internal ammonia levels make the animal more sensitive to elevated levels of ammonia in the environment. Despite this clear fact, ammonia toxicity tests are carried out on starved, resting, stress-free animals, conditions that minimize normal internal ammonia levels. That is, toxicity tests have been carried out when animals are least sensitive to ammonia. Recent studies indicate that coho salmon (*Oncorhynchus kisutch*) show a linear reduction in critical swimming velocity (U_{crit}, a measure of swimming performance) with increases in ambient ammonia concentration (Randall and Wicks, 2000). These swimming fish not only had a net influx of ammonia from the water, but also had to cope with ammonia accumulation in muscle due to deamination of adenylates during exercise (Mommsen and Hochachka, 1988). Because the increase in environmental ammonia resulted in increased plasma ammonia and this decreased the swimming ability, it is probable that elevated ammonia could reduce swimming performance and decrease the survival of migrating salmon. Taylor (2000) has shown that elevated ammonia levels in the body result in muscle depolarization, which will in turn reduce swimming performance. Increased acute ammonia toxicity is evident in swimming fish in a set of experiments testing the effect of total ambient ammonia on the mortality of rainbow trout under resting and exercise conditions (60% U_{crit}). The calculated LC_{50} for the resting fish under the conditions of these experiments (in the swimming respirometer but not swimming) is 207 mg N/liter total ammonia. This value is well above the EPA ammonia acute toxicity value of 48 mg N/liter total ammonia at pH 7.0. The LC_{50} for swimming fish of 32 mg N/liter total ammonia is below the EPA's acute value. Thus, the present EPA ammonia standards will not protect migrating salmon.

In fish farms, ammonia excreted from fish or released by bacterial decomposition of nitrogenous wastes results in an accumulation of ammonia. Thus, ammonia toxicity is a special concern in intensive fish culture, especially in farms where the stocking density is high. Accumulation of ammonia is also associated with demersal fish eggs and viviparity involving a prolonged maternal–fetal trophic relationship. Fed fish are more sensitive to environmental ammonia exposure than unfed fish (Wicks, unpublished observations). Feeding results in elevated plasma ammonia levels. Animals exposed to elevated environmental ammonia stop feeding (Wicks, unpublished observations), which presumably reduces the probability of ammonia intoxication. Plasma ammonia levels following feeding can approach those observed in fish exposed to the 96-h LC_{50} test. The extent of increase in ammonia production following feeding will undoubtedly be related to the protein content of the food and ration size (Leung *et al.*, 1999). Thus, diets could be designed to reduce the magnitude of the postprandial ammonia pulse and, therefore, reduce the impact of feeding on ammonia toxicity. Other stresses, such as crowding or rapid shifts in water temperature, can also exacerbate ammonia toxicity due to increased ammonia production by the fish following cortisol release in response to these stresses. Thus, permissible ammonia levels may need to be more restrictive in aquaculture because of the crowded conditions associated with fish culture.

III. STRATEGIES TO REDUCE, TOLERATE, OR AVOID AMMONIA TOXICITY

Fish exposed on land or living in high-pH waters may have difficulties maintaining ammonia excretion. Other fish can live in ammonia solutions, which will impair ammonia excretion by diffusion. All of these situations can lead to ammonia accumulation by the fish. In the case of exposure to alkaline pH or terrestrial conditions, the fish have difficulty excreting ammonia that is internally produced. A similar situation may occur when fish are exposed to low levels of ammonia. However, the fish will have to detoxify not only endogenously produced ammonia but also that penetrating inward into the fish when the external ammonia concentration reaches a level that reverses the pNH_3 gradient. Available data indicate that various fishes might have adopted different mechanisms to avoid ammonia toxicity depending on its origin and the conditions that leads to ammonia intoxication.

Ammonia toxicity can be ameliorated by preventing ammonia accumulation in the body by either decreasing production, maintaining or enhancing excretion, and/or converting ammonia to less toxic compounds for storage or excretion. A broad outline of ammonia production and excretion is presented in Fig. 4.

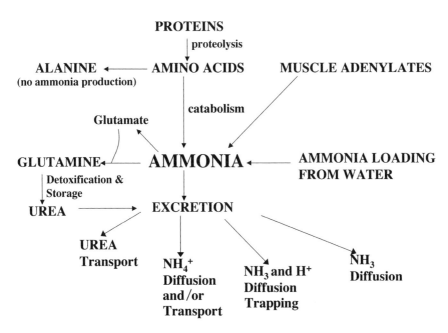

Fig. 4. Summary of strategies adopted by fish to ameliorate ammonia toxicity.

A. Reduction in Ammonia Production

After 24–48 h of starvation in mammals, liver glycogen is exhausted and lipolysis is enhanced to spare the remaining carbohydrate reserves and peripheral proteins; only after lipid depletion will extensive protein depletion occur, which is accompanied by an increase in urea production. On the other hand, muscle protein depletion is important in fish during periods of food deprivation (Mommsen et al., 1980). In fact, one of the primary sources of metabolic energy in carnivorous fishes is protein, rather than carbohydrate or lipids as in many mammals (Moon and Johnston, 1981). The main storage location of utilizable protein is white muscle. The amino acids released by proteolysis can be either oxidized by tissues for energy or converted to other utilizable forms by anabolic pathways.

If the rate of proteolysis remains the same, but the rate of amino acid catabolism decreases, ammonia production will decrease but the concentration of free amino acids will increase. Hence, increases in concentrations of free amino acids under an experimental condition in which fishes have difficulties in excreting ammonia may not indicate the detoxification of ammonia to amino acids as suggested by Iwata et al. (1981) and Iwata (1988) for the mudskipper *Periophthalmus cantonensis*. It may simply suggest a decrease in the rate of amino acid catabolism or a change in the catabolic pathways involved.

Although Wilkie and Wood (1995) suggested that there was no reduction of ammonia production in the rainbow trout (*O. mykiss*) exposed to high pH, a later study indicated that exposure of rainbow trout to pH 10 water does result in a reduction in ammoniagenesis (Wilson et al., 1998). The Lahontan cutthroat trout, *O. clarki henshawi*, appears to permanently lower its rates of nitrogenous waste production immediately following transfer from its juvenile freshwater habitat (pH 8.4) to Pyramid Lake, Nevada (pH 9.4) (Wilkie et al., 1994). The actual mechanism is not clear. Gordon et al. (1969, 1970, 1978) suggested that a common response to the lack of external water to carry ammonia away from the body of mudskippers that leave water for appreciable periods of time is a reduction in nitrogen metabolism. However, they did not give any data to support their proposition. The mudskipper, *Periophthalmodon schlosseri*, remains quiescent in total darkness (Ip, unpublished observation) and there is a decrease in tissue total free amino acids (TFAAs) when the mudskipper is exposed to terrestrial conditions under constant darkness (Fig. 5A). When *P. schlosseri* is exposed to 24 h of terrestrial conditions under a 12-h dark:12-h light regime, the fish can be very active and the levels of TFAA increase significantly in the tissues and plasma (Fig. 5B). An analysis of the balance sheet between the reduction in ammonia excretion and accumulation of ammonia, urea, alanine, glutamine, and glutamate in a 70-g *P. schlosseri* shows that catabolism of amino acids is suppressed in *P. schlosseri* exposed to the terrestrial condition under a dark:light regime, but there is still an accumulation of ammonia in the tissues (Lim et al., 2001). Reduction in both

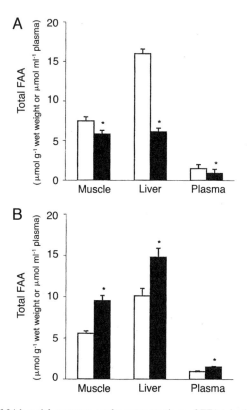

Fig. 5. Effect of 24-h aerial exposure on the concentrations of TFAA in the muscle, liver, and plasma of *P. schlosseri*; □ = submerged, ■ = exposed aerially (terrestrial condition). (A) Total darkness; (B) 12 h of light: 12 h of dark. * = Significantly ($P < 0.05$) different from submerged control.

proteolysis and amino acid catabolism is an effective strategy to slow the internal buildup of ammonia. However, it also reduces the utilization of amino acids as an energy source. Operating by itself, it may not be a good mechanism for fishes that are active during aerial exposure.

Unlike *P. schlosseri,* there is no increase in TFAA and no accumulation of glutamate or glutamine in the tissues of *Boleophthalmus boddaerti* during aerial exposure (Ip *et al.,* 1993). *B. boddaerti* also undergoes a reduction in the rate of amino acid catabolism under such a condition (Lim *et al.,* 2001). The absence of any significant accumulation of alanine, glutamine, or glutamate in the muscle supports this proposition. These data suggest that *B. boddaerti* does not rely on protein as an energy source during aerial exposure. Consequently, the decrease in the rate of ammonia excretion during 24-h aerial exposure leads to a small increase in ammonia in its tissues. Compared to the submerged, exercised controls, glycogen levels in the muscle of *B. boddaerti* decrease significantly after 24-h aerial exposure followed by exercise. Thus, *B. boddaerti* uses glycogen as metabolic fuel during aerial exposure. This strategy offers a limited amount of energy for a short period of time and, as a result, *B. boddaerti* does not stay away from water for long periods of time.

The capability of reducing amino acid catabolism during aerial exposure may not be a common phenomenon among fishes. For example, when the marble goby, *Oxyeleotris marmoratus,* is exposed to terrestrial conditions for 72 h, it does not undergo a reduction in amino acid catabolism (Jow *et al.*, 1999). Instead, protein or amino acid catabolism may have increased during aerial exposure (see Section III.C) because glutamine accumulates to levels far in excess of that needed to detoxify ammonia if production remains constant (Jow *et al.*, 1999).

B. Partial Amino Acid Catabolism Leading to the
Formation and Storage of Alanine

For mammals, alanine released from skeletal muscle plays an important role in the interorgan metabolism of carbon and nitrogen (Smith, 1986; Welbourne, 1987). Alanine is released by skeletal muscle and taken up by the liver where it is an important glucogenic precursor (Felig, 1975) and a regulator of protein synthesis (Peres-Sala *et al.*, 1987) and ketone body production (Nosadini *et al.*, 1980). Alanine is instrumental in supplying carbon backbones for *de novo* synthesis of glucose and, in fact, it is likely to be the most important gluconeogenic substrate *in vivo*. However, the existence of a glucose-alanine cycle, which tightly links extrahepatic alanine synthesis with muscular activity in fish, is doubtful (Mommsen and Walsh, 1991). First, rates of hepatic gluconeogenesis in fishes are generally insufficient to make the same important contribution to glucose turnover as in mammals. Second, it appears that ammonia and not alanine is the predominant product of muscular activity in fish muscle (Milligan and Wood, 1987; Mommsen and Hochachka, 1988; Wood, 1988). However, alanine production by extrahepatic tissues, especially the white muscle, is important during periods of extended starvation in fishes. Starvation leads to extensive muscular proteolysis (French *et al.*, 1983; Leech *et al.*, 1979; Mommsen *et al.*, 1980), and alanine appears to be the main amino acid exported from the muscles and available to the liver. Under such conditions, alanine serves as a gluconeogenic substrate for the fish liver.

In fish, alanine alone constitutes 20–30% of the total amino acid pool in the skeletal muscle (Hochachka and Guppy, 1987). Ip *et al.* (1993) reported increases in alanine content in the liver, muscle, and plasma (57, 22, and 56%, respectively) of *P. schlosseri* exposed to terrestrial conditions for 24 h. Similar phenomena were not observed in the muscle of aerially exposed *P. cantonensis* (Iwata *et al.*, 1981). Recently, Ip *et al.* (2001) verified again that alanine levels increased in the

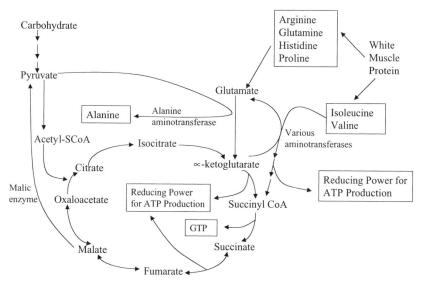

Fig. 6. Examples of partial amino acid catabolism leading to the production of ATP and the accumulation of alanine, without releasing ammonia.

tissues and plasma of *P. schlosseri* by at least twofold after 24 h of aerial exposure under a 12-h dark:12-h light regime.

Most of the free amino acids (FAAs) can be converted to alanine without releasing ammonia (Fig. 6). The overall quantitative energetics would appear to be quite favorable. The net conversion of glutamate to alanine would yield 10 mol of ATP per mole of alanine formed. This value would be higher for the conversion of proline or arginine to alanine (Hochachka and Guppy, 1987). Significant decreases in the levels of proline have been observed in the tissues of *P. schlosseri* after 24-h aerial exposure under a dark:light regime. Therefore, proline may also serve as an alternative substrate for alanine formation. This favorable ATP yield from amino acid catabolism is not accompanied by a net release of ammonia (Ip et al., 1993). Production and accumulation of alanine would therefore be advantageous to *P. schlosseri* when it is confronted with the problem of ammonia excretion during aerial exposure. In addition to the high levels of ATP generated during partial amino acid catabolism, alanine forms a useful carrier of amino acid carbon for further metabolism elsewhere. It can be converted to pyruvate for further oxidation in the heart, liver, and red muscle. Alternatively, it can serve as a glucose precursor in gluconeogenic tissues such as kidney or liver (Hochachka and Guppy, 1987). More importantly, it reduces the dependence on carbohydrate and spares the glycogen store till the need arises.

B. boddaerti does not rely on protein as an energy source but uses glycogen as metabolic fuel during aerial exposure (see above). Exercise during terrestrial exposure in *P. schlosseri*, however, does not affect the glycogen content of the muscle, but leads to the accumulation of alanine instead. This again verifies that *P. schlosseri* adopts the strategy that involves the formation of alanine, making ammonia excretion less of a problem during terrestrial conditions (Ip *et al.*, 2001).

According to Motimore and colleagues (Meijer *et al.*, 1990), a regulatory group of amino acids including glutamine, leucine, tyrosine, proline, histidine, tryptophan, and methionine that, together with the synergistic coregulator alanine (which does not inhibit proteolysis on its own), is responsible for the antiproteolytic response. For fish that have difficulty excreting endogenous ammonia, partial amino acid catabolism leading to the formation of alanine coupled with a decrease in the rates of proteolysis and amino acid catabolism would be the most cost-effective way to reduce the rate of ammonia buildup. This would allow amino acids to be used as an energy source during adverse conditions without polluting the internal environment. Two other obligatory air breathers, *Channa asiatica* (snake head) and *Monopterus albus* (eel), which can be quite active on land, adopt partial amino acid catabolism as a strategy to avoid ammonia toxicity during aerial exposure (Ip and Chew, unpublished results).

C. Ammonia Detoxification and Storage as Glutamine and/or Urea

1. Glutamine

Ammonia toxicity can be avoided by converting ammonia to less toxic compounds like glutamine and urea. Glutamine is produced from glutamate and NH_4^+, the reaction catalyzed by glutamine synthetase in the muscle and/or liver (Fig. 7). Glutamate may in turn be produced from α-ketoglutarate and NH_4^+, catalyzed by

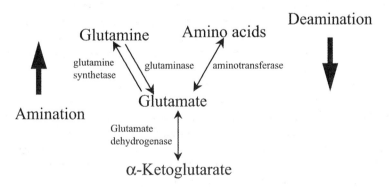

Fig. 7. Glutamine and glutamate metabolism.

glutamate dehydrogenase, or α-ketoglutarate and other amino acids catalyzed by various transaminases. In other words, formation of one glutamine molecule incorporates two ammonia molecules (Campbell, 1973). The marble goby, *O. marmoratus*, is a facultative air-breather capable of surviving under terrestrial conditions for up to several days, but ammonia levels in its muscle do not increase during aerial exposure (Jow et al., 1999). Aerially exposed *O. marmoratus* do not detoxify ammonia to urea, despite the fact that all five enzymes of the ornithine–urea cycle are present in the liver. The glutamine content in the muscle of *O. marmoratus* exposed to terrestrial conditions, however, increases threefold in 72 h (Jow et al., 1999). The hepatic glutamine content peaks at 24 h of aerial exposure, whereas that in the muscle peaks only after 24 h. This indicates that the glutamine formed in the liver is perhaps later shuttled to muscle, which acts as a reservoir for glutamine accumulation. Consequently, a steady-state glutamine level in the plasma is maintained. The glutamine level rises by about 0.7 μmol/g in the muscle after 72 h of aerial exposure. Because approximately 60% of a 500-g *O. marmoratus* is muscle, about 210 μmol of glutamine would have been accumulated (Jow et al., 1999). Because 2 mol of NH_4^+ are needed to produce 1 mol of glutamine, 420 μmol of ammonia would have been detoxified in this manner. This would be more than enough to account for the 161 μmol of ammonia produced during the entire period, assuming rates of production similar to that observed in water. It is unlikely, therefore, that amino acid catabolism has been reduced, but rather that the ammonia produced has been channeled toward the formation of glutamine instead. Protein or amino acid catabolism in *O. marmoratus* may in fact increase during aerial exposure, since glutamine accumulates to levels far in excess of that needed to detoxify ammonia predicted to be produced during this period. To our knowledge, glutamine has only been reported to play a role in ammonia detoxication in fish in response to high environmental ammonia concentrations (Levi et al., 1974; Arillo et al., 1981, Dabrowska and Wlasow, 1986; Mommsen and Walsh, 1992; Peng et al., 1998). Thus, *O. marmoratus* may represent the first fish reported to be able to detoxify internally generated ammonia by glutamine formation.

There is also an increase in the hepatic glutamine synthetase activity in *O. marmoratus* exposed to air. There is a good correlation between the peak of hepatic glutamine synthetase activity and the peak of glutamine accumulation in the muscle of the fish exposed to terrestrial conditions. Hepatic glutamate dehydrogenase activity also increases but there is no increase in glutamate. Hence, the glutamate produced must have been channeled into the formation of glutamine (Jow et al., 1999). In other cases where environmental ammonia is detoxified to glutamine (Levi et al., 1974; Arillo et al., 1981; Peng et al., 1998), neither cerebral nor hepatic glutamine synthetase activity is found to be significantly different from control values. This indicates that preexisting glutamine synthetase activity is sufficient to account for the increased glutamine formation (Walsh et al., 1993;

Wright et al., 1993). Although Peng et al. (1998) observed some increase in the specific activity of cerebral glutamine synthetase in mudskippers in response to ammonia loading, those from the liver remained unchanged. Hepatic glutamine synthetase activity increases several-fold in the gulf toadfish *Opsanus beta* kept under conditions of confinement/crowding, leading to an induction of ureotely (Walsh et al., 1994).

Contrary to the production of alanine, formation of glutamine is energetically expensive. One mole of ATP is required for the production of every amide group of glutamine catalyzed by glutamine synthetase. If the starting point is ammonia and α-ketoglutarate, every mole of ammonia detoxified formed would result in the hydrolysis of 2 mol of ATP equivalents. It is logical to utilize this pathway during critical situations, especially to protect the brain during ammonia loading. However, it is uncertain why *O. marmoratus* would adopt this strategy to reduce ammonia buildup during aerial exposure. Another fish that adopts the same strategy is *Bostrichthys sinensis* (Ip and Chew, unpublished results). Both fish are relatively inactive on land and the reduced energy demand for muscular activity may provide them with the opportunity to exploit glutamine formation as a means to detoxify ammonia. Alternatively, glutamine may be an important substrate for the synthesis of other compounds essential to their survival in adverse conditions.

Because ammonia exerts its toxic effects on the brain, it is important to have mechanisms to protect the brain by converting the toxic ammonia to nontoxic forms. The brain is often the organ undergoing the largest increase in glutamine concentration in fish exposed to ammonia. Glutamine synthetase activity occurs in high levels in fish brain with a binding constant for ammonia in the micromolar range (Webb and Brown 1980; Mommsen and Walsh, 1992). Rainbow trout exposed to un-ionized ammonia show an increase in glutamine levels in the brain (Arillo et al., 1981). There is a linear correlation between cerebral glutamine levels in goldfish and ambient ammonia concentrations up to 0.75 mM NH_4Cl (Levi et al., 1974). Plasma ammonia levels increase threefold in goldfish exposed to 0.75 mM NH_4Cl for 24–48 h. Brain glutamine increases 10-fold and glutamine shows the highest percentage increase compared to other amino acids studied. The common carp, *Cyprinus carpio*, exposed to 0.1 mg NH_3/liter has increased levels of amino acids in the brain and muscle, by 30 and 24%, respectively (Dabrowska and Wlasow, 1986). Glutamate increases significantly from 171 to 231 mg/100 g dry weight, providing a pool of substrate for glutamine synthesis, since the amount of glutamate supplied by the blood is not sufficient to maintain the rate of glutamine synthesis (Dabrowska and Wlasow, 1986).

Glutamine in the brain of ammonia-exposed teleosts may be converted to glutamate by glutaminase after the ammonia insult has subsided (Mommsen and Walsh, 1992). However, glutamine is more likely to be released from the brain under chronic ammonia exposure. Glutamine is then transported to the liver, kidney, and gills of ammoniotelic teleosts where it is deaminated by glutaminase, which is present in high activity in these tissues (Mommsen and Walsh, 1992). Glutamine can also be used for urea synthesis if the capability exists in that fish. The presence of glutamine synthetase activity in fish liver is variable (Campbell and Anderson, 1991). It is below the level of detection in many nonureosmotic fishes like the chinook salmon (*Onchorynchus tshawytscha*), starry flounder (*Platichthys stellatus*) and copper rock fish (*Ictalurus punctatus*). In contrast, in the liver of ureosmotic fishes (e.g., elasmobranch), the level of activity is higher than that in the brain (Campbell and Anderson, 1991). It is observed that the activity of glutamine synthetase in elasmobranch liver is at least 3–12 times higher than that in teleost liver (Webb and Brown, 1980). In agreement with the low level of glutamine synthetase in teleost liver, less than a 20% increase in hepatic glutamine is noted together with a corresponding decrease in glutamate in rainbow trout exposed to ammonia (Arillo *et al.*, 1981). For goldfish exposed to ammonia, hepatic ammonia levels increase from 3 to 7.8 μmol/g tissue, while glutamine increases only slightly (Levi *et al.*, 1974). Exposure of Lake Magadi tilapia to ammonia does not result in an activation of glutamine synthetase in the brain or liver (Walsh *et al.*, 1993), suggesting that preexisting glutamine synthetase levels may be sufficient to account for the increased glutamine formation in the liver of ammonia-exposed fish (Wright *et al.*, 1993).

For mudskippers [*P. schlosseri* and *B. boddaerti* (Peng *et al.*, 1998); *P. cantonensis* (Iwata, 1988)] exposed to high ambient ammonia concentrations, the TFAA concentrations in the brain, liver, and muscle increase, largely due to increases in glutamine. High glutamine synthetase activities are present in the brains of *P. schlosseri* and *B. boddaerti*. The activities of glutamate dehydrogenase (aminating) from the brains of these two mudskippers increase significantly when exposed to ammonia. Hence, *P. schlosseri* apparently utilizes different biochemical pathways to deal with being active during aerial exposure versus ammonia loading. During aerial exposure, it undergoes partial amino acid metabolism, forming alanine and reducing ammonia production (see Section III.B). When exposed to ammonia, the penetrating NH_3 is detoxified to glutamine. How these two biochemical pathways are regulated in response to increases in internal ammonia levels is uncertain at this moment.

In mammals, glutamine plays a major role as a transporter of nitrogen and carbon within the body and is the major precursor of urinary ammonia during metabolic acidosis (Welbourne, 1987; Welbourne and Joshi, 1990). Skeletal muscle is believed to be the major site of net glutamine synthesis in mammals, with smaller contributions being made by the liver (Welbourne and Joshi, 1990), lung (Ardawi, 1990), brain (Grill *et al.*, 1992), and adipose tissue (Frayan *et al.*, 1990). Upon release from the skeletal muscle, glutamine is extracted by the gut, where it serves as a major oxidative substrate (Windmueller and Spaeth, 1978), and by the kidneys, where it plays a vital role in the regulation of acid–base balance (Goldstein, 1986; Welbourne, 1987). It has been shown that glutamine is synthesized in rat skeletal muscle by the amidation of glutamate with NH_4^+ (Chang and Goldberg, 1978a,b; Rowe, 1985). The mammalian liver also contains considerable activities of the enzymes glutamine synthetase and glutaminase and has the capacity for either the net synthesis or net degradation of glutamine. For mitochondria isolated from the red muscle of *Salvelinus namaycush* or *Amia calva*, glutamine serves as the best oxidative amino acid substrate *in vitro* (Chamberlin et al., 1991). Glutamine-derived carbon appears to be channeled into the TCA cycle via glutamate dehydrogenase and complete oxidation of glutamine carbon in red muscle may also involve mitochondrial malic enzyme.

2. Urea

Urea forms an appreciable component of nitrogen output even when fish are in water (Campbell and Anderson, 1991; Wood, 1993; Saha and Ratha, 1998). The major pathways for urea synthesis in teleosts are arginolysis, uricolysis, and the urea cycle (Mommsen and Walsh, 1989, 1992; Anderson, 1995; Walsh, 1997; Wright, 1995; Chapter 7 of this volume). Although the urea cycle enzymes are probably present in all teleost fishes, perhaps expressed only during embryogenesis in most cases, and uricolysis is known to occur in many species, urea synthesis as the major mechanism of ammonia detoxification appears to occur in only a few species of teleost fish, usually as a unique adaptation to unusual environmental circumstances.

To date, three champions of ammonia tolerance have left their mark in the comparative literature. Both *Heteropneustes fossilis* (Saha and Ratha, 1998) and *Oreochromis alcalicus grahami* (Randall et al., 1989) detoxify ammonia to urea via the urea cycle as adaptations to air exposure and highly alkaline conditions, respectively. But, the third, the mudskipper *P. schlosseri*, survives 446 μM NH_3 by actively excreting NH_4^+ (see Section III.D) and detoxifying ammonia by converting it to glutamine.

Detailed information on the topics of the evolution, synthesis, and function of the urea cycle and of urea excretion in fishes can be found in Chapters 1 and 5–8 in this volume and will not be discussed further here.

D. Ammonia Excretion

1. Ammonia Diffusion

In general, ammonia is transferred from one compartment to another by diffusion of NH_3 down the NH_3 partial pressure gradient (Fig. 8). This requires the presence of water channels in the membrane and, if they are absent, a lipid membrane of high fluidity and, therefore, permeable to NH_3. Diffusion of NH_3 across the body surface, usually the gills, is the major pathway for excretion in most aquatic animals. Gills collapse in air and terrestrial exposure reduces ammonia

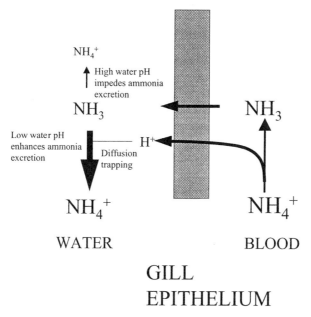

Fig. 8. Movement of NH_3 across the branchial epithelia of fish.

excretion in many fish. In a few fish, ammonia can be released into air by volatilization (Wilson, Randall, and Ip, unpublished results). The weather loach, *Misgurnus anguillicaudatus,* is capable of releasing 30% of the ammonia excreted as NH_3 during aerial exposure (Ip, Chew, Randall, and Wilson, unpublished results).

2. AMMONIA TRAPPING

In water at pH 7.0, less than 1% of any ammonia will be present as NH_3, and NH_3 excreted by the fish into water of this pH will be converted to NH_4^+. This conversion of NH_3 to NH_4^+ (ammonia trapping) will maintain the NH_3 gradient across the gills and enhance ammonia excretion (Fig. 8). Acidification of water next to the gills will, therefore, aid ammonia excretion (Wright *et al.,* 1989). Carbon dioxide and/or proton excretion via a gill epithelium apical proton pump will acidify gill water (Randall and Wright, 1989; Lin and Randall, 1990; Wilkie and Wood, 1991, 1994, 1995). There is no evidence, however, of manipulation of acid excretion to enhance ammonia excretion in fish, except possibly in the mudskipper, *P. schlosseri.* This fish has the capability of adjusting the pH of small volumes of seawater (Hong, 1997). When placed in 5 times its own volume of 50% seawater at pH 9 in the presence of 2 mM Tris buffer, it can lower the external pH by 1.30 units within 6 h. Stabilization of pH at a value close to 7 occurs after approximately 12 h. This water acidification occurs even if carbon dioxide from endogenous production is removed from the water by aeration. Under these conditions acidification is augmented by increased ammonium ion excretion (Ip, Chew, and Randall, unpublished data).

Mudskippers live in burrows formed in the soft mud of mangrove swamps and emerge from these burrows to feed during low tide (Ip *et al.*, 1990). In addition, they rear the developing embryos in this burrow. The growth of the mudskipper embryos requires the catabolism of yolk protein and this results in the production of ammonia. The burrow water is hypoxic and hypercapnic and has high ammonia levels. Because the burrow is poorly flushed, acidification of the small amount of burrow water by the fish is feasible. Acidification of the burrow water to pH 7.0 lowers the concentration of NH_3 in the external medium and therefore reduces the toxicity of ammonia to both the embryos and the fish themselves.

3. ACTIVE TRANSPORT OF AMMONIUM IONS

The mudskipper *P. schlosseri* lives on the mudflats of mangrove swamps in Singapore and Malaysia. This fish has the greatest reported ammonia (NH_3) tolerance among teleosts (Ip *et al.*, 1993; Peng *et al.*, 1998). Even though tissue levels are only around 0.2 mM, *P. schlosseri* exhibits no increase in tissue ammonia concentrations after 6 d of exposure to 8 mM NH_4Cl (36 μM NH_3) at pH 7.0. This indicates that either ammonia is not entering the fish or ammonia that entered is being removed. However, fish exposed to 8 mM NH_4Cl at pH 9 died within 2 h, presumably due to rapid NH_3 entry resulting from elevated NH_3 levels in the water. This, together with the fact that tissue ammonia levels are elevated in fish exposed to 100 mM NH_4Cl (446 μM NH_3), indicates that ambient ammonia is entering the fish. Unlike the Lake Magadi tilapia (Randall *et al.*, 1989), *P. schlosseri* does not convert accumulated ammonia to urea (Peng *et al.*, 1998). Thus, it would appear that ammonia entering the fish is either stored or excreted to keep tissue ammonia levels stable. Ammonia excretion is maintained when the fish are exposed to up to 30 mM NH_4Cl for 6 days, indicating that storage must play only a minor role and the majority must be excreted. Because the excretion rate of ammonia is constant and independent of the external ammonia concentration, it can be assumed that the efflux of ammonia increases to offset the increase in ammonia influx. Under such conditions, because both the NH_4^+ and NH_3 levels in the external media were higher than those in the blood, and because the transepithelial potential was not strong enough to maintain the NH_4^+ gradient (Randall *et al.*, 1999), excretion of ammonia by *P. schlosseri* must be active.

The gills of *P. schlosseri* are unique in having intrafilamentous interlamellar fusions (Low *et al.*, 1988, 1990; Wilson *et al.*, 1999), which help to prevent desiccation of the gill epithelium while the fish is out of water. However, these fusions also effectively extend the diffusion distance from the blood space to the apex of the lamellae. Hence, the gills of this mudskipper are not designed for gas exchange (Wilson *et al.*, 1999). Most of the large number of mitochondria-rich cells in the gills of *P. schlosseri* are isolated from one another by filament-rich cells of unknown function (Wilson *et al.*, 1999). Such an arrangement has not been observed in the branchial epithelium of other marine fishes. Also, the high density of mitochondria-rich cells within the tissue makes it an oxygen-demanding organ rather than a site for whole animal oxygen uptake. A study comparing enzyme profiles in the gills of *P. schlosseri* and *B. boddaerti* highlights the high activity of the gills of *P. schlosseri* (Low *et al.*, 1993). The former has (1) a greater potential for glycolytic flux (4-fold greater phosphofructose-kinase-1 activity), (2) a higher aerobic capacity (4-fold greater citrate synthase activity), (3) a greater capacity for utilizing exogenous glucose (four-fold higher hexokinase activity), and (4) a greater capacity for lactate production (high pyruvate kinase and lactate dehydrogenase activities, and higher pyruvate reductive capacity:lactate oxidative capacity ratio). Indeed, in contrast to gills of other fishes, which are lactate-consuming organs, the gill of *P. schlosseri* is a lactate-producing organ that accumulates lactate during environmental hypoxia (Low *et al.*, 1993; Ip and Low, 1990). Also, the branchial Na^+,K^+-ATPase activity is very high, 3-fold higher than that of *B. boddaerti* and approximately 10-fold higher than those of other fishes (Peng, 1994; Randall *et al.*, 1999). The abundance of mitochondria-rich cells within the lamellar epithelium of *P. schlosseri* is highly unusual for a marine fish (Wilson *et al.*, 1999). In addition, most of the mitochondria-rich cells are not associated with accessory cells and typical shallow tight junctions, which are of functional significance for NaCl elimination in other marine fish. It is possible that the gills of this mudskipper are involved in the elimination of ammonia against a concentration gradient.

The mammalian kidney tubule and collecting duct have several different mechanisms for the translocation of NH_4^+. For example, NH_4^+ will substitute for K^+ on the ouabain-sensitive Na^+,K^+-ATPase (Kurtz and Balaban, 1986; Wall, 1996) and on the furosemide (bimetanide)-sensitive $Na^+/K^+/2\ Cl^-$ cotransporter (Good, 1994). Recently, there has been a report of a specific verapamil-sensitive K^+-NH_4^+ antiporter (Amlal and Soleimani, 1997). These and other NH_4^+ transport systems have also been reported in tissues other than kidney. For example, transport of NH_4^+ takes place on the Na^+,K^+-ATPase of crab gill (Towle and Holleland, 1987) and insect rectal cells have an amiloride-sensitive system for the antiport of Na^+ and NH_4^+ (Thomson *et al.*, 1988). For the most part, these transport systems are involved in the "excretion" of NH_4^+ and H^+. Branchial ammonia excretion in dogfish pups is sensitive to the loop diuretic bumetanide, indicating $Na^+/2\ Cl^-/NH_4^+$ cotransport (Evans and More, 1988), via the substitution of NH_4^+ for K^+on a basolateral $Na^+/2\ Cl^-/K^+$ cotransporters (Good, 1988), may be present in most marine fish gills (Karnaky, 1986; Zadunaisky, 1984).

Substitution of NH_4^+ for K^+ in Na^+,K^+-ATPase has been proposed as a method of branchial ammonia excretion in fish (Claiborne *et al.*, 1982; Evans and Cameron, 1986). Because marine fishes have higher basolateral Na^+,K^+-

ATPase activities than their freshwater counterparts (Epstein et al., 1980), basolateral Na^+/NH_4^+ exchange could be important in seawater-adapted fishes. Indeed, NH_4^+ may even be a more effective counter ion than K^+ for the Na^+, K^+-ATPase in toadfish gills (Mallery, 1983). Because NH_4^+ extrusion across dogfish and scupin gills is inhibited when ouabain or K^+ are applied to the basolateral portion of the gills, it suggests that significant basolateral Na^+/NH_4^+ exchange takes place in the gills of some marine fishes (Claiborne et al., 1982). Activities of Na^+,NH_4^+-ATPase and Na^+,K^+-ATPase from P. schlosseri are comparable, and both are ouabain sensitive. Therefore, it is highly probable that there is a substitution of NH_4^+ for K^+ and that Na^+,K^+-ATPase and Na^+,NH_4^+-ATPase are the same transporter (Fig. 9). Inhibition of ammonia excretion and accumulation of ammonia in the plasma of P. schlosseri were observed when ouabain was added to seawater containing 2 mM NH_4Cl. A $Na^+/K^+/2\ Cl^-$ cotransporter as well as Na^+,K^+-ATPase have been localized by immunohistochemistry on the basolateral membrane of the mitochondrial rich cells of P. schlosseri gills (Wilson et al., 2000), so it is possible that ammonium ions enter the gill epithelium via a $Na^+/K^+/2\ Cl$ cotransporter as well (Fig. 9).

Substitution of NH_4^+ for H^+ has been proposed as a method of branchial ammonia excretion across the apical surfaces of chloride cells in fish (Claiborne et al., 1982). Addition of amiloride to seawater results in inhibition of ammonia excretion by P. schlosseri, indicating that a Na^+/H^+ exchanger is involved in ammonia excretion. The presence of NHE2 and NHE3 in the apical crypt of the mitochondrial-rich cells of P. schlosseri was indicated by immunoflourescence (Wilson et al., 2000). Potassium nitrate (100 mM), however, had no effect on ammonia excretion, indicating that v-ATPases (H^+ pump), although present (Wilson et al., 1999), may have little involvement in ammonia excretion in this fish.

In further support of Na^+/NH_4^+ exchange and Na^+,NH_4^+-ATPase being major transporters in branchial ammonium ion excretion during high environmental ammonia levels, there is an increased accumulation of Na^+ in the plasma of P. schlosseri in relation to an increase in the external ammonia concentration (Ip et al., 2000). Because Na^+,K^+-ATPase is pumping Na^+ into the blood in exchange for NH_4^+, a favorable Na^+ chemical gradient develops within the cell to draw Na^+ into the cell from the external environment in exchange for NH_4^+, leaving the cell at the apical surface. This mechanism would result in the accumulation of Na^+ in the blood and facilitate the excretion of ammonia out of the body across the gill epithelia (Fig. 9).

The branchial Na^+/NH_4^+ exchange hypothesis was originally proposed by August Krogh 60 years ago (Krogh, 1939). Numerous studies apparently supporting this hypothesis have been published in recent years (Maetz and Garcia-Romeu, 1964; Payan, 1978; Payan et al., 1975, Wright and Wood, 1985, McDonald and Milligan, 1988; McDonald and Prior, 1988; Yesaki and Iwama, 1992), leading to the widely held view that Na^+/NH_4^+ exchange is the major mode of ammonia

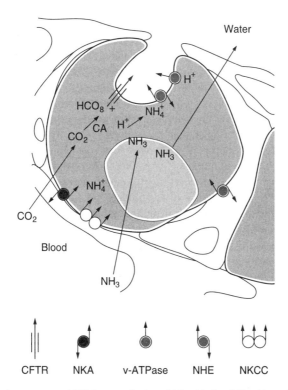

Fig. 9. Active transport of NH_4^+ across the branchial epithelia of *P. schlosseri*. Ammonia enters the mitochondria-rich (MR) cells from the blood either by NH_3 diffusion or active transport by $Na^+,K^+(NH_4^+)$-ATPase (NKA) and/or $Na^+/K^+(NH_4^+)/2\ Cl^-$ cotransporter (NKCC). Intracellular carbonic anhydrase (CA) provides H^+ for NH_3 protonation by catalyzing CO_2 hydration. The NH_4^+ formed is moved across the apical membrane in exhange for Na^+ by a Na^+/H^+ exchanger (NHE)-like carrier. The accumulated HCO_3^- exits the cell apically via a cystic fibrosis transmembrane receptor (CFTR)-like anion channel. The proton pump (v-ATPase) maintains a low pH for the external water, keeping the NH_4^+ outside the MR cells.

efflux across fish gills. After surveying the literature, however, Wilkie (1997) held the opinion that apical Na^+/NH_4^+ exchange is of minor to moderate importance in marine fishes and is likely to be entirely absent from freshwater fish species.

Early conclusions favoring Na^+/NH_4^+ exchange in marine fishes were based on data generated via manipulation of external ammonia or Na^+ concentrations. Evans (1977) reported that increased ambient NH_4^+ leads to declines in Na^+ uptake in four species of marine fishes, including the toadfish. Subsequent reports that ammonia flux was markedly depressed in toadfish, as well as in dogfish pups, exposed to Na^+-free water indicate that ammonia flux is partially dependent on Na^+ as a counterion (Evans, 1982). However, Na^+-free water has no effect on the ammonia flux in skates (*R. erinacea*) and hagfish (Evans, 1984b; Evans *et al.*, 1979). Evans and Cameron (1986) suggest that a Na^+-free effect, leading to reductions in the NH_4^+ permeability of branchial paracellular channels, may account for reduced ammonia flux in some of these Na^+-free water experiments. Whereas several studies demonstrate an amiloride-induced blockade of ammonia flux in marine invertebrates (Hunter and Kirschner, 1986; Pressley *et al.*, 1981), similar experiments in marine fishes are without effect. In the case where amiloride inhibits the ammonia flux in the toadfish, the effect may be due to an amiloride-induced inhibition of the basolateral Na^+,K^+-ATPase (Soltoff and Mandel, 1983) because ammonia flux patterns are unaltered when amiloride is added to the perfusate in the presence of ouabain (Evans *et al.*, 1989). Because NH_4^+ is likely to compete with K^+ for binding sites on Na^+,K^+-ATPase, these results suggest the presence of basolateral, not apical, Na^+/NH_4^+ in the toadfish. However, as Wilkie (1997) rightly points out: "Although significant apical Na^+/NH_4^+ exchange can probably be ruled out in freshwater fishes and at least four marine species, it should be kept in mind that only a tiny fraction of the world's total fish population has been studied. Since marine fish gills are likely to contain Na^+/H^+ antiporters, the possibility of apical Na^+/NH_4^+ exchange in these animals cannot be discounted." To date, *P. schlosseri* is one specimen that proves Wilkie's (1997) prediction correct.

During a terrestrial excursion, ammonia presumably accumulates in the water-filled interlamellar spaces in the gills of *P. schlosseri*. Passive diffusion of NH_3 down its partial pressure gradient into these restricted interlamellar water spaces would no longer function, and NH_4^+ would then be actively excreted into these spaces. Presumably, NH_4^+ enters the mitochondria-rich cells from the blood via the Na^+,K^+-ATPase and out into the interlamellar space by Na^+/H^+ exchanger. The rise in blood pH observed following air exposure (elimination of NH_4^+ represents the movement of acid equivalents; Kok *et al.*, 1998) and increased Na^+ uptake during ammonia loading (Randall *et al.*, 1999; Ip *et al.*, 2000) corroborate this hypothesis.

Active pumping of NH_4^+, in the absence of back diffusion of NH_3, is energetically more efficient than turning ammonia into urea or glutamine. Assuming NH_4^+ has an affinity similar to or greater than that of K^+ to the K^+-binding site in Na^+,K^+-ATPase (Mallery, 1983), 1 mol of ATP is utilized for every 2 mol of NH_4^+ eliminated. The question is why such a mechanism is not more widely adopted by fish. For such a mechanism to function, there must be a mechanism for the fish to prevent the back diffusion of NH_3 once the level of ammonia builds up in the environment. There are two possible solutions to such a problem. First, the membranes of cells, especially those of the gills and skin of *P. schlosseri*, may be relatively impermeable to NH_3 as has been suggested for the apical membrane of the renal tubule cells (Kikeri *et al.*, 1989). Second, if *P. schlosseri* has the capability of acidifying the external environment, the NH_4^+ excreted will stay in the ionized form and will be prevented from penetrating back to the tissues as NH_3. In conclusion, it appears that *P. schlosseri* is able to tolerate high environmental ammonia concentrations by actively excreting ammonium ions across the gills, thereby maintaining low tissue ammonia concentrations. Furthermore, it can excrete acid in response to high levels of ambient ammonia, which lowers the toxicity of ammonia to the fish and the embryos developing within the burrow.

IV. SUMMARY

Ammonia is an unusual toxicant in that it is produced by, as well as being poisonous to, animals. In aqueous solution ammonia has two species, NH_3 and NH_4^+, total ammonia is the sum of $[NH_3] + [NH_4^+]$ and the pK of this ammonia/ammonium ion reaction is around 9.5. The NH_3/NH_4^+ equilibrium both internally in animals and in ambient water depends on temperature, pressure, ionic strength, and pH; pH is most often of greatest significance to animals.

Elevated ammonia levels in the environment are toxic. Temperature has only minor effects on ammonia toxicity expressed as total ammonia in water, and ionic strength of the water can influence ammonia toxicity, but pH has a very marked effect on toxicity. Acid waters ameliorate, whereas alkaline waters exacerbate ammonia toxicity. The threshold concentration of total ammonia ($[NH_3] + [NH_4^+]$) resulting in unacceptable biological effects in freshwater, promulgated by the EPA (1998), is 3.48 mg N/liter at pH 6.5 and 0.25 mg N/liter at pH 9.0. There is only a relatively small saltwater data set, and a paucity of data on ammonia toxicity in marine environments, particularly chronic toxicity. The national criteria promulgated in the EPA (1989) saltwater document is a criterion continuous concentration (chronic value) of 0.99 mg N/liter total ammonia and a criterion maximum concentration (half the mean acute value) of 6.58 mg N/liter total ammonia, somewhat less than the equivalent freshwater pH 8.0 values of 1.27 and 8.4 mg N/liter total ammonia, respectively. This is consistent with marine species being somewhat more sensitive to ammonia than freshwater species.

Toxicity studies are usually carried out on unfed, resting fish in order to facilitate comparison of results. Based on recent studies, however, environmental stresses, including swimming, can have dramatic effects on ammonia toxicity. It is also clear that feeding results in elevated postprandial body ammonia levels. Thus, feeding will probably also exacerbate ammonia toxicity. Fish may be more susceptible to elevated ammonia levels during and following feeding or when swimming. Thus, present ammonia criteria may fail to protect migrating fish and may be inappropriate for fish fed on a regular basis.

Most teleost fish are ammonotelic, producing and excreting ammonia by diffusion of NH_3 across the gills. They are very susceptible to elevated tissue ammonia levels under adverse conditions. Some fish avoid ammonia toxicity by utilizing several physiologic mechanisms. Suppression of proteolysis and/or amino acid catabolism may be a general mechanism adopted by some fishes during aerial exposure or ammonia loading. Others, like the mudskipper, can undergo partial amino acid catabolism and use amino acids as an energy source, leading to the accumulation of alanine, while active on land. Some fish convert excess ammonia to less toxic compounds including glutamine and other amino acids for storage. A few species have active ornithine–urea cycles and convert ammonia to urea for both storage and excretion. Under conditions of elevated ambient ammonia, the mudskipper *P. schlosseri* can continue to excrete ammonia by active transport of ammonium ions. There are indications that some fish may be able to manipulate the pH of the body surface to facilitate NH_3 volatilization during aerial exposure, or that of the external medium to lower the toxicity of ammonia during ammonia loading. Future investigation of these aspects of "environmental ammonia detoxification" may produce new information on how fish avoid ammonia intoxication.

REFERENCES

Alderson, R. (1979). The effect of ammonia on the growth of juvenile Dover sole, *Solea solea* (L.) and turbot, *Seophthalmus maximus* (L.). *Aquaculture* **17**, 291–309.

American Petroleum Institute. (1981). "The Sources, Chemistry, Fate and Effects of Ammonia in Aquatic Environments," p. 145. American Petroleum Institute, Washington, DC.

Amlal, H., and Soleimani, M. (1997). K^+/NH_4^+ antiporter: a unique ammonium carrying transporter in the kidney inner medulla. *Biochim. Biophys. Acta* **1323**, 319–333.

Anderson, P. M. (1995). Urea cycle in fish: molecular and mitochondrial studies. In "Fish Physiology, Volume 14, Ionoregulation: Cellular and Molecular Approaches to Fish Ionic Regulation" (C. M. Wood and T. J. Shuttleworth, eds.), Chap. 3, pp. 57–83. Academic Press, New York.

Ardawi, M. S. M. (1990). Glutamine-synthesizing activity in lungs of fed, starved, acidosis, diabetic, injured and septic rats. *Biochem. J.* **270**, 829–832.

Arillo, A., Margiocco, C., Melodia, F., Mensi, P., and Schenone, G. (1981). Ammonia toxicity mechanisms in fish: studies on rainbow trout (*Salmo gairdneri* Rich). *Ecotoxicol. Environ. Safety* **5**, 316–325.

Bever, K., Chenoweth, M., and Dunn, A. (1981). Amino acid gluconeogenesis and glucose turnover in kelp bass (*Paralabax* sp.). *Am. J. Physiol.* **240**, R246–R252.

Binstock, L., and Lecar, H. (1969). Ammonium ion currents in the squid giant axon. *J. Gen. Physiol.* **53**, 342–361.

Bosman, D. K., Deutz, N. E. P., Maas, M. A. W., van Eijik, H. M. H., Smit, J. J. H., de Haan, J. G., and Chamuleau, R. A. F. M. (1992). Amino acid release from cerebral cortex in experimental acute liver failure, studied by *in vivo* cerebral cortex microdialysis. *J. Neurochem.* **59**, 591–599.

Boutilier, R. G., Heming, T. A., and Iwama, G. K. (1984). Physicochemical parameters for use in fish respiratory physiology. In "Fish Physiology" (W. S. Hoar and D. J. Randall, eds.), Vol. 10A, pp. 403–430. Academic Press, Orlando, FL.

Brett, J. R., and Zala, C. A. (1975). Daily pattern of nitrogen excretion and oxygen consumption of sockeye salmon (*Oncorhynchus nerka*) under controlled conditions. *J. Fish. Res. Bd. Can.* **32**, 2479–2486.

4. AMMONIA TOXICITY, TOLERANCE, AND EXCRETION

Brierly, G. P., and Stoner, C. D. (1970). Swelling and contraction of heart mitochondria suspended in ammonium chloride. *Biochemistry* **9**, 708–713.

Burrows, R. E. (1964). Effects of accumulated excretory products on hatchery-reared salmonids, Research Report 66. Fish and Wildlife Service, U. S. Dept. of Interior, Washington, DC.

Cameron, J. N., and Heisler, N. (1983). Studies of ammonia in the rainbow trout: physico-chemical parameters, acid-base behaviour and respiratory clearance. *J. Exp. Biol.* **105**, 107–126.

Cameron, J. N., and Heisler, N. (1985). Ammonia transfer across fish gills: A review. *In* "Circulation, Respiration and Metabolism" (R. Gilles, ed.), pp. 91–100. Springer-Verlag, Berlin.

Campbell, J. W. (1973). Nitrogen excretion. *In* "Comparative Animal Physiology" (C. L. Prosser, ed.), 3rd ed., pp. 279–316. W. B. Saunders, Philadelphia.

Campbell, J. W. (1991). Excretory nitrogen metabolism. *In* "Experimental and Metabolic Animal Physiology. Comparative Animal Physiology" (C. L. Prosser, ed.), 4th ed., pp. 277–324. Wiley-Interscience, New York.

Campbell, J. W. (1997). Mitochondrial ammonia metabolism and the proton-neutral theory of hepatic ammonia detoxication. *J. Exp. Zool.* **278**, 308–321.

Campbell, J. W., and Anderson, P. M. (1991). Evolution of mitochondrial enzyme systems in fish: the mitochondrial synthesis of glutamine and citrulline. *In* "Biochemistry and Molecular Biology of Fishes" (P. W. Hochachka and T. P. Mommsen, eds.), Vol. 1, pp. 43–76. Elsevier, Amsterdam.

Chamberlin, M. E., Glemet, H. C., and Ballantyne, J. S. (1991). Glutamine metabolism in a holostean fish (*Amia calva*) and a teleost (*Salvelinus namaycush*). *Am. J. Physiol.* **260**, R159–R166.

Chang, T. W., and Goldberg, A. L. (1978a). The origin of alanine production in skeletal muscle. *J. Biol. Chem.* **253**, 3677–3684.

Chang, T. W., and Goldberg, A. L. (1978b). The metabolic fates of amino acids and the formation of glutamine in skeletal muscle. *J. Biol. Chem.* **253**, 3685–3695.

Claiborne, J. B., Evans, D. H., and Goldstein, L. (1982). Fish branchial Na^+/H^+ exchange is via basolateral Na^+,K^+-activated ATPase. *J. Exp. Biol.* **96**, 431–434.

Cooper, J. L., and Plum, F. (1987). Biochemistry and physiology of brain ammonia. *Physiol. Rev.* **67**, 440–519.

Cowey, C. B., De La Higuera, M., and Adron, J. W. (1977). The effect of dietary composition and insulin on gluconeogenesis in rainbow trout (Salmo gairdneri). *Br. J. Nutr.* **38**, 385–395.

Dabrowska, H., and Wlasow, T. (1986). Sublethal effect of ammonia on certain biochemical and haematological indicators in common carp (*Cyprinus carpio L.*). *Comp. Biochem. Physiol.* **83C**, 179–184.

Driedzic, W. R., and Hochachka, P. W. (1976). Control of energy metabolism in fish white muscle. *Am. J. Physiol.* **230**, 579–582.

Emerson, K., Lund, R. E., Thurston, R. V., and Russo, R. C. (1975). Aqueous ammonia equilibrium calculations: effect of pH and temperature. *J. Fish. Res. Bd. Can.* **32**, 2379–2383.

Epstein, F. H., Silva, P., and Kormanik, G. (1980). Role of Na^+,K^+-ATPase in chloride cell function. *Am. J. Physiol.* **238**, R246–R250.

Evans, D. H. (1977). Further evidence for Na^+/NH_4^+ exchange in marine teleost fish. *J. Exp. Biol.* **70**, 213–220.

Evans, D. H. (1982). Mechanisms of acid extrusion by marine fishes: the teleost, *Opsanus beta*, and the elasmobranch, *Squalus acanthias*. *J. Exp. Biol.* **97**, 289–299.

Evans, D. H. (1984a). The roles of gill permeability and transport mechanisms in euryhalinity. *In* "Fish Physiology" (W. S. Hoar and D. J. Randall, eds.), Vol. 10B, pp. 239–283. Academic Press, New York.

Evans, D. H. (1984b). Gill Na^+/H^+ and Cl^-/HCO_3^- exchange systems evolved before the vertebrates entered freshwater. *J. Exp. Biol.* **113**, 465–469.

Evans, D. H., and Cameron, J. N. (1986). Gill ammonia transport. *J. Exp. Zool.* **239**, 17–23.

Evans, D. H., and More, K. J. (1988). Modes of ammonia transport across the gill epithelium of the dogfish pup (*Squalus acanthias*) *J. Exp. Biol.* **138**, 375–397.

Evans, D. H., Kormanik, G. A., and Krasny, E. J., Jr. (1979). Mechanisms of ammonia and acid excretion by the little skate, *Raja erinacea. J. Exp. Zool.* **208**, 431–437.

Evans, D. H., More, K. J., and Robbins, S. L. (1989). Modes of ammonia transport across the gill epithelium of the marine teleost fish *Opsanus beta. J. Exp. Biol.* **144**, 339–356.

Felig, P. (1975). Amino acid metabolism in man. *Ann. Rev. Biochem.* **44**, 939–955.

Felipo, V., Kosenko, E., Minana, M.-D., Marcaida, G., and Grisolia, S. (1994). Molecular mechanism of acute ammonia toxicity and of its prevention by L-carnitine. *In* "Hepatic Encephalopathy, Hyperammonemia and Ammonia Toxicity" (V. Felipo and S. Grisola, eds.), pp. 65–77. Plenum Press, New York.

Flessner, M. F., Wall, S. M., and Knepper, M. D. (1991). Permeability of rat collecting duct segments to NH_3 and NH_4^+. *Am. J. Phsyiol.* **260**, F264–F272.

Frayan, K. N., Khan, K., Coppack, S., and Elia, M. (1990). Amino acid metabolism in human subcutaneous adipose tissue. *Clin. Sci.* **80**, 471–474.

French, C. J., Hochachka, P. W., and Mommsen, T. P. (1983). Metabolic organization of liver during spawning migration of sockeye salmon. *Am. J. Physiol.* **245**, R827–R830.

Fromm, P. O., and Gillette, J. R. (1968). Effect of ambient ammonia on blood ammonia and nitrogen excretion of rainbow trout (*Salmo gairdneri*). *Comp. Biochem. Physiol.* **26**, 887–896.

Goldstein, L. (1986). Interorgan glutamine relationships. *Fedn. Proc.* **45**, 2186–2179.

Goldstein, L., Claiborne, J. B., and Evans, D. E. (1982). Ammonia excretion by the gills of two marine teleost fish: the importance of NH_4^+ permeance. *J. Exp. Zool.* **219**, 395–397.

Good, D. W. (1988). Active absorption of NH_4^+ by rat medullary thick ascending limb: inhibition by potassium. *Am. J. Physiol.* **255**, F78-F87.

Good, D. W. (1994). Ammonium transport by the thick ascending limb of Henle's loop. *Annu. Rev. Physiol.* **56**, 623–647.

Gordon, M. S., Boòtius, I., Evans, D. H., McCarthy, R., and Oglesby, L. C. (1969). Aspects of the physiology of terrestrial life in amphibious fishes. I. The mudskipper, *Periophthalmus sobrinus. J. Exp. Biol.* **50**, 141–149.

Gordon, M. S., Fischer, J., and Tarifeno, E. (1970). Aspects of the physiology of terrestrial life in amphibious fishes. II. The Chilean clingfish, *Sicyases sanguineus. J. Exp. Biol.* **53** 559.

Gordon, M. S., Ng, W. W. M., and Yip, A. Y. W. (1978). Aspects of the physiology of terrestrial life in amphibious fishes. III. The Chinese mudskipper *Periophthalmus cantonensis. J. Exp. Biol.* **72**, 57–75.

Grill, V., Bjorkman, O., Gutniak, M., and Lindquist, M. (1992). Brain uptake and release of amino acids in nondiabetic and insulin-dependent diabetic subjects. Importance of glutamine release for nitrogen balance. *Metabolism* **41**, 28–34.

Hampson, B. L. (1976). Ammonia concentration in relation to ammonia toxicity during a rainbow trout rearing experiment in a closed freshwater–seawater system. *Aquaculture* **9**, 61–70.

Handy, R. D., and Poxton, M. G. (1993). Nitrogen pollution in mariculture: toxicity and excretion of nitrogenous compounds by marine fish. *Rev. Fish Biol. Fish.* **3**, 205–241.

Harader, R. R., Jr., and Allen, G. H. (1983). Ammonia toxicity to chinook salmon parr: reduction in saline water. *Trans. Am. Fish. Soc.* **112**, 834–837.

Haywood, G. P. (1983). Ammonia toxicity in teleost fishes: a review. Canadian Technical Report of Fisheries and Aquatic Sciences/Rapport Technique Canadien des Sciences Halieutiques et Aquatiques.

Heisler, N. (1990). Mechanisms of ammonia elimination in fishes. *In* "Animal Nutrition and Transport Processes. 2. Transport, Respiration and Excretion: Comparative and Environmental Aspects" (J. P. Truchot and B. Lahlou, eds.), pp. 137–151. Karger, Basel.

Herbert, D. W. M., and Shurben, D. S. (1965). The susceptibility of salmonid fish to poisons under estuarine conditions. *J. Air Water Poll.* **9**, 89–91.

Hilgier, W., Haugvicova, R., and Albrecht, J. (1991). Decreased potassium-stimulated release of 3HDaspartate from hippocampal slices distinguishes encephalopathy related to acute liver failure from that induced by simple hyperammonemia. *Brain Res.* **567,** 165–168.

Hillaby, B. A., and Randall, D. J. (1979). Acute ammonia toxicity and ammonia excretion in rainbow trout (*Salmo gairdneri*). *J. Fish. Res. Bd. Can.* **36,** 621–629.

Hindfelt, B., Plum, F., and Duffy, T. E. (1977). Effect of acute ammonia intoxication on cerebral metabolism in rats with portacaval shunts. *J. Clin. Invest.* **59,** 386–396.

Hochachka, P. W., and Guppy, M. (eds.). (1987). "Metabolic Arrest and the Control of Biological Time," pp. 101–119. Harvard University Press, London.

Hong, L. N. (1997). A comparative study of the effects of environmental pH on two mudskippers. Honors thesis, Department of Biological Sciences, National University of Singapore, p. 87.

Hunter, K. C., and Kirschner, L. B. (1986). Sodium absorption coupled to ammonia excretion in osmoconforming marine invertebrates. *Am. J. Physiol.* **251,** R957–R962.

Ip, Y. K., and Low, W. P. (1990). Lactate production in the gills of the mudskipper *Periophthalmodon schlosseri* exposed to hypoxia. *J. Exp. Zool.* **253,** 99–101.

Ip, Y. K., Chew, S. F., Lim, A. L. L., and Low, W. P. (1990). The Mudskipper. *In* "Essays in Zoology, Papers Commemorating the 40th Anniversary of Department of Zoology" (L. M. Chou and P. K. L. Ng, eds.), pp. 83–95. National University of Singapore Press, Singapore.

Ip, Y. K., Lee, C. Y., Chew, S. F., Low, W. P., and Peng, K. W. (1993) Differences in the responses of two mudskippers to terrestrial exposure. *Zool. Sci.* **10,** 511–519.

Ip, Y. K., Peng, K. W., Chew, S. F., Kok, W. K., Wilson, J., and Randall, D. J. (2000). The mudskippers: ammonia toxicity and tolerance. Proceedings of the Fifth International Symposium on Fish Physiology, Toxicology and Water Quality, 10–13 November, 1998, City University of Hong Kong. p. 69–86.

Ip, Y. K., Lim C. B., Chew, S. F., Wilson, J. M. and Randall, D. J. (2001). Production and accumulation of alanine through partial amino acid catabolism facilitate *Periophthalmodon schlosseri* (mudskipper) to use amino acids as an energy source while active on land. *J. Exp. Biol.* **204,** 1615–1624.

Iwama, G. K., McGeer, J. C., Wright, P. A., Wilkie, M. P., and Wood, C. M. (1997). Divalent cations enhance ammonia excretion in Lahontan cutthroat trout in highly alkaline water. *J. Fish Biol.* **50,** 1061–1073.

Iwata, K. (1988). Nitrogen metabolism in the mudskipper, *Periophthalmus cantonensis:* changes in free amino acids and related compounds in carious tissues under conditions of ammonia loading with reference to its high ammonia tolerance. *Comp. Biochem. Physiol.* **91A,** 499–508.

Iwata, K., Kakuta, M., Ikeda, G., Kimoto, S., and Wada, N. (1981). Nitrogen metabolism in the mudskipper, *Periophthalmus cantonensis:* a role of free amino acids in detoxification of ammonia produced during its terrestrial life. *Comp. Biochem. Physiol.* **68A,** 589–596.

Jow, L. Y., Chew, S. F., Lim, C. B., Anderson, P. M., and Ip, Y. K. (1999). The marble goby *Oxyeleotris marmoratus* activates hepatic glutamine synthetase and detoxifies ammonia to glutamine during air exposure. *J. Exp. Biol.* **202,** 237–245.

Karnaky, K. J. (1986). Structure and function of the chloride cell of *Fundulus heteroclitus* and other teleosts. *Am. Zool.* **26,** 209–224.

Khoo, K. H., Culberson, C. H., and Bates, R. G. (1977). Thermodynamics of the dissociation of ammonium ion in seawater from 5 to 40C. *J. Solution Chem.* **6,** 281–290.

Kikeri, D., Sun, A., Zeidel, M. L., and Hebert, S. C. (1989). Cell membranes impermeable to NH_3. *Nature* **339,** 478–480.

Kloppick, E., Jacobasch, G., and Rapoport, S. (1967). Steigerung der glykolyse durch den einfluss von ammoniumionen auf die phosphofruktokinaseaktivitat. (Enhancement of the glycolytic rate by action of ammonium ion on phosphofructokinase activity). *Acta. Biol. Med. Ger.* **18,** 37–42.

Kok, W. K., Lim, C. B., Lam, T. J., and Ip, Y. K. (1998). The mudskipper *Periophthalmodon schlos-*

seri respires more efficiently on land than in water and vice versa for *Boleophthalmus boddaerti.* *J. Exp. Zool.* **280,** 86–90.

Krogh, A. (1939, reprinted 1965). "Osmotic Regulation in Aquatic Animals." Dover Publications, New York.

Kurtz, I., and Balaban, R. S. (1986). Ammonium as a substrate for Na^+,K^+-ATPase in rabbit proximal tubules. *Am. J. Physiol.* **250,** F497-F502.

Lande, M. B., Donovan, J. M., and Zeidel, M. L. (1995). The relationship between membrane fluidity and permeabilities to water, solute, ammonia, and protons. *J. Gen. Physiol.* **106,** 67–84.

Lang, T., Peters, G., Hoffmann, R., and Meyer, E. (1987). Experimental investigations on the toxicity of ammonia: effects on ventilation frequency, growth, epidermal mucous cells, and gill structure of rainbow trout *Salmo gairdneri. Dis. Aquat. Org.* **3,** 159–165.

Leech, A. R., Goldstein, L., Cha, C. J., and Goldstein, J. M. (1979). Alanine biosynthesis during starvation in skeletal muscle of the spiny dogfish, *Squalus acantias. J. Exp. Zool.* **207,** 73–80.

Leung, K. M. Y., Chu, J. C. W., and Wu, R. S. S. (1999). Effects of body weight, water temperature and ration size on ammonia excretion by the areolated grouper (*Epinephelus areolatus*) and mangrove snapper (*Lutjanus argntimaculatus*). *Aquaculture* **170,** 215–227.

Levi, G., Morisi, G, Coletti, A., and Catanzaro, R. (1974). Free amino acids in fish brain: normal levels and changes upon exposure to high ammonia concentrations *in vivo* and upon incubation of brain slices. *Comp. Biochem. Physiol.* **49A,** 623–636.

Lim C. B., Chew, S. F., Anderson, P. M., and Ip, Y. K. (2001). Mudskippers (*Periophthalmodon schlosseri* and *Boleophthalmus boddaerti*) reduce the rates of protein and amino acid catabolism to slow down internal ammonia build up during aerial exposure in constant darkness. *J. Exp. Biol.* **204,** 1605–1614.

Lin, H., and Randall, D. J. (1990). The effect of varying water pH on the acidification of expired water in rainbow trout. *J. Exp. Biol.* **149,** 149–160.

Lin, H., and Randall, D. J. (1991). Evidence for the presence of an electrogenic proton pump on the trout gill epithelium. *J. Exp. Biol.* **161,** 119–134.

Lin, H., and Randall, D. J. (1995). Proton pumps in fish gills. Cellular and molecular approaches to fish ionic regulation. *In* "Fish Physiology" (C. M. Wood and T. J. Shuttleworth, eds.), Vol. 14, pp. 229–255. Academic Press, Orlando, FL.

Low, W. P., Lane, D. J. W., and Ip, Y. K. (1988). A comparative study of terrestrial adaptations in three mudskippers—*Periophthalmus chrysospilos, Boleophthalmus boddaerti* and *Periophthalmus schlosseri. Biol. Bull.* **175,** 434–438.

Low, W. P., Lane, D. J. W., and Ip, Y. K. (1990). A comparative study of the gill morphometry in three mudskippers—*Periophthalmus chrysospilos, Boleophthalmus boddaerti* and *Periophthalmus schlosseri. Zool. Sci.* **7,** 29–38.

Low, W. P., Peng, K. W., Phuan, S. K., Lee, C. Y., and Ip, Y. K. (1993). A comparative study on the responses of the gills of two mudskippers to hypoxia and anoxia. *J. Comp. Physiol.* **163B,** 487–494.

Maetz, J., and Garcia-Romeu, F. (1964). The mechanisms of sodium and chloride uptake by the gills of a freshwater fish, *Carassius auratus.* II. Evidence for NH_4^+/Na^+ and HCO_3^-/Cl^- exchanges. *J. Gen. Physiol.* **47,** 1209–1227.

Mallery, C. H. (1983). A carrier enzyme basis for ammonium excretion in teleost gill-NH_4^+-stimulated Na^+-dependent ATPase activity in *Opsanus beta. Comp. Biochem. Physiol.* **74,** 889–897.

Mans, A. M., Biebuyck, J. F., and Hawkins, R. A. (1983). Ammonia selectively stimulates neutral amino acid transport across blood–brain barrier. *Am. J. Physiol.* **245,** C74-C77.

Marcaida, G., Felipo, V., Hermenegildo, C., Minana, M. D., and Grisolia, S. (1992). Acute ammonia toxicity is mediated by NMDA type of glutamate receptors. *FEBS Lett.* **296,** 67–68.

McDonald, D. G., and Milligan, C. L. (1988). Sodium transport in the brook trout, *Salvelinus fontinalis:* effects of prolonged low pH exposure in the presence and absence of aluminum. *Can. J.*

Fish. Aquat. Sci. **45,** 1606–1613.
McDonald, D. G., and Prior, E. T. (1988). Branchial mechanisms of ion and acid–base regulation in the freshwater rainbow trout, *Salmo gairdneri. Can. J. Zool.* **66,** 2699–2708.
McDonald, D. G., Tang, Y., and Boutilier, R. G. (1989). Acid and ion transfer across the gills of fish: mechanisms and regulation. *Can. J. Zool.* **67,** 3046–3054.
McKenzie, D. J., Randall, D. J., Lin, H., and Aota, S. (1993). Effects of changes in plasma pH, CO_2 and ammonia on ventilation in trout. *Fish Physiol. Biochem.* **10,** 507–515.
Meijer, A. J., Lamers, W. H., and Chamuleau, R. A. F. M. (1990). Nitrogen metabolism and ornithine cycle formation. *Physiol. Rev.* **70,** 701–748.
Merkens, J. C., and Downing, K. M. (1957). The effect of tension of dissolved oxygen on the toxicity of un-ionized ammonia to several species of fish. *Ann. Appl. Biol.* **45,** 521–527.
Milligan, C. L., and Wood, C. M. (1987). Muscle and liver intracellular acid–base and metabolite status after strenuous activity in the inactive benthic starry flounder *Platichthys stellatus. Physiol. Zool.* **60,** 54–68.
Mommsen, T. P., and Hochachka, P. W. (1988). The purine nucleotide cycle as two temporally separated metabolic units: a study on trout muscle. *Metabolism* **37,** 552–556.
Mommsen, T. P., and Walsh, P. J. (1989). Evolution of urea synthesis in vertebrates: the piscine connection. *Science* **243,** 72–75.
Mommsen, T. P., and Walsh, P. J. (1991). Urea synthesis in fishes: evolutionary and biochemical perspectives. *In* "Biochemistry and Molecular Biology of Fishes, Volume 1, Phylogenetic and Biochemical Perspectives" (P. W. Hochachka and T. P. Mommsen, eds.), pp. 137–163. Elsevier, New York.
Mommsen, T. P., and Walsh, P. J. (1992). Biochemical and environmental perspectives on nitrogen metabolism in fishes. *Experientia* **48,** 583–593.
Mommsen, T. P., French, C. J., and Hochachka, P. W. (1980). Sites and patterns of protein and amino acid utilization during the spawning migration of salmon. *Can. J. Zool.* **58,**1785–1799.
Moon, T. W., and Johnston, I. A. (1981). Amino acid transport and interconversions in tissues of freshly caught and food-deprived plaice, *Pleuronectes platessa* L. *J. Exp. Biol.* **19,** 653–663.
Nosadini, R., Datta, H., Hodson, A., and Alberti, K. G. M. M. (1980). A possible mechanism for the antiketogenic action of alanine in the rat. *Biochem. J.* **190,** 323–332.
Payan, P. (1978). A study of the Na^+/NH_4^+ exchange across the gill of the perfused head of the trout (*Salmo gairdneri*). *J. Comp. Physiol.* **124,** 181–188.
Payan, P., Matty, A. J., and Maetz, J. (1975). A study of the sodium pump in the perfused head preparation of the trout *Salmo gairdneri* in freshwater. *J. Comp. Physiol.* **104,** 33–48.
Peng, K. W. (1994). A comparative study on the environmental ammonia tolerance of two mudskippers. M.Sc. thesis, National University of Singapore.
Peng, K. W., Chew, S. F., Lim, C. B., Kuah, S. S. L., Kok, W. K., and Ip, Y. K. (1998). The mudskippers *Periophthalmodon schlosseri* and *Boleophthalmus boddaerti* can tolerate environmental NH_3 concentrations of 446 and 36 μM, respectively. *Fish Physiol. Biochem.* **19,** 59–69.
Pequin, L., and Serfaty, A. (1963). L'excretion ammoniacale chez un Teleosteen dulcicole *Cyprinius carpio* L. *Comp. Biochem. Physiol.* **10,** 315–324.
Peres-Sala, M. D., Parrilla, R., and Ayuso, M. S. (1987). Key role of L-alanine in the control of hepatic protein synthesis. *Biochem. J.* **241,** 491–498.
Person-Le Ruyet, J., Boeuf, G., Zambonino-Infant, J., Helgason, S. and Le Roux, A. (1998). Short-term physiological changes in turbot and seabream juveniles exposed to exogenous ammonia. *Comp. Biochem. Physiol.* **119A,** 511–518.
Pressley, T. A., Graves, J. S., and Krall, A. R. (1981). Amiloride-sensitive ammonium and sodium ion transport in the blue crab. *Am. J. Physiol.* **241,** R370–R378.
Randall, D. J., and Wicks, B. J. (2000). Fish: ammonia production, excretion and toxicity. Proceedings of the Fifth International Symposium on Fish Physiology, Toxicology and Water Quality, 10–13 November, 1998, City University of Hong Kong. p. 41–50.

Randall, D. J., and Wright, P. A. (1989). The interaction between carbon dioxide and ammonia excretion and water pH in fish. *Can. J. Zool.* **67**, 2936–2942.
Randall, D. J., Wood, C. M., Perry, S. F., Bergman, H., Maloiy, G. M., Mommsen, T. P., and Wright, P. A. (1989). Urea excretion as a strategy for survival in a fish living in a very alkaline environment. *Nature* **337**, 165–166.
Randall, D. J., Wilson, J. M., Peng, K. W., Kok, T. W. K., Kuah, S. S. L., Chew, S. F., Lam, T. J., and Ip. Y. K. (1999). The mudskipper, *Periophthalmodon schlosseri*, actively transports NH_4^+ against a concentration gradient. *Am. J. Physiol.* **46**, R1562–R1567.
Rao, V. L. R., Murthy, C. R. K., and Butterworth, R. F. (1992). Glutamatergic synaptic dysfunction in hyperammonemic syndromes. *Metab. Brain Dis.* **7**, 1–20.
Reichenbach-Klinke, H. H. (1967). Untersuchungen uber die Einwirkung des Ammoniakgehalts auf den Fischorganismus. *Archiv Fischereiwissenschaft* **17**, 122–132.
Rowe, W. B. (1985). Glutamine synthetase from muscle. *In* "Methods in Enzymology" (A. Meister, ed.), pp. 199–212. Academic Press, New York.
Saha, N., and Ratha, B. K. (1998). Ureogenesis in Indian air-breathing teleosts: adaptation to environmental constraints. *Comp. Biochem. Physiol.* **120A**, 195–208.
Sayer, M. D. J., and Davenport, J. (1987a). Ammonia and urea excretion in the amphibious teleost *Blennius pholis* exposed to fluctuating salinity and pH. *Comp. Biochem. Physiol.* **87**, 851–857.
Sayer, M. D. J., and Davenport, J. (1987b). The relative importance of the gills to ammonia and urea excretion in five seawater and one freshwater teleost species. *J. Fish Biol.* **31**, 561–570.
Schenker, S., McCandless, D. W., Brophy, E., and Lewis, M. S. (1967). Studies on the intracerebral toxicity of ammonia. *J. Clin. Invest.* **46**, 838–848.
Schmidt, W., Wolf, G., Grungreiff, K., and Linke, K. (1993). Adenosine influences the high-affinity-uptake of transmitter glutamate and aspartate under conditions of hepatic encephalopathy. *Metab. Brain Dis.* **8**, 73–80.
Sears, E. S., McCandless D. W., and Chandler, M. D. (1985). Disruption of the blood–brain barrier in hyperammonemic coma and the pharmocologic effects of dexaethasone and difluoromethyl ornithine. *J. Neurosci. Res.* **14**, 255–261.
Smart, G. (1976). The effect of ammonia exposure on gill structure of the rainbow trout (*Salmo gairdneri*). *J. Fish Biol.* **8**, 471–475.
Smart, G. R. (1978). Investigations of the toxic mechanisms of ammonia to fish-gas exchange in rainbow trout (*Salmo gairdneri*) exposed to acutely lethal concentrations. *J. Fish Biol.* **12**, 93–104.
Smith, R. J. (1986). Role of skeletal muscle in interorgan amino acid exchange. *Fedn. Proc.* **45**, 2172–2176.
Soderberg, R. W., and Meade, J. W. (1992). Effects of sodium and calcium on acute toxicity of unionized ammonia to Atlantic salmon and lake trout. *J. Appl. Aquat.* **1**, 83–92.
Soltoff, S. P., and Mandel, L. J. (1983). Amiloride directly inhibits the Na^+,K^+-ATPase activity of rabbit kidney proximal tubules. *Science* **220**, 957–959.
Stephan, C. E., Mount, D. I., Hansen, D. J., Gentile, J. H., Chapman, G. A., and Brungs, W. A. (1985). Guidelines for deriving numerical national water quality criteria for the protection of aquatic organisms and their uses, NTIS #PB85-227049. National Technical Information Service, Springfield, VA.
Taylor, E. (2000). Effects of exposure to sublethal levels of copper on brown trout: Mechanisms of ammonia toxicity. Proceedings of the Fifth International Symposium on Fish Physiology, Toxicology and Water Quality, 10–13 November, 1998, City University of Hong Kong. p. 51–68.
Thomson, R. B., Thomson, J. M., and Phillips, J. E. (1988). NH_4^+ transport in acid-secreting insect epithelium. *Am. J. Physiol.* **254**, R348–R356.
Thurston, R. U., Russo, R. C., and Smith, C. E. (1978). Acute toxicity of ammonia and nitrite to cutthroat trout fry. *Trans. Am. Fish. Soc.* **107**, 361–368.

Thurston, R. V., Russo, R. C., and Vinogradov, G. A. (1981). Ammonia toxicity to fishes. The effect of pH on the toxicity of the un-ionized ammonia species. *Environ. Sci. Tech.* **15,** 837–840.

Tommaso, J. R., Goudie, C. A., Simco, B. A., and Davis, K. B. (1980). Effects of environmental pH and calcium on ammonia toxicity in channel catfish. *Trans. Am. Fish. Soc.* **109,** 229–234.

Towle, D. W., and Holleland, T. (1987). Ammonium ion substitutes for K^+ in ATP-dependent Na^+ transport by basolateral membrane vesicles. *Am. J. Physiol.* **252,** R479–R489.

U.S. Environmental Protection Agency. (1984). Ambient Water Quality Criteria for Ammonia—1984. National Technical Information Service, Springfield, VA.

U.S. Environmental Protection Agency. (1989). Ambient Water Quality Criteria for Ammonia (Saltwater). National Technical Information Service, Springfield, VA.

U.S. Environmental Protection Agency. (1998). Addendum to "Ambient Water Quality Criteria for Ammonia—1984." National Technical Information Service, Springfield, VA.

U.S. National Research Council Committee on Medical and Biologic Effects of Environmental Pollutants (1979). Ammonia. University Park Press, Baltimore, MD.

Wajsbrot, N., Gasith, A., Krom, M. D., and Popper, D. M. (1991). Acute toxicity of ammonia to juvenile gilthead seabream *Sparus aurata* under reduced oxygen levels. *Aquaculture* **92,** 277–288.

Wall, S. (1996). NH_4^+ augments net acid secretion by a ouabain-sensitive mechanism in isolated perfused inner medullary collecting ducts. *Am. J. Physiol.* **270,** F432–F439.

Walsh, P. J. (1997). Evolution and regulation of ureogenesis and ureotely in (batrachoidid) fishes. *Ann. Rev. Physiol.* **59,** 299–323.

Walsh, P. J., Bergman, H. L., Narahara, A., Wood, C. M., Wright, P. A., Randall, D. J., Maira, J. N., and Laurent, P. (1993). Effects of ammonia on survival, swimming, and activities of enzymes of nitrogen metabolism in the Lake Magadi tilapia *Oreochromis alcalicus grahamii*. *J. Exp. Biol* **180,** 323–387.

Walsh, P. J., Tucker, B. C., and Hopkins, T. E. (1994). Effects of confinement/crowding on ureogenesis in the Gulf toadfish *Opsanus beta*. *J. Exp. Biol.* **191,** 195–206.

Walton, M. J., and Cowey, C. B. (1977). Aspects of ammoniogenesis in rainbow trout, *Salmo gairdneri*. *Comp. Biochem. Physiol.* **57,** 143–149.

Webb, J. T., and Brown, G. W. (1980). Glutamine synthetase: assimilatory role in liver as related to urea retention in marine chondrichthyes. *Science* **208,** 293–295.

Welbourne, T. C. (1987). Interorgan glutamine flow in metabolic acidosis. *Am. J. Physiol.* **253,** F1069-F1076.

Welbourne, T. C., and Joshi, S. (1990). Interorgan glutamine metabolism during acidosis. *J. Parenter. Enter. Nutr.* **14,** 77S–85S.

Whitfield, M. (1974). The hydrolysis of ammonium ions in sea water—a theoretical study. *J. Mar. Biol. Ass. U.K.* **54,** 565–580.

WHO. (1986). Ammonia, Environmental Health Criteria 54. World Health Organization, Geneva.

Wilkie, M. P. (1997). Mechanisms of ammonia excretion across fish gills. *Comp. Biochem. Physiol.* **118,** 39–50.

Wilkie, M. P., and Wood, C. M. (1991). Nitrogenous waste excretion, acid–base regulation and ionoregulation in rainbow trout (*Oncorhynchus mykiss*) exposed to extremely alkaline water. *Physiol. Zool.* **64,** 1069–1086.

Wilkie, M. P., and Wood, C. M. (1994). The effects of extremely alkaline water (pH 9.5) on rainbow trout gill function and morphology. *J. Fish Biol.* **45,** 87–98.

Wilkie, M. P., and Wood, C. M. (1995). Recovery from high pH exposure in rainbow trout: white muscle ammonia storage, ammonia washout and the restoration of blood chemistry. *Physiol. Zool.* **68,** 379–401.

Wilkie, M. P., and Wood, C. M. (1996). The adaptations of fish to extremely alkaline environments. *Comp. Biochem. Physiol.* **113,** 665–673.

Wilkie, M. P., Wright, P. A., Iwama, G. K., and Wood, C. M (1993). The physiological responses of the Lahontan cutthroat trout (*Oncorhynchus clarki henshawi*), a resident of highly alkaline Pyramid Lake (pH 9.4), to challenge at pH 10. *J. Exp. Biol.* **175,** 173–194.

Wilkie, M. P., Wright, P. A., Iwama, G. K., and Wood, C. M. (1994). The physiological adaptations of the Lahontan cutthroat trout (*Oncorhynchus clarki henshawi*) following transfer from well water to the highly alkaline waters of Pyramid Lake, Nevada (pH 9.4). *Physiol. Zool.* **67,** 355–380.

Wilson, J. M., Iwata, K., Iwama, G. K., and Randall, D. J. (1998). Inhibition of ammonia excretion and production in rainbow trout during severe alkaline exposure. *Comp. Biochem. Physiol.* **121,** 99–109.

Wilson, J. M., Randall, D. J., Kok, T. W. K., Vogl, W. A., and Ip, Y. K. (1999). Fine structure of the gill epithelium of the terrestrial mudskipper, *Periophthalmodon schlosseri*. *Cell Tissue Res.* **298,** 345–356.

Wilson, J. M, Randall, D. J., Donowitz, M., Bogl, A. W., and Ip, Y. K. (2000). Immunolocalization of ion transport proteins to the mudskipper (*Periophthalmodon schlosseri*) branchial epithelium mitochondria-rich cells. *J. Exp. Biol.* **203,** 2297–2310.

Wilson, R. W., Wright, P. M., Munger, S., and Wood, C. M. (1994). Ammonia excretion in freshwater rainbow trout (*Oncorhynchus mykiss*) and the importance of gill boundary layer acidification: lack of evidence for Na^+-NH_4^+ exchange. *J. Exp. Biol.* **191,** 37–58.

Windmueller, H. G., and Spaeth, A. E. (1978). Identification of ketone bodies and glutamine as the major respiratory fuels *in vivo* in postabsorptive rat small intestine. *J. Biol. Chem.* **253,** 69–76.

Wood, C. M. (1988). Acid–base and ionic exchanges at gills and kidney after exhaustive exercise in the rainbow trout. *J. Exp. Biol.* **136,** 461–481.

Wood, C. M. (1993). Ammonia and urea metabolism and excretion. *In* "The Physiology of Fishes" (D. H. Evans, ed.), pp. 379–425. CRC Press, Boca Raton, FL.

Wright, P. A. (1995). Nitrogen excretion: three end products, many physiological roles. *J. Exp. Biol.* **198,** 273–281.

Wright, P. A., and Wood, C. M. (1985). An analysis of branchial ammonia excretion in the freshwater rainbow trout: effects of environmental pH change and sodium uptake blockade. *J. Exp. Biol.* **114,** 329–353.

Wright, P. A., Heming, T., and Randall, D. J. (1986). Downstream pH changes in water flowing over the gills of rainbow trout. *J. Exp. Biol.* **126,** 499–512.

Wright, P. A., Randall, D. J., and Perry, S. F. (1989). Fish gill boundary layer: a site of linkage between carbon dioxide and ammonia excretion. *J. Comp. Physiol.* **158,** 627–635.

Wright, P. A., Iwama, G. K., and Wood, C. M. (1993). Ammonia and urea excretion in Lahontan cutthroat trout (*Oncorhynchus clarki henshawi*) adapted to the highly alkaline Pyramid Lake (pH 9.4). *J. Exp. Biol.* **175,** 153–172.

Yesaki, T. Y., and Iwama, G. K. (1992). Some effects of water hardness on survival, acid–base regulation, ion regulation and ammonia excretion in rainbow trout in highly alkaline water. *Physiol. Zool.* **65,** 763–787.

Zadunaisky, J. A. (1984). The chloride cell: the active transport of chloride and the paracellular pathways. *In* "Fish Physiology" (W. S. Hoar and D. J. Randall, eds.), Vol. 10B, pp. 129–176. Academic Press, New York.

Zischke, J. A., and Arthur, J. W. (1987). Effects of elevated ammonia levels on the fingernail clam, *Musculium transversum,* in outdoor experimental streams. *Arch. Environ. Contam. Toxicol.* **16,** 225–231.

5

ONTOGENY OF NITROGEN METABOLISM AND EXCRETION

P. A. WRIGHT AND H. J. FYHN

  I. Early Development of Fishes
    A. Stages of Development
 II. Free Amino Acids
    A. The Egg Pool
    B. Changes during Development
    C. Oocyte Hydration in Marine Teleosts
    D. Buoyancy of Pelagic Teleost Eggs
    E. Phylogenetic Aspects
III. Nutritional Demands
    A. Amino Acids as Fuel
    B. The Fish Larvae at First Feeding
IV. Nitrogen End Products and Their Chemistry in Solution
    A. Ammonia
    B. Urea
    C. Uric Acid
 V. Nitrogen Metabolism
    A. Ammonia Metabolism
    B. Urea Metabolism
VI. Nitrogen Excretion
    A. Developmental Changes in the Pattern of Nitrogen Excretion
    B. Tolerance to Elevated Environmental Ammonia and pH
    C. Development of Respiratory and Ion-Regulatory Structures
    D. Site and Mechanisms of Ammonia and Urea Transport
VII. Summary
    References

I. EARLY DEVELOPMENT OF FISHES

Nitrogen metabolism and excretion during fish development are critical aspects of early physiology because the major fuel source is obtained endogenously through absorption of the yolk amino acids. Catabolism of proteins and amino acids results in the formation of ammonia, a potentially toxic nitrogenous end product that must be either eliminated or modified to prevent damage to the developing embryonic tissue. Alternatively, if ammonia is sequestered or retained, this may occur in compartments (e.g., yolk) separate from the vulnerable developing tissue. Of the nitrogen excretory products, ammonia primarily, but also urea, is excreted in early fish stages.

Ontogeny is the entire sequence of events involved in the development of an individual organism. Nitrogen excretion during early life stages is not static but is influenced by both developmental stage and environmental conditions. Thus, many different factors unique to early development may influence nitrogen metabolism and excretion. For example, early in embryogenesis many species feed endogenously on yolk nutrients, whereas in later stages, food is acquired exogenously. Also, the physical properties of the embryo (e.g., presence of an egg capsule) influence the rate of nitrogen waste excretion to the environment. The microenvironment in which the embryos or larvae live may also determine the patterns of nitrogen excretion. For example, embryos of viviparous fishes are in intimate contact with maternal fluids and tissues, whereas in oviparous species, embryos may be benthic or pelagic and are completely dependent on nutrients endowed on them at spawning. Hence, these and related factors are important considerations when discussing the ontogeny of nitrogen metabolism and excretion.

Early fish development has been studied in three extant classes: the primitive jawless fishes (Agnatha), the cartilaginous fishes (Chondrichthyes), and the bony fishes (Osteichthyes). This chapter focuses mostly on the Osteichthyes; where data are available, Chondrichthyes and Agnatha are included. It is beyond the scope of this chapter to review fish embryogenesis and early life history; for comprehensive treatises, see Blaxter (1988), Balon (1990), Groot and Margolis (1991), Kamler (1992), Helfman et al. (1997), and Langeland and Kimmel (1997). For an excellent review of maternal–embryonic relationships in viviparous fishes, see Wourms et al. (1988) and for nitrogen excretion in prenatal elasmobranchs, Kormanik (1992, 1993, 1995).

A. Stages of Development

Several classification systems have been used to describe early stages of fish development. For salmonids, we adopt the system used by Balon (1975, 1990) (Fig. 1). The cleavage egg or simply *egg* is unfertilized, the *embryo* refers to the stage between fertilization and hatching, the *eleutheroembryo* or free embryo is the endogenously feeding stage (i.e., yolk sac still present), whereas the *alevin* (larval vestige) is the actively feeding stage (i.e., yolk sac is completely absorbed).

In many marine species, there is a true larval stage with special larval organs that are later replaced or lost during metamorphosis. We adopt the commonly used terminology of an *egg stage* until hatching, *yolk-sac stage* after hatch when yolk

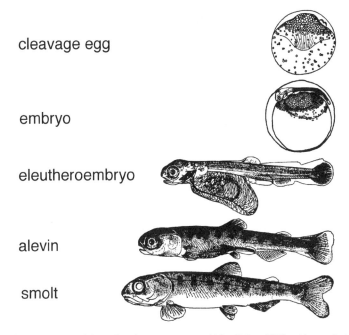

Fig. 1. Early stages of *Oncorhynchus* sp. ontogeny. (After Balon, 1975, with permission.)

still remains, *larval stage* until metamorphosis, and *juvenile stage* after metamorphosis until the reproductive *adult stage* (Kendall *et al.*, 1984).

The terminology of Fig. 2 for the teleost egg and embryo is used throughout. The *chorion* encloses the developing embryo, which floats in the *perivitelline space* (PVS) filled with the *perivitelline fluid* (PVF). The embryo is divided into the *body* and the *yolk* compartment enclosed in the *yolk sac*. Before epiboly (gastrulation) the yolk is covered by the *perivitelline membrane,* whereas after epiboly and closure of the blastopore, the yolk is enclosed in the cellular yolk sac with the innermost layer being the *yolk syncytial layer* (Heming and Buddington, 1988; Trinkhaus, 1992, 1993, 1996).

II. FREE AMINO ACIDS

A. The Egg Pool

Amino acids are quite soluble and occur free in solution in the intra- and extracellular body fluids. This pool of dissolved amino acids is termed *free amino*

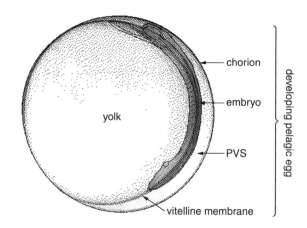

Fig. 2. Terminology of the teleost egg and embryo. PVS, perivitelline space.

acids (FAA) as opposed to the polymerized amino acids. The two pools are dynamically interconnected through cellular metabolism. In mammals, the total FAA concentration is maintained at levels of 2–5 mM in blood plasma and 20–30 mM in muscle tissue (Harper, 1983; Crim and Munro, 1994), whereas in fishes the muscle concentrations are somewhat higher at 50–70 mM (Lyndon *et al.*, 1993; Carter *et al.*, 1995; Yamamoto *et al.*, 2000). Such values are often taken as typical for blood plasma and tissues, but this is not necessarily always the case. One exception is the pelagic egg of marine teleosts where concentrations of 130–200 mM may be found (Fyhn, 1993; Rønnestad and Fyhn, 1993; Mellinger, 1994; Rønnestad *et al.*, 1999a). This large pool of FAA has significant implications, as discussed below. Another example is marine invertebrates, which use FAA as intracellular osmolytes to counterbalance the high salinity of the environment and thus may have FAA levels up to 200–500 mM (e.g., Yancey *et al.*, 1982; Chamberlin and Strange, 1989). The high FAA concentrations of marine invertebrates include also marine zooplankton (Fyhn *et al.*, 1993, 1995; Næss *et al.*, 1995) and this is relevant for the developing marine fish larvae at the time when they start exogenous feeding by catching and ingesting prey organisms in the ocean or in mariculture (see Section III.B).

The large FAA pool of the pelagic eggs of marine teleosts is mainly localized to the yolk compartment. For example, in microdissected, lyophilized larvae of Atlantic halibut (*Hippoglossus hippoglossus*) ∼98% of the total FAA content is found in the yolk by day 2 posthatch (Rønnestad *et al.*, 1993) as verified by Finn *et al.* (1995b) on larvae 4 days posthatch. Direct measurements of the yolk FAA concentration based on samples taken by micropuncture agree well with the calculated concentration from yolk morphometry and analyses of the dissected lyophilized larvae (Rønnestad et al., 1993).

The total amount of FAA of pelagic marine teleost eggs correlates with the size of the egg. For example, the large (3 mm in diameter) egg of the Atlantic halibut (Fyhn, 1989; Finn et al., 1991) contains more than 15 times the FAA content of the small (1.2 mm in diameter) egg of the lemon sole, *Microstomus kitt* (Rønnestad et al., 1992a). Calculations of the average FAA concentration in the egg underestimate the actual yolk concentration if correction is not made for the PVS or the embryonic body. The PVS as a percentage of total egg volume varies between fish species (Laale, 1980) and has values of 10–15% for Atlantic halibut (Mangor-Jensen and Waiwood, 1995), 18–23% for Atlantic cod (Mangor-Jensen, 1987), 50–60% for striped bass, *Morone saxatilis* (Mangor-Jensen et al., 1993), and up to 85% for the long rough dab, *Hippoglossoides platessoides limandoides* (Lønning and Davenport, 1980). Estimates of the relative size of the embryo compared to the yolk volume in marine fish eggs are lacking in the literature. When the calculations of yolk FAA concentration are based on yolk volumes determined by morphometry, values of 130–180 mM can be estimated for eggs without oil globules (Rønnestad et al., 1992a, 1993; Finn et al., 1995a,b) and 80–160 mM for eggs with oil globules (Rønnestad et al., 1992b, 1994, 1998a; Fyhn and Govoni, 1995; Finn et al., 1996; Sivaloganathan et al., 1998).

The profile of the FAA pool of the pelagic eggs is remarkably similar among the tested fishes whether they are taken from the boreal (Thorsen et al., 1993) or the tropical region (Rønnestad et al., 1996). The FAA pool consists of about 55% indispensable amino acids dominated by leucine, isoleucine, lysine, and valine, and about 45% dispensable amino acids dominated by alanine and serine. [We follow the terminology of Harper (1974, 1983) of dispensable and indispenable amino acids rather than using the terms nonessential and essential amino acids. This is appropriate for embryonic and larval stages due to their high growth rates, which may make even dispensable amino acids growth limiting owing to the high synthetic demands of the larvae.]

The egg FAA content (moles per egg) may vary between females and between egg batches for a multiple batch spawner [e.g., Atlantic cod (Mårstøl et al., 1993)]. A progressive decrease of 30–45% in the mean content of FAA occurrs from the early to the late egg batches. Within each batch, however, the variation in the egg FAA content is small (about 5% S.D., matching the analytical variability of the amino acid analysis). A similar interbatch variability in the egg size has been found for the Atlantic cod (Kjesbu et al., 1991; Kjesbu and Kryvi, 1993). Thus, although the amount of FAA invested in each egg differs between batches and throughout the spawning season, the amount per egg within each batch is well controlled. As a consequence, use of single-batch material is highly recommended in studies of embryonic and larval performance in order to minimize the data variation that otherwise may conceal an ontogenetic pattern in the process under investigation.

In marine teleosts the egg FAA pool makes up a large fraction of the total amino acid content (free + protein bound). In eggs without oil globule(s) like those of Atlantic cod, Atlantic halibut, or the barfin flounder, *Verasper moseri,* the FAA content typically amounts to 25–40% (w/w) of the total amino acid content (Finn, 1994; Thorsen, 1995; Matsubara and Koya, 1997; Rønnestad *et al.,* 1999a). In pelagic eggs with oil globule(s) like those of European sea bass (*Dicentrarchus labrax*), gilthead seabream (*Sparus aurata*), turbot, Asian sea bass (*Lates calcarifer*), or spot (*Leiostomus xanthurus*) the percentages are smaller, about 20–25%, but still quite large (Rønnestad *et al.,* 1992b, 1994, 1998a; Finn *et al.,* 1996; Fyhn and Govoni, 1995; Sivaloganathan *et al.,* 1998; Parra *et al.,* 1999). Thus, estimates of the protein content in marine pelagic fish eggs or yolk-sac larvae based on the Kjeldahl or Nessler procedures must be treated with care due to a likely overestimation.

The water salinity at spawning affects the protein content of marine teleost pelagic eggs, but not their FAA content. Thus, for acclimatized Atlantic cod that spawned at salinities of 7 ppt in the Baltic Sea or 34 ppt in the oceanic water off the Lofoten Islands, Thorsen *et al.* (1996) reported egg protein contents of 17 and 51 μg/egg, respectively, while the egg FAA content was about 240 nmol/egg in both cases. The FAA pool of the brackish water cod eggs accounted for about 65% of the total amino acid content. This high value, however, was the result of the low protein content and not because of an increased FAA content as might be surmized from the percentage value alone. Similarly, the brackish water cod eggs had lower FAA concentration than their marine counterparts (131 versus 179 mM) because of a larger egg volume (Thorsen *et al.,* 1996). These cases illustrate the danger of drawing conclusions about body content changes based solely on data expressed as ratios or percentages (Rønnestad, 1995).

Compared to the pelagic eggs, the FAA pool of marine demersal eggs is much lower, typically 2–5% of the total amino acid content (Thorsen *et al.,* 1993, Rønnestad *et al.,* 1999a). In a comparative study of newly spawned eggs of Panama fishes, Rønnestad *et al.* (1996) showed that the FAA pool of pelagic eggs (12 species from the families Labridae, Scaridae, Acanthuridae, and Chaetodontidae) accounted for 35% of the total amino acid content, whereas the value was only 2.6% in demersal eggs (10 species from the families of Pomacentridae, Gobiidae, and Blennidae). Furthermore, the profile of the FAA pool of the demersal eggs is quite different from that of the pelagic eggs and is strongly dominated by taurine (Suzuki and Suyama, 1983; Hølleland and Fyhn, 1986; Thorsen *et al.,* 1993; Rønnestad *et al.,* 1996). Taurine, a degradation product of methionine and cysteine in fishes (Yokoyama and Nakazoe, 1996, 1998), is an amino acid analog with several biological functions including cellular osmoregulation (Huxtable, 1992;

Thoroed and Fugelli, 1994). In eggs of the lumpsucker (*Cyclopterus lumpus*), taurine accounts for 67% of the FAA pool (Thorsen *et al.*, 1993), and in a puffer fish *(Takifugu niphobles)*, the percentage is 74% (Suzuki and Suyama, 1983). This contrasts the situation in marine pelagic eggs where no individual amino acid comprises more than 15% of the total with taurine making up less than 5%.

Analyses of newly spawned eggs of freshwater fishes show that the FAA pool is small compared to the total amino acid content, usually less than 1.5% (Suzuki and Suyama, 1983; Dabrowski *et al.*, 1985; Srivastava and Brown, 1992; Srivastava *et al.*, 1995; Gunasekera *et al.*, 1996a, 1999; Terjesen *et al.*, 1997; Thompsen, 2000). Unlike the situation in the pelagic eggs of marine fish, the FAA profile in eggs of freshwater fishes is highly variable between species. For example, the percentage of indispensable FAA is ~25% in Atlantic salmon, *Salmo salar* (Srivastava *et al.*, 1995), but ~60% in Murray cod, *Maccullochella peelii peelii,* and trout cod, *M. macquariensis* (Gunasekera *et al.*, 1999). Similarly, taurine may be dominating in some species [e.g., sweet smelt, *Plecoglossus altivelis,* pond smelt, *Hypomesus olidus,* and carp, *Cyprinus carpio* (Suzuki and Suyama, 1983)] or apparently be lacking [Nile tilapia, *Oreochromis niloticus* (Gunasekera *et al.*, 1996a), Murray cod, and trout cod (Gunasekera *et al.*, 1999)].

B. Changes during Development

The yolk FAA pool in pelagic eggs of marine teleosts decreases during development, although differences exist between species in terms of *when,* relative to hatch, the decrease occurs. The time of hatching is not fixed in relation to developmental stage. The presence of an oil globule(s) also influences the developmental dynamics of the yolk FAA pool. In marine teleost eggs with an oil globule(s), such as turbot (Rønnestad *et al.*, 1993; Finn *et al.*, 1996; Conceição *et al.*, 1997a), gilthead sea bream, *S. aurata (*Rønnestad *et al.*, 1994), red sea bream, *Pagrus major* (Seoka *et al.*, 1997), European sea bass, *D. labrax* (Gatesoupe, 1986; Rønnestad *et al.*, 1998a), Asian seabass, *L. calcarifer* (Sivaloganathan *et al.*, 1998), and Senegal sole, *Solea senegalensis* (Parra *et al.*, 1999), the FAA decrease occurs mostly during the egg stage, whereas in eggs lacking oil globules such as lemon sole (Rønnestad *et al.*, 1992a), plaice, *P. platessa* (Rainuzzo *et al.*, 1993), and Atlantic cod (Fyhn *et al.*, 1987; Finn *et al.*, 1995a), the depletion continues beyond hatch and up to the time of first feeding. In Atlantic halibut, which lacks oil globules and hatches at an immature stage (Haug, 1990; Pittman, 1991; Kjørsvik and Reiersen, 1992), there is hardly any decrease in the egg FAA pool before hatch (Fyhn, 1989; Finn *et al.,* 1991).

The decrease in the FAA pool does not result from a diffusive loss to the ambient seawater. This conclusion is supported by experiments on yolk-sac larvae of the Atlantic halibut which showed no significant loss of FAA to the incubation medium (Rønnestad, 1993). In these experiments, the medium FAA concentration remained at about 0.3 μM over 94–116 h of incubation despite a decline in the FAA content of the incubated larvae that could have increased the medium concentration several hundred-fold. The conclusion agrees with previous findings for teleost eggs and yolk-sac larvae demonstrating that the perivitelline membrane has an exceptionally low permeability for water and ions (Potts and Rudy, 1969; Mangor-Jensen, 1987; Riis-Vestergaard, 1987; Tytler et al., 1993; Mangor-Jensen et al., 1993).

In the pelagic eggs of marine fish, all yolk FAA except taurine and phosphoserine decrease during development. Both indispensable and dispensable amino acids participate in the decline, and there seems to be no sparing effect for the former (Finn et al., 1995a,b, 1996; Rønnestad et al., 1999a). Taurine is maintained at a constant level while the content of phosphoserine actually increases (Rønnestad and Fyhn, 1993; Rønnestad et al., 1999a). By the end of yolk resorption, taurine dominates the FAA pool (Rønnestad et al., 1993; Finn et al., 1995a,b, 1996; Sivaloganathan et al., 1998). Its function during fish development remains to be established, but in mammals it has been shown to be crucial for early development of various organs, including brain, eye, retina, and heart, as well as for bile salt conjugation (Sturman, 1993). Conceição et al. (1997a) found a correlation between growth rate and taurine content for turbot larvae suggestive of a dietary requirement for taurine during early development in this species. Phosphoserine, the major component of yolk phosvitins (Murakami et al., 1991; Reith et al., 2001), exists at trace levels in the egg but accumulates progressively in the larval body compartment (Rønnestad et al., 1992a, 1993; Finn et al., 1995a,b, 1996). The mechanism of yolk phosvitin degradation may result from the joint action of proteases and acid phosphatases in the medaka, Oryzias latipes (Murakami et al., 1991).

The decline in the FAA pool correlates with the decrease in yolk volume (Rønnestad et al., 1992a,b, 1993; Fyhn and Govoni, 1995; Finn et al., 1995a,b, 1996; Sivaloganathan et al., 1998). The vitelline syncytium layer, with its abundance of mitochondria, endocytotic vesicles, and endoplasmatic reticulum, is the likely organ responsible for the transfer of yolk nutrients to the embryonic tissues (Heming and Buddington, 1988; Kjørsvik and Reiersen, 1992; Sire et al., 1994; Mani-Ponset et al., 1996; Poupard et al., 2000). This view is supported by developmental data from several fish species which show that the yolk reservoirs are not exploited before epiboly (Rønnestad et al., 1992a,b, 1993; Finn et al., 1995a,b, 1996). In yolk-sac larvae of Atlantic halibut, water, protein, and FAA do not decrease in synchrony during development (Rønnestad et al., 1993; Finn et al., 1995b): FAA decrease first, followed by water, and then proteins. Thus, during the period of maximal yolk resorption in Atlantic halibut larva, the FAA concentration decreases by 60–67% (i.e., from 150 to 50–60 mM), even though the yolk volume decreases by ~80%. Concomitantly, the yolk protein concentration increases by a factor of 2–3 [65–160 mg/ml (Finn et al., 1995b); 90–290 mg/ml

(Rønnestad et al., 1993)]. Based on the changes in the yolk FAA pool profile during development of the Atlantic halibut larvae, Rønnestad et al. (1993) suggested that selective uptake mechanisms were present for the different amino acids in the syncytium. The data of Finn et al. (1995b) support this conclusion. Conceição et al. (1998a) found that yolk proteins were taken up in bulk by nonselective endocytosis in the freshwater African catfish. New studies, specifically addressing uptake of the various yolk nutrients during the endogenous feeding phase of the fish embryo, are necessary to clarify the mechanisms involved.

In contrast to the situation for marine teleosts, the few studies of the FAA pool in freshwater teleost eggs and larvae show increasing rather than decreasing contents during yolk-sac development (Suzuki et al., 1991; Srivastava and Brown, 1992; Srivastava et al., 1995; Gunasekera et al., 1996a,b, 1999; Terjesen et al., 1997; Thompsen, 2000). For example, the African catfish (*Clarias gariepinus*) has a small FAA pool at spawning [about 9 nmol/individual (ind), or $\sim$0.5% the total amino acid content] increasing to a maximum (82.0 $\pm$ 5.1 nmol/ind; 6% of total content) at final yolk resorption (Terjesen et al., 1997). Microdissection of newly hatched larvae revealed that $\sim$75% of the FAA pool is localized to the yolk compartment and that the subsequent increase occurred in the body of the African catfish. Thus, the average body tissue FAA concentration reached a value of $\sim$47 mM during maximal growth but decreased to $\sim$11 mM in the starving larvae past yolk resorption (Terjesen et al., 1997). The changes for taurine differed from that of other FAA (Terjesen, 1995). In the starving larvae, after final yolk resorption, the taurine content did not decrease but remained at a level of about 4.5 nmol/ind or a mean body concentration of about 2 mM. In fed larvae, taurine increased most of all the FAA and reached a mean body concentration of $\sim$12 mM after about 10 days when it made up a third of the total FAA pool. Taking into account the methionine and cysteine content of the proteins of the yolk-sac larvae of the African catfish (Conceição et al., 1998a), the decrease in their protein content (Terjesen et al., 1997) was more than enough to accommodate the observed increase in taurine and methionine levels during the yolk-sac stage (Terjesen, 1995). An unanswered question is why taurine is sequestered to this high degree in the developing larvae of freshwater and marine teleost larvae.

C. Oocyte Hydration in Marine Teleosts

More than 100 years ago Fulton (1891, 1898) and Milroy (1898) pointed out the remarkable volume increase during final oocyte maturation in marine teleosts and suggested an osmotic mechanism whereby a watery fluid was secreted into the oocytes by the enclosing follicle. Based on these early findings and ideas, more recent studies have revealed further details of the oocyte swelling mechanism in teleosts (see reviews by Guraya, 1986; Mommsen and Walsh, 1988; Selman and Wallace, 1989; Tyler and Sumpter, 1996). The involvement of FAA as osmolytes in the oocyte swelling mechanism of marine teleosts with pelagic eggs has recently attracted much interest (Fyhn, 1993; Mellinger, 1994; Thorsen and Fyhn, 1991, 1996; Thorsen et al., 1993, 1996; Matsubara and Koya, 1997; Matsubara et al., 1999; Rønnestad et al., 1999a; Fyhn et al., 1999; Finn et al., 2000; Reith et al., 2001).

When teleost eggs spawn in freshwater versus seawater they meet quite different osmotic challenges. In the former, the eggs are hyperosmotic and face a passive water influx, whereas in the latter they are hyposmotic and experience water efflux. Moreover, early embryos lack the organs responsible for osmoregulation found in the adult fish (e.g., kidney, gills) and must rely on compensatory means included with the egg before spawning. This is true even though the osmotic problems are minimized by the exceedingly low permeability of the egg vitelline membrane (Potts and Rudy, 1969; Mangor-Jensen, 1987; Riis-Vestergaard, 1987; Tytler et al., 1993; Mangor-Jensen et al., 1993). A volume decrease (i.e., net water loss) in developing marine teleost embryos [e.g., Atlantic cod (Mangor-Jensen, 1987), Atlantic halibut (Mangor-Jensen and Waiwood, 1995), and striped bass (Mangor-Jensen et al., 1993)], as well as a volume increase (i.e., net water influx) in freshwater teleost embryos [e.g., Atlantic salmon (Li et al., 1989)], has been reported based on measurements of changes in PVS volume. As a counteraction, the water content of the eggs at spawning seems to be adjusted according to the environmental osmolality. Eggs of freshwater teleosts have lower relative water contents [63 ± 7%, average of 25 species (Kamler, 1992)] than eggs of marine teleosts [benthic eggs: 70–80% (Craik and Harvey, 1986; Suzuki and Suyama, 1983); pelagic eggs: 86–93% (Craik and Harvey, 1986, 1987; Matsubara and Koya, 1997; Finn et al., 2000)]. The especially high water content of the marine pelagic eggs is related to their buoyancy (see Section II.D).

The increase in FAA content of the oocyte during final maturation causes osmotic water influx and swelling of the yolk compartment, transforming the yolk from an opaque mass of crystalline yolk platelets into the typical hyaline and homogeneous yolk of the mature egg (Selman and Wallace, 1989; Kjesbu and Kryvi, 1989, 1993; Selman et al., 1994; Wallace et al., 1993; Cerda et al., 1996). A temporal hyperosmolality of the oocytes relative to the blood and ovarian fluid during the actual swelling phase has recently been noted in Atlantic halibut (Østby et al., 2000). Through this preovulatory hydration, the egg acquires the necessary water to overcome osmotic problems during early development in seawater.

Several studies have demonstrated the involvement of FAA in oocyte hydration in the pelagic eggs of marine fishes (Thorsen, 1995; Thorsen and Fyhn, 1991, 1996; Thorsen et al., 1993). These studies utilized the *in vitro* technique for hormonally induced, vitellogenic oocytes as developed by the group of Wallace and Selman (Wallace and Selman, 1978, 1981, 1985, 1990; Selman and Wallace, 1989; Cerda et al., 1996). In oocytes and eggs taken directly from spawning plaice and lemon sole, a 10- to 15-fold increase in total FAA correlated with a 4- to 7-

Table I
Protein and Free Amino Acid (FAA) Content in Prehydrating Oocytes and Eggs[a]

			Volume (μL/ind)	Protein (μg/ind)	FAA (μg/ind)	FAA (%)
Plaice						
	In vivo	Oocyte	0.94 ± 0.03	227 ± 9.0	6.3 ± 0.3	2.7 ± 0.1
	"	Egg	3.29 ± 0.13	172 ± 8.6	57.9 ± 1.8	25.2 ± 1.3
	In vitro (L-15)	Egg	3.72 ± 0.12	164 ± 6.0	53.9 ± 1.4	24.9 ± 0.8
	" (FO)	Egg	3.67 ± 0.09	159 ± 7.0	54.9 ± 2.0	25.7 ± 0.8
Lemon sole						
	In vivo	Oocyte	0.16 ± 0.02	40.6 ± 12.0	1.2 ± 0.1	2.9 ± 0.5
	"	Egg	1.01 ± 0.09	36.9 ± 3.3	16.6 ± 1.4	31.1 ± 1.7
	In vitro (L-15)	Egg	1.16 ± 0.11	31.9 ± 3.1	13.6 ± 1.3	29.8 ± 0.0
	" (FO)	Egg	0.98 ± 0.11	29.0 ± 1.9	12.9 ± 1.1	30.8 ± 2.3

After Thorsen and Fyhn, 1996, with permission from The Fisheries Society of the British Isles and Academic Press.

[a] Protein and FAA content in prehydrating oocytes and eggs of plaice (*Pleuronectes platessa*) and lemon sole (*Microstomus kitt*) *in vivo*, and in eggs matured *in vitro* in culture medium with (L-15 medium) and without (FO medium) the addition of 17 amino acids (each at about 1.5 mM). The FAA contents are also expressed as a percentage of total amino acid content (sum of FAA and proteinic amino acids).

fold increase in oocyte volume, while simultaneously, the protein content decreased by 10–25% (Table I). When the oocytes were hormonally induced to swell *in vitro*, with or without FAA in the medium, they attained a FAA profile almost identical to that of *in vivo* swollen oocytes (Fig. 3), although the FAA content (amount/egg) was lower than in the *in vivo* eggs (Table I). This probably results from protein uptake during the swelling process *in vivo*, which is prevented during the *in vitro* experiments. An increase in the FAA pool that is larger than the decrease in the protein content is often found during oocyte swelling *in vivo* in marine fishes with pelagic eggs (Matsubara and Koya, 1997; Finn *et al.*, 2000). The lower content of taurine in the *in vitro* eggs compared to the *in vivo* condition (Fig. 3) undoubtedly results from the lack of taurine in the culture medium. Recent biopsy data from spawning Atlantic halibut, Atlantic cod, haddock (*Melanogrammus aeglefinus*), and yellowtail flounder (*Pleuronectes ferrugineus*) show a volume increase of 5–8 times for the swelling oocytes correlated with an increase in their ninhydrin positive substances (NPS) contents of 8–10 times and a decrease in the soluble protein content of 15–35% (Finn *et al.*, 2000). [Measurement of NPS is used as an estimate of total soluble -amino acids in a protein precipitated solution, although the reaction detects other amino compounds such as ammonia and soluble peptides but not proline (Moore and Stein, 1954)].

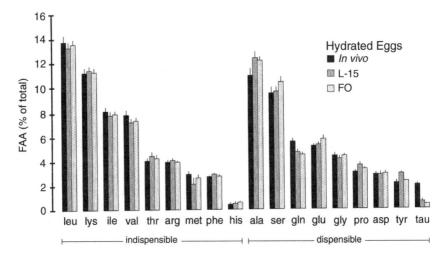

Fig. 3. Free amino acid profiles of eggs of plaice (*P. platessa*) taken from a spawning fish *in vivo* and from eggs matured *in vitro* in culture medium with (L-15 medium) and without (FO medium) the addition of 17 amino acids (each at about 1.5 mM). (After Thorsen and Fyhn, 1996, with permission from the The Fisheries Society of the British Isles and Academic Press.)

The FAA of the pelagic fish eggs appears to be derived from hydrolysis of specific yolk proteins (Thorsen and Fyhn, 1991, 1996; Thorsen *et al.*, 1993; Matsubara and Sawano, 1995; Matsubara and Koya, 1997; Matsubara *et al.*, 1999; Finn *et al.*, 2000). The similarity of the egg FAA profile for 7 species of boreal marine fishes and 12 species of tropical marine fishes (Thorsen *et al.*, 1993; Rønnestad *et al.*, 1996) also points to a common yolk protein source for the FAA pool. Greeley *et al.* (1986) showed that extensive proteolysis of a yolk protein with a molecular mass of 104–108 kDa and of the highly phosphorylated phosvitins (15–60 kDa) occurred during final oocyte maturation in several marine fishes with pelagic eggs. The gel electrophoretic studies conducted by Thorsen and coworkers with plaice, lemon sole, and Atlantic cod focused on an ~100-kDa lipovitellin as the source for the FAA pool (Thorsen *et al.*, 1993; Thorsen and Fyhn, 1996; Thorsen *et al.*, 1996). Similarly, distinct changes in the electrophoretic yolk protein banding patterns during oocyte maturation have been shown for gilthead sea bream and the European seabass (Carnevali *et al.*, 1992, 1993) and for Atlantic halibut, Atlantic cod, haddock, and yellowtail flounder (Finn *et al.*, 2000). In the latter study, oocyte protein hydrolysis occurred concomitant with an increase in the FAA pool. Oocyte hydration correlated with breakdown of yolk lipovitellins has also been found in the Japanese eel, *Anguilla japonica*, after injections with salmon pituitary homogenates (Okumura *et al.*, 1995).

The detailed biochemical studies of Matsubara and coworkers with the barfin flounder indicate that full hydrolysis of a heavy-chain lipovitellin (107 kDa) occurs during oocyte final maturation concomitant with partial hydrolysis of phosphovitin and another small yolk protein, the β-component (Matsubara and Sawano, 1995; Matsubara and Koya, 1997; Matsubara et al., 1999). More recently in haddock, Reith et al., (2001) reported that there are two distinct lipovitellins in the oocytes (Had1 and Had2, both ~110 kDa, 54% identical amino acid sequences) and that Had2 is fully degraded while only a part of Had1 (108 amino acids at the C-terminal end) is degraded during final oocyte maturation. Thus, the available evidence supports the idea that the FAA pool of marine pelagic eggs originates from the proteolysis of specific yolk proteins, mainly a lipovitellin of 105–110 kDa. The enzymatic mechanism responsible for this selective yolk protein hydrolysis, probably lysosomal cathepsin L (Carnevali et al., 1999), and the control function of this mechanism should be rewarding fields for further study.

Other osmolytes, especially K^+, but also NH_4^+, Cl^-, and inorganic phosphate increase several fold during oocyte swelling, thereby implicating these solutes in the water uptake process (Craik and Harvey, 1984, 1986, 1987; Thorsen and Fyhn, 1991; Østby et al., 2000). The relative importance of inorganic ions and FAA as osmoeffectors in oocyte swelling differ between pelagic and demersal eggs (Thorsen et al., 1993; Rønnestad et al., 1996). In pelagic eggs, the FAA make up ~50% of the solutes responsible for the osmotic water influx, whereas in the demersal eggs the percentage is much less, ~15% with a dominance of taurine (Thorsen and Fyhn, 1991; Thorsen, 1995). A marine teleost egg sampled at random, therefore, may be categorized as demersal or pelagic based on the percentage of solutes as FAA and the taurine content. The recent study of Hartling and Kunkel (1999) on the demersal eggs of the winter flounder (*Pleuronectes americanus*) is interesting in this regard because the yolk lipovitellin is partly degraded during the transition from oocyte to egg, while there is no corresponding increase in the FAA pool (N. Finn, personal communication).

D. Buoyancy of Pelagic Teleost Eggs

The majority of marine teleosts have pelagic eggs (Kendall et al., 1984). Egg buoyancy is maintained during development and promotes dispersal by facilitating drift within ocean currents. At the same time, buoyancy increases the chances of survival of the newly hatched larvae by positioning them in the upper water layers where planktonic food organisms abound. The egg buoyancy of marine teleosts is mainly due to the large volume of dilute yolk, although buoyancy is partly assisted by the presence of oil globules (Shelbourne, 1965; Craik & Harvey, 1987; Finn et al., 1996). The increased water content of Atlantic cod eggs when spawned in the brackish water of the Baltic Sea compared to eggs in full strength seawater (97 versus 93%) supports the hypothesis that the yolk promotes buoyancy (Thorsen et al., 1996). The heavy components of the fish egg are primarily the chorion, but also body proteins and other dry materials with the exception of lipids. However, the volume of the embryo in marine teleosts decreases during early development due to osmotic water loss, and buoyancy should therefore decrease with time, making the egg prone to sinking. Since this does not occur, volume changes in the developing embryo must somehow be counteracted so that the forces promoting flotation or sinking are balanced, that is, the egg remains neutrally buoyant.

Increased buoyancy can be achieved by exchanging a light solute with a heavier one, for example, NH_4^+ for K^+. In marine teleost embryos an accumulation of ammonia occurs concomitant with a loss of K^+ (Fyhn, 1993; Finn et al., 1995a,b, 1996; Terjesen et al., 1998). In vivo NMR studies indicate that ammonia is trapped as NH_4^+ in an acidic yolk (pH = 5.3–5.6) in yolk-sac larvae of plaice and Atlantic halibut (Grasdalen and Jørgensen, 1987; Jørgensen and Grasdalen, 1990). Other ions must also be involved in this exchange mechanism since the stoichiometric match between accumulated NH_4^+ and K^+ is not 1:1 as it should be to maintain electroneutrality of the body fluids. Hence, future measurements should focus on other possible light solutes that assist in providing lift.

Stoichiometric calculations of the changes in FAA and protein contents of developing eggs and yolk-sac larvae of lemon sole, turbot, Atlantic cod, and Atlantic halibut suggest that ammonia formation results from FAA rather than from protein catabolism (Rønnestad et al., 1992a,b; Finn et al., 1995a,b, 1996; Terjesen et al., 1998). Further, the proportion of total ammonia production that is accumulated during early development varies between species: Atlantic cod, ~25% (Finn et al., 1995a); turbot, ~30% (Rønnestad et al., 1992b) or ~25% (Finn et al., 1996); lemon sole, ~5% (Rønnestad et al., 1992a); and Atlantic halibut, ~40% (Terjesen et al., 1998). Obviously, the accumulated ammonia must be included in the overall nitrogen budget when estimating metabolic rates of teleost embryos and yolk-sac larvae. Furthermore, the high levels of ammonia in the oocytes and eggs of marine teleosts add to the difficulty of interpreting measurements of FAA based only on NPS measurements (e.g., Craik and Harvey, 1987). More studies are necessary before the biological significance of the variation in ammonia accumulation can be understood.

Accumulated ammonia is quickly excreted on hatching (e.g., Finn et al., 1995a,b, 1996; Sivaloganathan et al., 1998) when the embryo is freed from the heavy chorion and may no longer need the additional buoyancy. This supports the idea that ammonia is sequestered to assist in buoyancy regulation (Fig. 4). The slow start of ammonia excretion in Atlantic halibut after hatch may relate to its deep water spawning and the immature status of the larvae at hatch (Haug, 1990). The yolk-sac larvae may need additional flotation to reach the surface layer at the commencement of exogenous feeding (Terjesen et al., 1998). It would be interesting to learn whether other teleosts with deep water spawning (e.g., the Pacific black cod, *Anoplopoma fimbria*) show a similar delay in ammonia excretion after hatch.

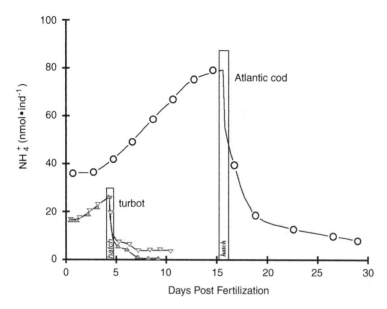

Fig. 4. Changes in total ammonium content in developing eggs and larvae of Atlantic cod, *Gadus morhua* (6°C), and turbot, *Scophthalmus maximus* (15°C). (After Finn *et al.* 1995a, 1996 with permission from Springer-Verlag and Elsevier Science.)

Further support for the buoyancy function of accumulated NH_4^+ in teleost eggs is provided by comparing Atlantic cod eggs spawned in full-strength seawater or brackish water of the Baltic Sea (Thorsen *et al.*, 1996). The brackish water-spawned eggs had a higher ammonia content relative to the seawater-spawned eggs (51 versus 30 mM), but a reduced K^+ concentration (30 versus 42 mM). Similar studies on other species, including those with demersal marine eggs and freshwater eggs, would provide a clearer picture of the ionic aspects of buoyancy regulation.

E. Phylogenetic Aspects

The fossil record of the early Actinopterygians (teleost ancestors) suggests evolution in freshwater for more than 200 million years before the teleosts returned to the sea during the Jurassic period, apparently with a rapid differentiation of new species in the Cretaceous period about 100 million years ago (Halstead, 1985; Moyle, 1993; Long, 1995). Indeed, the hyposmotic condition of extant marine teleosts—a condition that is quite unusual among aquatic animals—is assumed to be an evolutionary vestige of their freshwater ancestry (Smith, 1932, 1953; Evans, 1993). The evolutionary success of teleosts has been related to the special yolk structure and the meroblastic cleavage of the teleost egg (Collazo *et al.,* 1994; Collazo, 1996). Their rapid differentiation in the marine environment during the Cretaceous period has been related to the large FAA pool of their eggs at spawning (Fyhn *et al.,* 1999).

In extant teleosts, blood plasma osmolality is somewhat higher in marine species compared to freshwater species, but in both cases it is homeostatically maintained by physiological means involving several organs and specialized cells (Evans, 1993). The problem of maintaining the water balance of the early embryo, however, must be solved by different means since the necessary organs are not yet differentiated. Measurements of the osmolalities of teleosts embryos and larvae show that they are comparable to that of the adults (reviewed by Riis-Vestergaard, 1987; Tytler *et al.,* 1993). The first forays of teleosts that led to colonization of the oceans probably involved excursions into the sea for feeding and return to the freshwater environment for spawning, as occurs today with anadromic salmonids. This assured that the embryos could rely on mechanisms and adaptations that were already suited for a life in freshwater. A complete marine life cycle for teleosts could not occur until the egg and embryo had acquired the necessary seawater adaptations, particularly, maintenance of embryonic water balance.

Fyhn *et al.* (1999) proposed that the key step in teleost evolution that allowed their successful radiation in the oceans about 100 million years ago was yolk protein hydrolysis at final oocyte maturation with a resulting increase in the FAA pool and ensuing oocyte hydration. Both the osmotic challenges faced by the eggs and the pelagic distribution of buoyant eggs would benefit from oocyte hydration. The similarity of the FAA pool in marine pelagic teleost eggs studied so far (Thorsen *et al.,* 1993, Rønnestad *et al.,* 1996) suggests that the phenomenon is of ancient origin and central to the adaptation of teleosts to life in a saline environment, as well as that the hydrolyzed yolk protein(s) are derived from a common origin (see Section II.C).

III. NUTRITIONAL DEMANDS

Fish larvae may grow exceedingly fast with growth rates of $30-100\% \cdot \text{day}^{-1}$ (Kamler, 1992). Because growth in fishes is primarily protein accretion, larvae have a large demand for amino acids. In addition, amino acids are heavily used as substrates in the energy dissipation of fish larvae (see Section III.A) and are precursors in the synthesis of many other important biomolecules involved in the growth process, for example, purines and pyrimidines of the nucleic acids, the porphyrine nucleus of hemoglobin and cytochromes, phosphocreatine, as well as in peptide- or amino acid-derived hormones. Together, this multitude of functions results in an exceptional demand for amino acids by growing fish larvae. However, most studies of the nutritional requirements of fish larvae during the last decades have focused on lipids and especially on the indispensible nature of the polyunsaturated fatty acids (e.g., Koven *et al.*, 1992; Watanabe, 1993; Mellinger, 1995; Rønnestad *et al.*, 1995; Finn *et al.*, 1995d; Izquierdo, 1996; Shields *et al.*, 1999; Bell *et al.*, 1999). A wider nutritional approach is now recommended (Watanabe and Kiron, 1994) and the larval demands for vitamins (Dabrowski, 1992; Mangor-Jensen *et al.*, 1994; Merchie, 1995; Takeuchi *et al.*, 1995; Rønnestad *et al.*, 1997, 1998b,c, 1999b), amino acids (Yùfera *et al.*, 1999; Lopez-Alvarado and Kanazawa, 1994; Lopez-Alvarado *et al.*, 1994), and specific fuels (Rønnestad, 1992; Rønnestad and Fyhn, 1993; Finn *et al.*, 1995a,b,c, 1996; Rønnestad *et al.*, 1999a) have been emphasized.

The anabolic function of amino acids in fish larvae is given in Chapter 2. Models of amino acid metabolism and the estimates of cost of protein growth in fish larvae provide increased nutritional insight (Conceição *et al.*, 1995, 1997a,b, 1998b; Houlihan *et al.*, 1995). A critical review of experimental design and statistical data treatment in studies of amino acid requirements of fishes has recently been published (Shearer, 2000). The following section focuses on the use of amino acids as fuel in fish embryos and larvae.

A. Amino Acids as Fuel

The teleost egg is essentially a closed system with regard to nutrients, and all precursors necessary for growth of the developing embryo must be included at spawning (Needham, 1931, 1942; Finn, 1994). Based on this idea, the biochemical composition of the newly spawned fish eggs has been reviewed in search of egg quality criteria (Kjørsvik *et al.*, 1990). In addition to their function as precursors in the growth process, amino acids are also excellent energy substrates and are readily accepted, after deamination, into the citric acid cycle for aerobic production of ATP. Their use as a dominating fuel in marine teleost embryos and yolk-sac larvae is now firmly documented (reviewed by Rønnestad and Fyhn, 1993; Finn, 1994; Rønnestad *et al.*, 1999a).

A possible anaerobic contribution to routine metabolism of teleost larvae was addressed by Finn *et al.* (1995c) using calorespirometry on turbot eggs and yolk-sac larvae from 0 to 8 days postfertilization. The data suggest that anaerobiosis contributes insignificantly to routine metabolism because the values were within the theoretical oxyenthalpic scope for fully aerobic metabolism. This conclusion is supported by low lactic acid levels in egg and yolk-sac larvae of Atlantic halibut, Atlantic cod, and turbot (Finn *et al.*, 1995a,b, 1996) and by low activities of glycolytic enzymes, but high levels of oxidative enzymes of the citric acid cycle in turbot embryos (Segner *et al.*, 1994).

Quantitative metabolic studies of body composition (FAA, protein, carbohydrates, and lipids) concurrently with determinations of ammonia production and O_2 consumption allow an assessment of the relative importance of amino acids as a fuel in aerobic metabolism. From the molar amounts of specific amino acids catabolized, one can calculate the moles of ammonia produced as well the moles of O_2 consumed during a given period. Further, if a certain molar amount of known FAA is polymerized into proteins, one can calculate the mass of proteins produced from the molecular weights of the amino acids involved. The calculated values can then be compared with those obtained experimentally and analytically, and the agreement statistically tested. This quantitative strategy has been the basis for the advances in fish larval energetics, which have demonstrated that FAA are an important fuel during the egg and yolk-sac larval stages of marine teleosts (Fyhn, 1989, 1990; Rønnestad, 1992; Finn, 1994; Rønnestad and Fyhn, 1993; Rønnestad et al., 1999a).

The catabolism of FAA in marine pelagic fish larvae account for a large share of the oxygen consumed during yolk-sac development (Fig. 5). Based on data from studies of several marine fish species, Finn (1994) calculated that the embryos and yolk-sac larvae of eggs without oil globules (Type I egg) derived an average of 70% of their metabolic energy from amino acids (20–40% from FAA and 20–50% from protein), whereas the average figure for eggs with oil globules (Type II egg) was 50%, equally shared between FAA and protein. The balance was made up mainly by lipids while glycogen contributed 1% or less to the total energy dissipation. Thus, the FAA pool needs to be taken into consideration when studying energy dissipation and yolk absorption in embryonic and larval fishes.

For older larvae of Atlantic halibut, after first feeding when growth relies on ingested external feed, Rønnestad and Naas (1993) showed that approximately 60% of the metabolic energy continued to be derived from the catabolism of amino acids. In the warm water Asian seabass (Type II egg) only 14% of the aerobic energy dissipation is derived from amino acid catabolism during the yolk-sac stage (Sivaloganathan et al., 1998). These authors suggest that a correlation exists between habitat temperature and the amino acid dependency in that larvae of cold water teleosts rely more on amino acid fuels than warm water species.

Of note is the increased catabolism of proteins when the FAA pool is exhausted at final yolk resorption (Fig. 5). The polymerization of some of the FAA during the earlier embryonic period appears to provide a labile protein pool that can be accessed when FAA are no longer available later in development. This observation is interesting since it apparently departs from the general idea that amino acids in animals are not stored in deposits for energy purpose, as is the case for glycogen and lipids.

The current estimates of the use of amino acids as an embryonic and larval fuel in marine fishes have been based on the assumption that ammonia is the only excretory product resulting from amino acid catabolism. It has been assumed that any urea excreted was derived from purine and pyrimidine catabolism via nucleic acid degradation (van Waarde, 1988). The calculated total production of ammonia of the developing embryo (sum of accumulated and excreted ammonia) is, however, often less than what could be accounted for by catabolism of the disappear-

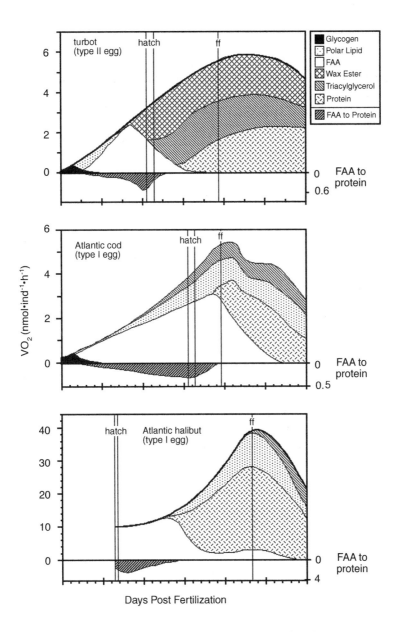

Fig. 5. Sequence of catabolic substrate oxidation in developing embryos and larvae of turbot (15°C), Atlantic cod (6°C), and Atlantic halibut (6°C). The contribution of each substrate to energy metabolism is expressed as a fraction of the total aerobic metabolic rate as measured by oxygen uptake. Also shown is the synthesis rate of body protein from free amino acids (FAA, nmol·ind^{-1} h^{-1}). Ff, first feeding. (After Finn et al. 1995a,b, 1996, with permission from Springer-Verlag and from Elsevier Science.)

ing FAA (Rønnestad and Fyhn, 1993; Finn et al., 1995a,b, 1996). It is, therefore, interesting that recent studies have shown that at least part of the urea synthesized is derived from amino acids in yolk-sac larvae of Atlantic cod (Chadwick and Wright, 1999) and of Atlantic halibut (Terjesen et al., 2000; see Section V.B). Taking urea excretion into account when calculating total nitrogen excretion will improve the correspondence between the estimated energy budgets of these larvae and their catabolism of amino acids.

B. Fish Larvae at First Feeding

When the availability of nutritional reserves in the yolk falls below that required for growth and development, the larvae must commence exogenous feeding by catching and ingesting suitable prey. Long ago, the Norwegian marine biologist Johan Hjort (1869–1948) observed large annual variations in commercial fish stocks and the dominance of strong year-classes in the catches (Hjort, 1914). He advanced the hypothesis that the availability of the right kind of plankton prey during a critical period when the larvae started external feeding was crucial for their survival and, thus, for the year-class strength. Small zooplankton (mainly nauplii of copepods but also rotifers, ciliates, and various invertebrate larvae) are generally assumed to be the main food organisms of fish larvae. Marine zooplankton contain a large pool of FAA (Båmstedt, 1986; Fyhn et al., 1993, 1995), as do most marine invertebrates where intracellular FAA largely contribute to osmoregulation in a saline environment (Yancey et al., 1982; Chamberlin and Strange, 1989). Marine phytoplankton, which also contain a large pool of FAA (e.g., Fyhn et al., 1993), are ingested by marine fish larvae during a short period prior to the consumption of zooplankton (Van der Meeren, 1991; Naas et al., 1992; Reitan et al., 1993, 1994; Lazo et al., 2000). Because marine teleosts have likely always fed on planktonic organisms, one might wonder whether they have become adapted to—or even dependent on—an initial prey containing a large content of FAA.

Rust et al. (1993) examined nutrient assimilation in 15-day posthatch larvae of walleye, *Stizostedion vitreum,* using their method of controlled tube feeding of radiolabeled nutrients. Walleye larvae assimilated FAA better than polypeptides, which again was assimilated better than proteins. The differences became less as the larvae grew older (Rust, 1995). Similar results have been obtained for marine fish larvae of Japanese flounder (Rønnestad et al., 2000a) and Senegal sole (Rønnestad et al., 2000b). These findings imply that the assimilation of amino acids from protein digestion in early fish larvae may be insufficient to cover metabolic demands for energetic and synthetic purposes. A supply of *free* amino acids in the diet, therefore, may be beneficial for survival and optimal growth of the marine fish larvae during a developmental window from the time of first feeding until the intestine is differentiated and has acquired full capacity to digest ingested proteins

(Fyhn, 1989, 1990, 1993; Rønnestad and Fyhn, 1993; Rønnestad et al., 1999a, 2000a,b). *Artemia* nauplii, which are the most commonly used live first-feed organism for fish larvae in mariculture, contain less FAA than do natural marine plankton such as copepods (Næss et al., 1995; Helland et al., 2000). Procedures have been recently developed to enrich first-feed organisms with water soluble nutrients (Tonheim et al., 2000).

Fish larvae at the start of external feeding may employ other mechanisms to increase the assimilation of amino acids from the intestine. Indeed, the addition of exogenous enzymes improves assimilation of pelleted feed (Walford et al., 1991; Kolkovski et al., 1993, 1997). Endocytotic uptake of whole proteins occurs in the hind gut mucosa for some fish larvae (Watanabe, 1984; Deplano et al., 1991; Kurokawa et al., 1996). However, this mechanism needs to be quantified in order to assess its importance for the amino acid budget of the growing larvae.

IV. NITROGEN END PRODUCTS AND THEIR CHEMISTRY IN SOLUTION

A. Ammonia

Similar to adult fish, ammonia is the most important nitrogenous waste product during early stages of development in teleost species. Ammonia (NH_3) is a small polar substance that binds a proton to form the ammonium ion NH_4^+ in aqueous solution (see Chapter 4). The term *ammonia* will be used to refer to the total of the un-ionized and ionized forms ($NH_3 + NH_4^+$). The pK of the NH_3/NH_4^+ reaction is between 9.0 and 9.5 and, therefore, at physiologic pH, most ammonia exists as NH_4^+. Ammonia (NH_3) is highly permeant (~ 0.03 cm·s^{-1}; Lande et al., 1995), whereas the permeability of NH_4^+ is typically much lower. Taking the pK and permeability characteristics into account, NH_3 levels are higher in alkaline compartments, but total ammonia levels are higher in more acidic compartments. The yolk of the freshwater rainbow trout is relatively acidic (pH ~ 6.35) and yolk ammonia levels are elevated compared to the PVF and embryonic tissue (Fig. 6; Rahaman-Noronha et al., 1996). Similarly, in the freshwater African catfish the yolk ammonia concentration was 25 mM, whereas the larval body was only 8.5 mM (Terjesen et al., 1997). Yolk ammonia concentrations in marine teleosts may be even higher (see Section II.D.).

The distribution of ammonia may be influenced by the electrical potential across the cell membrane if the membrane is relatively permeable to NH_4^+. In rainbow trout embryos, the calculated NH_4^+ equilibrium potential (Em NH_4^+) is approximately equal to the measured electrical potential between the PVF and yolk, providing evidence for a NH_4^+ permeable yolk-sac membrane (Rahaman-Noronha et al., 1996). These findings, however, do not agree with other reports of

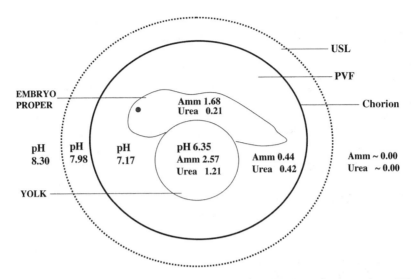

Fig. 6. Schematic diagram of a rainbow trout embryo depicting the unstirred water layer (USL) next to the chorion, enclosing the perivitelline fluid (PVF), embryo proper, and yolk. Ammonia (Amm) and urea concentrations in the PVF (mM) and pH values are taken from Rahmahan-Noronha (1995) and Rahmahan-Noronha et al. (1996), whereas the ammonia and urea concentrations in the embryo proper and yolk were measured by Steele et al. (2001). Means of 4–8 measurements. Yolk urea content was significantly higher than embryo urea content (Student's t test, $p < 0.05$).

a yolk-sac membrane relatively impermeable to ions and water (Potts and Rudy, 1969; Alderdice, 1988) and possibly urea (see Section IV.B). Hence, the concentration of ammonia in the yolk may be influenced by a number of interacting factors, and our understanding would be greatly enhanced by the development of microtechniques to study ammonia dynamics *in vivo*.

Ammonia freely diffuses across the egg case of chondrichthyes (Kormanik, 1989; Kormanik *et al.*, 1992), the chorion of teleosts (Rahaman-Noronha *et al.*, 1996), and integuments of young fish (Paley *et al.*, 1993; Wright and Land, 1998). In Atlantic halibut embryos, Terjesen *et al.* (1998) reported that 2 days of exposure to monumental ammonia levels (up to 27 mM NH$_4$Cl in seawater, pH ~ 8.2) had no influence on total ammonia concentrations (the ammonia of the PVF was rinsed away before analysis). Only after 6 days of exposure did ammonia levels rise substantially (Terjesen *et al.*, 1998). Thus, halibut embryos may have an unusually low permeability to NH$_3$ similar to *Xenopus* oocytes (Burckhardt and Frömter, 1992).

Diffusion of ammonia out of the encapsulated embryo may be hindered by the lack of respiratory convection (i.e., gill ventilation) and indirect contact with bulk water. Indeed, ammonia levels steadily rise in both freshwater [e.g., rainbow trout

(Wright et al., 1995), African catfish (Terjesen et al., 1997)] and marine species [e.g., Atlantic cod (Fyhn and Serigstad, 1987; Finn et al., 1995a; Chadwick and Wright, 1999) and turbot (Finn et al., 1996)] during embryogenesis. After hatch, the gills are initially undeveloped (Rombough and Moroz, 1990, 1997; Rombough, 1999), but the eleutheroembryo or yolk-sac larvae typically have intimate contact with a large volume of water and freedom for whole-body movement. For these reasons ammonia content might be expected to decline after hatching. In marine species, ammonia levels peak prior to hatch and then fall precipitously shortly afterward (by 2.5- to 6.8-fold), with the exception of the Atlantic halibut (Terjesen et al., 1998), in which the decline occurs more slowly (Table II; see also Section II.D.). In the two freshwater species for which data are available, ammonia levels continue to rise after hatch (Table II). Clearly, developmental changes in ammonia content are not simply related to the respiratory factors discussed above. Changes in the permeability of membranes or the rate of ammonia production during and/or after hatching may be critical as well.

Is it coincidental that ammonia levels peak in a number of species prior to hatch? If waste excretion is limited when the embryo reaches a certain size, as is probably the case for oxygen (Rombough, 1988), then elevated levels of the potentially toxic ammonia may be one of the triggers for hatching. The physiological cues for hatching are not well understood, but environmental factors, as well as developmental stage are important (for review, see Yamagami, 1988). Extrinsic factors (e.g., hypoxia, temperature, light, pH) that affect the rate of hatching (Hagenmaier, 1974; DiMichele and Powers, 1984; Oppen-Bernsten et al., 1990a; Helvik et al., 1991) are thought to suppress or stimulate the release of the hatching enzyme (Yamagami, 1988). Although considerable information has been recently published on the biochemical and molecular characteristics of the hatching proteins (e.g., Yasumasu et al., 1992, 1996), there is virtually no information on the physiological regulation of these proteins. Understanding the physiological triggers for hatching and how the hatching process is regulated by intrinsic and extrinsic factors is an important goal.

B. Urea

Urea excretion is considered to be of minor importance to overall nitrogenous excretion in most adult teleosts (for reviews, see Wood, 1993; Korsgaard et al., 1995; Wright and Land, 1998; Chapters 1 and 7, this volume), but it may be more significant in some species during early stages of development. Urea ($CO(NH_2)_2$) is a small, dipolar, uncharged molecule. Urea has a permeability through artificial lipid bilayers that is relatively low (4×10^{-6} cm·s^{-1}; Gallucci et al., 1971; Lande et al., 1995) compared to other nonelectrolytes.

The egg capsule of the viviparous shark *Mustelus canis* (Lombardi and Files, 1993), the oviparous shark *Scyliorhinus canicula* (Hornsey, 1978), and two skate

Table II
Concentration of Ammonia and Urea (nmol·N·ind^{-1}) in Early Stages of Teleost Development

Species	Common name	Prehatch[a]	Posthatch[b]	Reference
Seawater				
Microstomus kitt	Lemon sole			
Ammonia		41.1	23.3	Rønnestad et al. 1992a
Sparus aurata	Gilthead sea bream			
Ammonia		22.3	8.5	Rønnestad et al., 1994
Gadus morhua	Atlantic cod			
Ammonia		45.6	18.1	Fyhn and Serigstad, 1987
Ammonia		52.2	31.3	Fyhn et al., 1987
Ammonia		78.9	18.4	Finn et al., 1995a
Ammonia		25.1	3.7	Chadwick and Wright, 1999
Urea		4.9	1.8	
Scophthalmus maximus	Turbot			
Ammonia		24.8	3.3	Rønnestad et al., 1992b
Ammonia		25.0	8.8	Rønnestad and Fyhn, 1993
Ammonia		26.4	5.1	Finn et al., 1996
Hippoglossus hippoglossus	Atlantic halibut			
Ammonia		626	470	Rønnestad and Fyhn, 1993
Ammonia-a[c]		291	352	Terjesen et al., 1998
Ammonia-b		357	486	
Freshwater				
Oncorhynchus mykiss	Rainbow trout			
Ammonia[d]		2.3	3.5	Rice and Stokes, 1974
Urea[d]		1.5	1.5	
Ammonia[d]		1.6	2.6	Wright et al., 1995
Urea[d]		2.5	2.3	
Clarias gariepinus	African catfish			
Ammonia		4.4	14.4	Terjesen et al., 1997

[a] 0–2 days prior to hatching.
[b] Approx. 2 days after hatching.
[c] Series a and series b represent two different groups of fish; see reference for more information.
[d] In mM.

species *Raja erinacea* and *R. ocellata* (Kormanik et al., 1992) does not present a barrier to urea diffusion. Urea uptake from the environment to the PVF is also relatively rapid in trout embryos when they are placed in a urea solution, indicating that the chorion is highly permeable (Rahaman-Noronha, 1995).

Data are sparse on the urea content of embryos (Table II). In rainbow trout, urea levels increase progressively during embryogenesis (Dépêche et al., 1979) in a pattern that parallels changes in ammonia content (Table II; Wright et al., 1995).

Urea content declines in the trout alevin stage again in tandem with ammonia levels (Wright et al., 1995). In contrast, urea content does not reflect ammonia accumulation in marine cod embryos and remains relatively constant prior to hatch (Chadwick and Wright, 1999). At any given time, the level of ammonia or urea in the animal is a balance between synthesis, storage, and elimination. The rate of urea excretion in cod embryos is considerably higher than in rainbow trout (Table III) and this may be one of many possible factors that might explain the observed differences in urea content between the two species.

The distribution of urea between the embryo and yolk in rainbow trout is puzzling (Fig. 6). The higher level of urea in the yolk compared to the embryo is difficult to explain because (1) urea is not known to complex with proteins and (2) the two compartments are intimately connected by the vascular system. Why would the yolk retain approximately sixfold higher levels of urea than the embryonic tissues of trout? It is undoubtedly a complicated question, involving the site of urea synthesis, blood flow between the embryo and the yolk, the relative permeability of the cutaneous and yolk-sac membranes, and the possible involvement of specialized urea transport proteins (see Section VI.D). These considerations are only speculative and much will be gained by investigations into the handling (i.e., production, retention, excretion) of urea by early embryos. In a review article, Kormanik (1993) hinted that urea reservoirs in the yolk of elasmobranch embryos were 10–60 times as large as the embryonic body concentrations. Clearly, important questions remain to be answered on the early development of urea retention in elasmobranches.

C. Uric Acid

To our knowledge, there is no evidence that uric acid is excreted during early fish development. Uric acid is much less soluble in water than ammonia or urea. It is a moderately strong organic acid with a pK_1 of 5.5 and its anion, urate, is normally complexed to K^+, Na^+, or NH_4^+ ions at physiological pH (Becker, 1993). Rice and Stokes (1974) reported low (0.1–0.5 mM) and unchanging levels of uric acid in whole trout embryos from fertlization to yolk absorption. Excretion of uric acid cannot be detected in eyed-up trout embryos (Rahaman-Noronha, 1995).

Urate precipitates from solution in avian embryos during later stages of incubation and exerts no influence on the osmotic properties of water within the allantoic sac (e.g., Packard and Packard, 1986). Considering the use of urate as a nitrogen waste product in bird embryos, it might be interesting to examine nitrogen excretion in the terrestrial embryos of the marine grunion, *Leuresthes tenuis* (Darken et al., 1998). Although grunion embryos are protected from harsh external conditions by several centimeters of sand, ammonia may accumulate in the absence of abundant water. Under these unusual developmental circumstances, is ammonia converted to urate in early embryos of *L. tenuis*? Further work on this interesting terrestrial species is bound to be rewarding.

Table III
Nitrogen Excretion Rates (nmol·N·ind^{-1}·h^{-1}), % Urea (= Urea-N Excretion/(Urea-N Excretion + Ammonia-N Excretion) and Ammonia Quotient (AQ=Ammonia Excretion/Oxygen Consumption) in Early Life Stages of Fishes

Species	Common name	Embryo[a]	Yolk-sac larvae[b]	Late larval stage[c]	Reference
Seawater					
Scophthalmus maximus	Turbot				
Ammonia		0.11	0.28	0.74	Finn et al., 1996
AQ		0.06	0.05	0.18	
Ammonia		0.24	0.16	0.68	Rønnestad and Fyhn, 1993
AQ		0.19	0.03	0.16	
Gadus morhua	Atlantic cod				
Ammonia		0.21	0.71	0.60	Finn et al., 1995a
AQ		0.22	0.14	0.16	
Ammonia		0.40–0.00[d]	1.18	—	Chadwick and Wright 1999
Urea		0.07–0.65	0.32	—	
% Urea		15–100	21	—	
Hippoglossus hippoglossus	Atlantic halibut				
Ammonia		—	2.80	3.00	Rønnestad and Fyhn, 1993
AQ		—	0.38	0.12	
Ammonia		—	3.20	2.70	Finn et al., 1995b
AQ		—	0.24	0.12	
Ammonia		0.70	2.60	—	Finn et al., 1991
AQ		0.13	0.11	—	
Sciaenops ocellatus	Red drum				
Ammonia		—	0.33	3.92	Torres et al., 1996
AQ		—	0.10	0.15	
Urea		—	0.16	0.84	
% Urea		—	33	18	
Sparus aurata	Gilthead sea bream				
Ammonia		0.03	—	0.15	Rønnestad et al., 1994
AQ		0.02	—	0.02	
Blennius pavo	Blenny				
Ammonia		—	2.94	7.06[e]	Klumpp and von Westernhagen, 1986

Species		Value 1	Value 2	Value 3	Reference
Microstomus kitt	Lemon sole				Rønnestad et al., 1992a
	Ammonia	0.01	1.04	0.33	
	AQ	0.01	0.45	0.16	
Cyclopterus lumpus	Lump-sucker				Davenport et al., 1983
	Ammonia	0.54	0.81	3.26	
Pleuronectes platessa	Plaice				Davenport et al., 1983
	Ammonia	0.54	0.49	—	
Freshwater					
Oncorhynchus mykiss	Rainbow trout				Smith, 1947
	Ammonia	4.58	9.29	43.82	
	Urea	0.08	4.70	9.48	
	% Urea	2	34	18	
	Ammonia	0.07[f]	0.23[f]	0.52[f]	Rice and Stokes, 1974
	Ammonia	0.14[f,g]	0.28[f]	0.51[f]	Wright et al., 1995
	Urea	0.01[f,g]	0.02	0.06	
	% Urea	7	7	11	
Clarias gariepinus	African catfish				Terjesen et al., 1997
	Ammonia	0.1–0.2	6.21	9.86	
	Urea	bld	1.29	1.71	
	% Urea	—	17	15	
Coregonus lavaretus	Whitefish				Dabrowski and Kaushik, 1984
	Ammonia	3.35	5.04	5.99	
	Urea	0.36	0.77	0.83	
	% Urea	10	13	12	
Cyprinus carpio	Common carp				Kaushik et al., 1982
	Ammonia	0.68	2.92	3.58	
	Urea	0.17	0.16	0.32	
	% Urea	20	5	8	

[a] Embryos were approx. midway between fertilization and hatching.
[b] Yolk-sac larvae or eleutheroembryo were approx. 2 days after hatching.
[c] Late larval or alevin stages were approx. 1–2 days after the yolk had been absorbed.
[d] The range indicates the data for the two different cod populations that were studied.
[e] Larvae were 10 days old.
[f] μmol·N·g^{-1} wet weight·h^{-1}.
[g] Pilley and Wright (2000).
bld, below level of detection

Urate has been heralded as an effective antioxidant and repair agent of oxidative damage to DNA bases (Simic and Jovanovic, 1989; Becker, 1993). Urate inhibits lipid autoxidation and protects thymine, uracil, and guanine from oxidation by ozone, forming urea and allantoin as by-products (Meadows *et al.*, 1986). The possible role of urate as an antioxidant during fish embryogenesis might prove to be an interesting area for future investigation.

V. NITROGEN METABOLISM

A. Ammonia Metabolism

In all teleost fishes studied to date, ammonia is produced and excreted very early after fertilization. The release of ammonia from the newly fertilized egg [e.g., lumpsucker, *Cyclopterus lumpus* (Davenport *et al.*, 1983), Atlantic halibut (Finn *et al.*, 1991)] seems not to be of metabolic origin but, rather, related to the formation of cross-linking isopeptide bonds by a transglutaminase during chorion hardening (Hagenmaier *et al.*, 1976; Oppen-Berntsen *et al.*, 1990b). Information is scant, however, on the pattern of expression of metabolic enzymes, let alone those related to protein metabolism, during early fish development. Presumably the metabolic pathways are the same as those found in adult fish liver, namely, the transamination of various amino acids (see Chapters 3 and 4). The level of glutamate oxaloacetate transaminase and glutamate pyruvate transaminase activities in African catfish were relatively high compared to glycolytic or gluconeogenic enzymes at very early developmental stages (0–20 mg; Segner and Verreth, 1995). Glutamate dehydrogenase is present in very early embryos of rainbow trout (before the eyed stage) and activity is located mostly in the embryonic tissue, with only ~3% found in the yolk fraction (Shahsavarani, 2000).

During exhaustive exercise deamination of adenylates in fish muscle is also a significant source of ammonia (Mommsen and Hochachka, 1988). One might expect ammonia synthesis from adenylate breakdown to be important around the time of hatching when muscular activity assists the embryo in escaping the egg capsule. To our knowledge, this additional source of ammonia has not been explored in hatching embryos, although Yamamoto (1984) ascribed this to be the source of the ammonia released in newly fertilized eggs of chum salmon, *Oncorhynchus keta*.

B. Urea Metabolism

There is a growing literature on urea synthesis during early stages of fish development (for reviews, see Korsgaard *et al.*, 1995; Wright and Land, 1998). In adult fish, urea is derived from either dietary arginine (catalyzed by arginase), uric acid degradation (catalyzed by uricase, allantoinase, and allantoicase), or synthesized *de novo* from glutamine and bicarbonate via the ornithine–urea cycle (see Chapter 7). The fish-type urea cycle consists of carbamoyl phosphate synthetase III (CPSase III), ornithine transcarbamylase (OTCase), argininosuccinate (AS), argininolyase (AL), and arginase, as well as the accessory enzyme, glutamine synthetase (GSase).

Griffith (1991) proposed that urea synthesis via the urea cycle probably evolved in early gnathostome fishes in order to detoxify ammonia produced from the catabolism of yolk amino acids as the main energy source during prolonged embryonic development. He speculated that the urea cycle was repressed and no longer functional in adults except in those ancestral fish that invaded seawater. This would explain the retention of the urea cycle in extant marine ureosmotic fishes and the extremely rare occurrence of a functional cycle in extant teleosts (Anderson, 1980; Mommsen and Walsh, 1989, 1991; Randall *et al.*, 1989; Saha and Ratha, 1989; Walsh, 1997; see also Chapter 1, this volume). This scenario fits with emerging data on the distribution of the urea cycle in elasmobranchs and teleosts.

Read (1968a) suggested that ureosmotic regulation and a functional urea cycle were present very early in the development in shark (*Squalus suckleyi*) and skate (*Raja binoculata*) based on the activities of two urea cycle enzymes, OTCase and arginase. GSase activity is present in the brain, liver, and kidney tissue of late-term dogfish embryos (Kormanik *et al.*, 1998). Embryos of dogfish (*Squalus acanthias*) have plasma urea levels comparable to adult levels ($\sim$340 mmol/liter; Evans *et al.*, 1982; Kormanik and Evans, 1986). Although data are limited, it does appear that ureoosmotic regulation develops early in elasmobranchs, but measurement of the complete suite of urea cycle enzyme activities, including the key enzyme, CPSase III, is an important step.

A functional urea cycle was thought to be absent in rainbow trout because OTCase and arginase, but not CPSase and AS, were detected (Rice and Stokes, 1974). The incorporation of radiolabeled bicarbonate into urea in embryos of the rainbow trout and guppy (*Poecilia reticulata*) suggested that the urea cycle was operational in these species (Dépêche *et al.*, 1979). There is now solid evidence that CPSase III and other urea cycle enzymes are induced early in embryogenesis and expressed at relatively high levels of activity compared to adult fishes. In the rainbow trout, CPSase III, OTCase, arginase and GSase activities are first detected just after hatching [42 days postfertilization (dpf); Wright *et al.*, 1995], but mRNA levels for CPSase III peak well before that time (10–14 dpf; Korte *et al.*, 1997). In marine Atlantic cod, the activities of these four urea cycle enzymes can be detected as early as 3 dpf (Chadwick and Wright, 1999). In the cod and trout studies, only arginase activity is expressed at higher levels in adult tissues compared to early life stages. Isolation and characterization of Atlantic halibut CPSase III in yolk-sac larvae were recently carried out by Terjesen *et al.* (2000). As in the

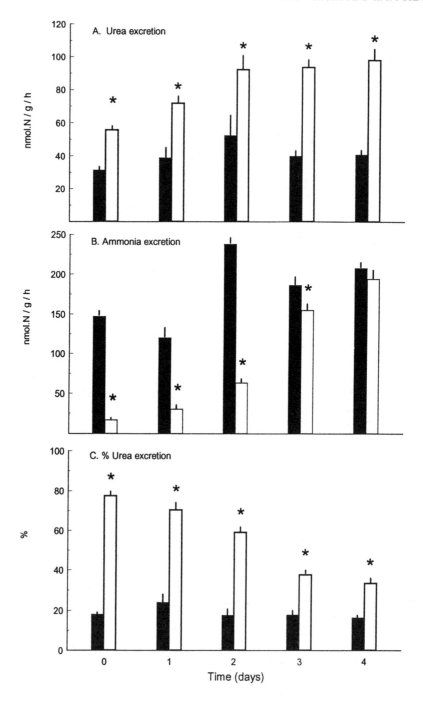

trout and cod, CPSase III activity was clearly expressed at relatively high levels early in development, but the enzyme was much less sensitive to the activator N-acetyl-L-glutamate than observed in other species. Thus far, AS and AL have not been measured in teleost embryos simply because of the difficulty of detecting very low activities in such small samples. It is assumed, however, that both enzymes are expressed in a developmental pattern similar to the other urea cycle enzymes. Taken together, the fact that phylogenetically distant teleosts with dissimilar life histories induce CPSase III, OTCase, arginase, and GSase during embryogenesis provides initial support for the hypothesis that early urea cycle enzyme expression is universal among teleost embryos.

Are urea cycle enzymes induced to detoxify excess ammonia during early life stages? When rainbow trout eleutheroembryo or alevin were acutely exposed (4 h) to treatments that inhibit ammonia elimination (i.e., alkaline water or elevated external ammonia), urea excretion significantly increased (Wright and Land, 1998), but urea cycle enzymes were not measured. In a separate study, eleutheroembryos were chronically exposed (4 days) to either control (pH 8.2) or alkaline water (pH 9.5) (P. A. Wright, A. K. Felskie, and P. M. Anderson, unpublished data). The rates of urea excretion were consistently elevated 1.7- to 2.4-fold (Fig. 7A), whereas ammonia excretion was significantly lower on the first 3 days of the exposure, but recovered by day 4 (Fig. 7B). Consequently, the proportion of nitrogen excreted as urea was more than 50% over the first 48 h and remained significantly elevated relative to control fish for the duration of the experiment (Fig. 7C). At the end of the 4-day exposure period, tissue ammonia levels were significantly elevated (fourfold), whereas tissue urea levels were unchanged (Table IV). The increase in urea excretion coupled with unchanging tissue urea concentrations indicates that urea synthesis probably was enhanced in the alkaline-exposed fish. There were no corresponding changes, however, in the levels of urea cycle enzyme activity (Table IV). Two possible interpretations for these results are (1) flux through the urea cycle pathway increases, but the constitutive level of activity was sufficient to accommodate the higher flux rate and/or (2) other pathways for urea synthesis are employed, for example, uricolysis (see Section V.B.3).

Urea cycle enzyme activity in liver tissue of larval sea lamprey (*Petromyzon marinus*) also did not respond to chronic elevation of environmental ammonia

---

Fig. 7. Nitrogen excretion in the eleutheroembryo stage (31–35 dpf, 12°C) of rainbow trout exposed to either control water (pH 8.2; black bars) or water of pH 9.5 (open bars) for 4 days. Urea and ammonia excretion rates were measured over a 3-h flux period each day. % Urea excretion is the urea-N excretion rate/ (urea-N excretion rate + ammonia-N excretion rate). A one-way analysis of variance was performed to evaluate differences. Where significant differences were found, the Bonferoni or Tukey's pair-wise comparison was applied ($p \leq 0.05$). *, significantly different from control measurement.

levels (2 mM NH$_4$Cl, pH 7.9) (Wilkie *et al.*, 1999). This was not unexpected because the enzyme CPSase III could not be detected in control fish, and the increase in urea excretion after 5 days of exposure to NH$_4$Cl was marginal (Wilkie *et al.*, 1999).

Another environmental stress that may stimulate urea synthesis is increased salinity. Acclimation of the ovoviviparous guppy to a saline-concentrated medium results in an increase in embryonic urea content (Dépêche and Schoffeniels, 1975). In osmotically stressed, separated guppy embryos, urea content is significantly elevated, but there is little change in the rate of urea formation (Dépêche *et al.*, 1979). As a comparison, very early trout embryos (2 dpf) were exposed to dilute seawater, and urea content and rate of synthesis were also increased (Dépêche *et al.*, 1979). These findings are difficult to interpret because (1) both species normally inhabit freshwaters and (2) the measured urea content values were low relative to ureosmotic regulators, which diminishes the possibility of urea as an important osmolyte.

1. REGULATION OF THE UREA CYCLE

The regulation of urea cycle enzyme expression by hormones in early development has been studied in mammals (Dingemanse and Lamers, 1995) and amphibians (Atkinson, 1995), but there is no comparable information in fish. For example, thyroid hormone treatment induces urea cycle enzyme expression, urea excretion, and metamorphosis in premetamorphic tadpoles (e.g., Helbing *et al.*, 1992). Metamorphosis in the Japanese flounder (*Paralichthys olivaceus*) and grouper (*Epinephelus coioides*) is regulated by the thyroid hormones (Inui and Miwa, 1985; Miwa and Inui, 1987; de Jesus *et al.*, 1991, 1993), but whether urea cycle enzymes are expressed in flounder and, if so, respond to thyroid hormone, is unknown and awaits further exploration.

2. LOCALIZATION OF UREA CYCLE ENZYMES

The localization of urea cycle enzymes in embryos is of interest because of recent reports of CPSase III expression in extrahepatic tissues, such as skeletal muscle, in various fish species (Korte *et al.*, 1997; Kong *et al.*, 1998; Felskie *et al.*, 1998; Lindley *et al.*, 1999; see also Chapter 7, this volume). In microdissected rainbow trout eleutheroembryo, CPSase III activity was exclusive to the embryonic body, and was not found in yolk or liver tissue (Steele *et al.*, 2001). It is difficult and time consuming to manually dissect yolk from the embryo proper, especially very soon after fertilization when the embryo is poorly formed or in species with relatively small embryos (<2 mm). We recently developed a centrifugation method that separates yolk from the embryonic tissue, with minimal cross-contamination (Shahsavarani, 2000). This technique should unlock avenues of investigation previously thought impossible, for example, the analysis of the

Table IV
Urea and Ammonia Content and Urea Cycle Enzyme Activities[a]

	Control water	Alkaline water
Ammonia (mM)	3.37 ± 0.01	13.02[b] ± 0.30
Urea, mM	1.17 ± 0.04	1.17 ± 0.12
CPSase III[c]		
μmol·min^{-1}·mg protein^{-1}	0.000016 ± 0.000004	0.000014 ± 0.000003
μmol·min^{-1}·g tissue^{-1}	0.0014 ± 0.0002	0.0015 ± 0.0003
OTCase		
μmol·min^{-1}·mg protein^{-1}	0.0033 ± 0.0005	0.0030 ± 0.0006
μmol·min^{-1}·g tissue^{-1}	0.289 ± 0.024	0.323 ± 0.059
GSase		
μmol·min^{-1}·mg protein^{-1}	0.020 ± 0.004	0.020 ± 0.003
μmol·min^{-1}·g tissue^{-1}	1.78 ± 0.20	2.08 ± 0.22

[a] In whole eleutheroembryo of rainbow trout exposed to control (pH 8.2) or alkaline water (pH 9.5) for 4 days (means ± S.E., n=6).
[b] Significantly different from control, unpaired t-test ($p<0.05$).
[c] Assayed in the presence of glutamine, N-acetylglutamate, and UTP.

complete ontogenic pattern of expression of enzymes or nitrogen end products in embryonic tissues of small marine embryos or very early stages of salmonid development.

3. URICOLYSIS

Less attention has been devoted to the role of uricolysis during early life stages in fish. The four enzymes, uricase, allantoinase, allantoicase, and ureidoglycollate lyase, that are typically found in relatively high levels in adult fish liver (Noguchi et al., 1979; Takada and Noguchi, 1986; Hayashi et al., 1989; Wright, 1993) are also present in agnathan ammocoetes (*P. marinus*) (Wilkie et al., 1999) and in embryos of the African catfish (Terjesen et al., 2001). Future studies should focus on the relative importance of uricolysis versus the urea cycle in urea synthesis during early stages of development where there is no information available.

VI. NITROGEN EXCRETION

A. Developmental Changes in the Pattern of Nitrogen Excretion

Smith (1947) was the first to examine nitrogen excretion in early stages of teleost development. He wrote "(t)he percentage of ammonia excreted by the egg is very high and there is no certain evidence of urea excretion by diffusion through the chorion before hatching" in rainbow trout. More recent work, however, clearly demonstrates that urea is excreted by a number of species and can contribute significantly to total nitrogen excretion (Table II; Yarzhombek and Maslennikova, 1971; Wright and Land, 1998).

Ammonia and urea excretion rates are typically low in embryos but rise with overall increases in metabolic rate and developmental progress. This is particularly true for freshwater species, such as trout, catfish, whitefish, and carp, where nitrogen excretion rates steadily increase from embryo to alevin, with ammonia excretion dominating (Table III). If the data of Smith (1947) are excluded (the urea values were thought to be suspect), the % urea excretion remains relatively constant in trout, catfish, and whitefish during early life stages (Table III). In the carp, however, urea excretion is proportionally higher in the embryos (20%) relative to later stages (8%). In contrast, Terjesen *et al.* (1998) found that urea accounted for about 20% of the total nitrogen excretion during the yolk-sac stage of the African catfish, increasing to about 45% for the exogenously feeding larvae.

In marine species, the pattern of nitrogen excretion is not as simple as the steady rise observed in many freshwater species since the ammonia produced may be temporarily sequestered for buoyancy purposes as well as be related to metabolic fuel selection (see Sections II.D and III.A). In addition, the efficiency by which food is utilized and converted to growth in different developmental stages may be a consideration in the percentage of nitrogen released as metabolic waste products (Klumpp and von Westernhagen, 1986).

Data on urea excretion rates in marine fishes are limited, but it does appear that urea is an excretory product during early development. In Atlantic cod embryos, the estimated urea excretion may account for up to 100% of the total nitrogen excretion in very early stages of development (Chadwick and Wright, 1999), although this conclusion is at variance with the balanced nitrogen and energy budgets of previous studies (Finn *et al.*, 1995a,d). In the red drum, *Sciaenops ocellatus*, urea excretion is also substantial, contributing 33% to total nitrogen excretion in yolk-sac larvae (Table III; Torres *et al.*, 1996). Perhaps the significant level of urea cycle enzyme activities in Atlantic cod embryos is a factor in the prodigious output of urea (Chadwick and Wright, 1999). To compare rates of urea excretion and synthesis, a more complete analysis of both urea cycle and uricolytic enzymes, as well as nitrogen excretion rates in various freshwater and marine species, is required.

The embryos of some viviparous fishes, such as the embiotocidae, are not attached to maternal tissue, contain little or no yolk, and have relatively long periods of gestation. Given these developmental conditions, it is interesting that pregnant female *Cymatogaster aggregata* ovarian fluid had a relatively high level of urea (10–12 mM), but maternal plasma urea levels were relatively low (De-

Vlaming *et al.,* 1983). These findings suggest that *C. aggregata* embryos synthesize and excrete urea to the surrounding ovarian fluid. During early stages *in ovario,* the eelpout *(Zoarces viviparus)* excretes predominantly urea [% urea ~65%, where % urea = urea excretion/(urea excretion + ammonia excretion) and, not surprisingly, urea concentration in the ovarian fluid is also moderately high (4–6 mM; Korsgaard, 1994). Later in the larval stage, the rate of urea excretion is attenuated when *Z. viviparus* are acclimated to seawater (Korsgaard, 1994) and they do not make more urea when exposed to elevated external ammonia (Korsgaard, 1997). The developmental pattern of urea cycle or uricolytic enzymes has not been measured in this species, but might provide an insight into the relatively high rates of urea excretion.

The landlocked and anadromous forms of the lamprey ammocoetes *(Petromyzon marinus)* both excrete 15–20% of nitrogenous wastes as urea (Wilkie *et al.,* 1999). Studies of the freshwater-adapted adult Pacific lamprey *(Entosphenus tridentatus)* revealed much lower rates of urea excretion (~1%) (Read, 1968b). Because a complete developmental pattern of nitrogen excretion in any one species of agnathan has not been studied, it is difficult to interpret these discrepancies.

B. Tolerance to Elevated Environmental Ammonia and pH

The toxicity of ammonia to adult fish is well established in the literature (see Chapter 4). It is surprising, therefore, that embryos of some species are tolerant to relatively high concentrations of ammonia, but after hatch they generally become progressively more sensitive (Rice and Stokes, 1975; McCormick *et al.,* 1984; Daniels *et al.,* 1987; Terjesen *et al.,* 1998). The channel catfish appears to be an exception because age is not a factor in its sensitivity to ammonia (Bader and Grizzle, 1992). It would be interesting to compare the development of pathways for ammonia detoxification (e.g., urea cycle) between species where age plays a role in ammonia toxicity and in those species, such as the channel catfish, where age appears not to be a factor.

Environmental ammonia can be of anthropogenic or natural sources. In regard to the latter, female salmonids bury their embryos and poor water circulation may be the cause of mortalities (Montgomery *et al.,* 1996). The depth of chum salmon (*O. keta*) embryos ranges between 10 and 49 cm within the stream gravel bed (Montgomery *et al.,* 1996). Under these conditions, high ammonia concentrations and low oxygen tensions in the surrounding medium may be a problem for some individuals. In hatcheries, ammonia may also accumulate in incubation troughs depending on the seeding density of the embryos (Bailey *et al.,* 1980). Rice and Bailey (1980) tested the sensitivity of pink salmon (*O. gorbuscha*) embryos to similar ammonia concentrations, however, and found no ill effects.

C. Development of Respiratory and Ion-Regulatory Structures

The development of gas and ion exchange surfaces is of prime importance to the excretion of nitrogenous wastes. In most adult fish, nitrogen excretion occurs principally at the gill, with minor involvement of the skin and kidney. Nitrogen excretion in embryos, however, occurs in the absence of functional gill filaments. Gas exchange between the teleost embryo and the environment is largely cutaneous and may include the head and trunk, fins, and yolk sac (Rombough and Moroz, 1990, 1997; Rombough, 1999). After hatch, the relative size of the skin surface decreases as the fish grows, but still accounts for a significant portion of the total potential respiratory surface area in larval and even juvenile fish (Rombough and Moroz, 1990, 1997). The gills replace the skin as the dominant site of respiratory exchange (>50%) when teleost larvae are ~100 mg (Rombough and Ure, 1991; Wells and Pinder, 1996; Rombough and Moroz, 1997). In many species, extrabranchial chloride cells are seen very early in development and may be important in ion regulation (Shen and Leatherland, 1978; Guggino, 1980; Alderdice, 1988; Hwang, 1989, 1990; Ayson *et al.*, 1994; Rombough, 1999). Rombough (1999) suggested that the appearance of chloride cells on the embryonic gill of rainbow trout (*Onchorhynchus mykiss*) before hatching when the skin surface area is adequate to meet the respiratory requirements indicates that gill development may be initially more important for ion regulation. Morphological and Na^+,K^+-ATPase activity data in tilapia (*O. mossambicus*) also indicate that the larval gill forms a functional ion regulatory organ prior to its role in gas exchange (Li *et al.*, 1995).

D. Site and Mechanisms of Ammonia and Urea Transport

The site(s) of nitrogenous waste excretion in embryonic and larval fishes has not been determined. This is a very intriguing question, because evidence suggests that respiration and ion regulation are mostly cutaneous early in development, with the gill becoming progressively more important later in development (see Section VI.C). As a dissolved gas, NH_3 should follow the pathway of other respiratory gases (e.g., CO_2, O_2). If NH_4^+ is also excreted (e.g., Na^+/NH_4^+ exchange), it may be released by the mitochondrial-rich or chloride cells present on the yolk-sac membrane and gills. After the yolk is absorbed, the site of NH_4^+ excretion may convert solely to the gill. The fact that ammonia concentrations in the water near the eleutheroembryo rainbow trout gill are significantly higher relative to water near the yolk sac and skin indicates that the gill may be the most important site of nitrogen excretion even prior to yolk-sac absoprtion (Misiaszek, 1996).

The current model of gill ammonia excretion in adult freshwater fish involves NH_3 passive diffusion under normal conditions without an additional role for Na^+/NH_4^+ exchange, as previously proposed (see Chapter 4). Instead of Na^+ uptake directly linked to NH_4^+ excretion, it is thought that Na^+ uptake via apical conductive channels is driven by the apical H^+-ATPase pump. Recent studies in young freshwater tilapia (32–64 mg; *O. mossambicus*) also provide evidence for the role of H^+-ATPase in Na^+ uptake (Fenwick *et al.*, 1999). Moreover, when young tilapia were exposed to external NH_4Cl, a vehicle to alter gill cell H^+ concentration, Na^+ uptake was significantly inhibited (Fenwick *et al.*, 1999). These authors hypothesize that during NH_4Cl exposure, NH_3 entry "mops up" gill cell protons, thereby reducing H^+ excretion via H^+-ATPase and retarding the influx of Na^+ via conductive apical channels. Furthermore, there appears to be no direct link between NH_4^+ efflux and Na^+ influx in trout larvae, because ammonia excretion is unaltered by acute changes in external Na^+ levels (Misiaszek, 1996).

In rainbow trout embryos, ammonia excretion is dependent on the P_{NH_3} gradient from the embryo to the water (Rahmahan-Noronha *et al.*, 1996). There is an acidic unstirred layer (USL) of water immediately adjacent to the chorion surface (Fig. 6). It is presumed that the acidic USL is maintained by H^+ and CO_2 excretion (CO_2 hydrates to form HCO_3^- and H^+ in the water; Wright *et al.*, 1986). Elimination of the acidic USL by the addition of buffer to the water decreases ammonia excretion (Rahmahan-Noronha *et al.*, 1996), as occurs in adult rainbow trout (Wright *et al.*, 1989). There have been no comparable studies in marine embryos, but the relatively high pH and buffering capacity of seawater may mean that NH_4^+ excretion plays a more important role similar to the situation in some adult marine species (for review, see Wood, 1993). To determine the site and mode of ammonia excretion (NH_3 versus NH_4^+) in early life stages of a broader number of species, the use of microelectrodes within the boundary water layer might be a reasonable approach (Rombough, 1998).

Diffusion of urea from fish tissues to the external environment may occur by simple or nonmediated transport, as well as by facilitated transport (see also Chapter 8). Urea excretion in trout eyed-up embryos (16–32 dpf) is dependent, in part, on a bidirectional facilitated urea transporter with a K_m of 2 mM, a value similar to the concentration of urea in the whole embryo (Pilley and Wright, 2000). Urea analogs (i.e., thiourea, acetamide) and inhibitors (e.g., phloretin) added to the external water reversibly inhibit urea excretion from the embryo. The tissue localization of this putative urea transporter in trout is unknown, but with the recent cloning of other fish urea transporters (Smith and Wright, 1999; Walsh *et al.*, 2000; see also Chapter 8, this volume), the molecular characterization and expression are within grasp.

VII. SUMMARY

In many fish species, the early pattern of nitrogen metabolism and excretion is influenced by the dominance of amino acids as the primary fuel, and by the resulting toxic end-product, ammonia. Pelagic eggs of marine teleosts at spawning contain a large yolk pool of FAA. During final oocyte maturation the FAA originate by hydrolysis of specific yolk proteins, and they assist in egg hydration. At the start of exogenous feeding, marine fish larvae will inevitably ingest a new supply of FAA with their zooplankton prey and they can thus continue an amino acid-based energy dissipation despite low intestinal proteolytic capacity. Ammonia excretion rates increase with developmental stage, but ammonia also accumulates to substantial levels around the time of hatching. Historically, urea excretion has been considered of minor importance in embryos and larvae of teleosts, but significant levels of urea cycle enzymes have recently been detected during embryogenesis in some species, despite low or nondetectable levels in the adults. Mechanisms of nitrogen excretion have been studied only in freshwater rainbow trout embryos, where ammonia excretion is dependent on the partial pressure gradient of NH_3. Future research on a diverse array of species with different life histories and environments will broaden our understanding of the ontogeny of nitrogen excretion.

REFERENCES

Alderdice, D. F. (1988). Osmotic and ionic regulation in teleost eggs and larvae. In "Fish Physiology" (W. S. Hoar and D. J. Randall, eds.), Vol. 11A, pp. 163–251. Academic Press, San Diego.

Anderson, P. M. (1980). Glutamine- and N-acetylglutamate-dependent carbamoyl phosphate synthetase in elasmobranchs. Science 208, 291–293.

Atkinson, B. G. (1995). Molecular aspects of ureogenesis in amphibians. In "Nitrogen Metabolism and Excretion" (P. J. Walsh and P. A. Wright, eds). pp. 133–146. CRC Press, Boca Raton, FL.

Ayson, F. G., Kaneko, T., Hasegawa, S., and Hirano, T. (1994). Development of mitochondrion-rich cells in the yolk-sac membrane of embryos and larvae of tilapia, Oreochromis mossambicus, in fresh water and seawater. J. Exp. Zool. 270, 129–135.

Bader, J. A., and Grizzle, J. M. (1992). Effects of ammonia on growth and survival of recently hatched channel catfish. J. Aquat. Anim. Health 4, 17–23.

Bailey, J. E., Rice, S. D., Pella, J. J., and Taylor, S. G. (1980). Effects of seeding density of pink salmon, Onchorhynchus gorbuscha, eggs on water chemistry, fry characteristics, and fry survival in gravel incubators. Fish. Bull. 78, 649–658.

Balon, E. K. (1975). Terminology of intervals in fish development. J. Fish. Res. Bd. Can. 32, 1663–1670.

Balon, E. K. (1990). Epigenesis of an epigeneticist: the development of some alternative concepts on the early ontogeny and evolution of fishes. Guelph Ichthyol. Rev. 1, 1–48.

Båmstedt, U. (1986). Chemical composition and energy content. In "The Biological Chemistry of Marine Copepods" (E. D. S. Corner, and S. C. M. O'Hara, eds.), pp. 1–58. Clarendon Press, Oxford.

Becker, B. F. (1993). Towards the physiological function of uric acid. Free Radic. Biol. Med. 14, 615–631.

Bell, J. G., Tocher, D. R., Farndale, B. M., McVicar, A. H., and Sargent, J. R. (1999). Effects of essential fatty acid-deficient diets on growth, mortality, tissue histopathology and fatty acid composition in juvenile turbot (Scophthalmus maximus). Fish Physiol. Biochem. 20, 263–277.

Blaxter, J. H. S. (1988). Development: eggs and larvae. *In* "Fish Physiology" (W. S. Hoar and D. J. Randall, eds.), Vol. 11A, pp. 177–252. Academic Press, San Diego.

Burckhardt, B.-C., and Frömter, E. (1992). Pathways of NH_3/NH_4+ permeation across *Xenopus laevis* oocyte cell membrane. *Pflügers Arch.* **420**, 83–86.

Carnevali, O., Mosconi, G., Roncarati, A., Belvedere, P., Romano, M., and Limatola, E. (1992). Changes in the electrophoretic pattern of yolk proteins during vitellogenesis in the gilthead sea bream, *Sparus aurata* L. *Comp. Biochem. Physiol.* **103B**, 955–962.

Carnevali, O., Mosconi, G., Roncarati, A., Belvedere, P., Romano, M., and Polzonetti-Magni, A. M. (1993). Yolk protein changes during oocyte growth in European sea bass *Dicentrarchus labrax* L. *J. Appl. Ichthyol.* **9**, 175–184.

Carnevali, O., Carletta, R., Cambi, A., Vita, A., and Bromage, N. (1999). Yolk formation and degradation during oocyte maturation in seabream, *Sparus aurata:* involvement of two lysosmal proteases. *Biol. Reprod.* **60**, 140–6.

Carter, C. G., He, Z-Y., Houlihan, D. F., McCarthy, I. D., and Davidson, I. (1995). Effect of feeding on the tissue free amino acid concentrations in rainbow trout (*Oncorhynchus mykiss* Waldbaum). *Fish Physiol. Biochem.* **14**, 153–164.

Cerda, J., Selman, K., and Wallace, R. A. (1996). Observations on oocyte maturation and hydration *in vitro* in the black sea bass, *Centropristis striata* (Serranidae). *Aquat. Living Resour.* **9**, 325–335.

Chadwick, T. D., and Wright, P. A. (1999). Nitrogen excretion and expression of urea cycle enzymes in the Atlantic cod (*Gadus morhua* L.): a comparison of early life stages with adults. *J. Exp. Biol.* **202**, 2653–2662.

Chamberlin, M. E., and Strange, K. (1989). Anisosmotic cell volume regulation: a comparative view. *Am. J. Physiol.* **257**, C159-C173.

Collazo, A. (1996). Evolutionary correlations between early development and life history in Plethodontid salamanders and teleost fishes. *Am. Zool.* **36**, 116–131.

Collazo, A., Bolker, J. A., and Keller, R. (1994). A phylogenetic perspective on teleosts gastrulation. *Am. Naturalist.* **144**, 133–152.

Conceição, L., Polat, A., Rønnestad, I., Machiels, M., and Verreth, J. A. J. (1995). A first attempt to estimate protein turnover using a simulation model for amino acid metabolism in yolk-sac larvae of *Clarias gariepinus* (Burcell) and *Hippoglossus hippoglossus* (L.). *Fish Physiol. Biochem.* **19**, 43–57.

Conceição, L. E. C., vanderMeeren, T., Verreth, J. A. J., Evjen, M. S., Houlihan, D. F., and Fyhn, H. J. (1997a). Amino acid metabolism and protein turnover in larval turbot (*Scophthalmus maximus*) fed natural zooplankton or *Artemia. Mar. Biol.* **129**, 255–265.

Conceição, L. E. C., Houlihan, D. F., and Verreth, J. A. J. (1997b). Fast growth, protein turnover and costs of protein metabolism in yolk-sac larvae of the African catfish (*Clarias gariepinus*). *Fish Physiol. Biochem.* **16**, 291–302.

Conceição, L, Ozòria, E. A., Suurd, E. A., and Verreth, J. A. J. (1998a). Amino acid profiles and amino acid utilization in larval African catfish (*Clarias gariepinus*): effects of ontogeny and temperature. *ICES Mar. Sci. Symp.* **201**, 80–86.

Conceição, L. E. C., Verreth, J. A. J., Verstegen, M. W. A., and Huisman, E. A. (1998b). A preliminary model for dynamic simulation of growth in fish larvae: application to the African catfish (*Clarias gariepinus*) and turbot (*Scophthalmus maximus*). *Aquaculture* **163**, 215–235.

Craik, J. C. A., and Harvey, S. M. (1984). Biochemical changes occurring during final maturation of eggs of some marine and freshwater teleosts. *J. Fish Biol.* **24**, 599–610.

Craik, J. C. A., and Harvey, S. M. (1986). Phosphorus metabolism and water uptake during final maturation of ovaries of teleosts with pelagic and demersal eggs. *Mar. Biol.* **90**, 285–289.

Craik, J. C. A., and Harvey, S. M. (1987). The causes of buoyancy in eggs of marine fishes. *J. Mar. Biol. Ass. U.K.* **67**, 169–182.

Crim, M. C., and Munro, H. N. (1994). Proetins and amino acids. *In* "Modern Nutrition in Health and Disease" (M. E. Shils, J. A. Olson, and M. Shike, eds.), pp. 3–35. Williams & Wilkins, Baltimore, MD.

Dabrowski, K. (1992). Ascorbate concentration in fish ontogeny. *J. Fish Biol.* **40,** 273–279.

Dabrowski, K., and Kaushik, S. J. (1984). Rearing of coregonid (*Coregonus schinzi* palea cuv. et val.) larvae using dry and live food. II. Oxygen consumption and nitrogen excretion. *Aquaculture* **41,** 333–344.

Dabrowski, K., Luczynski, M., and Rusiecki, M. (1985). Free amino acids in the late embryogenesis and prehatching stage of two coregonid fishes. *Biochem. Syst. Ecol.* **13,** 249–356.

Daniels, H. V., Boyd, C. E., and Minton, R. V. (1987). Acute toxicity of ammonia and nitrite to spotted seatrout. *Progr. Fish Cul.* **49,** 260–263.

Darken, R. S., Martin, K. L. M., and Fisher, M. C. (1998). Metabolism during delayed hatching in terrestrial eggs of a marine fish, the Grunion *Leuresthes tenius. Physiol. Zool.* **71,** 400–406.

Davenport, J., Lønning, S., and Kjørsvik, E. (1983). Ammonia output by eggs and larvae of the lumpsucker, *Cyclopterus lumpus,* the cod, *Gadus morhua* and the plaice, *Pleuronectes platessa. J. Mar. Biol. Ass. U.K.* **63,** 713–723.

de Jesus, E. G., Hirano, T., and Inui, Y. (1991). Changes in cortisol and thyroid hormone concentrations during early development and metamorphosis in the Japanese flounder, *Paralichthys olivaceus. Gen. Comp. Endocrinol.* **82,** 369–376.

de Jesus, E. G., Hirano, T., and Inui, Y. (1993). Flounder metamorphosis: its regulation by various hormones. *Fish Physiol. Biochem.* **11,** 323–328.

Dépêche, J., and Schoffeniels, E. (1975). Changes in electrolytes, urea and free amino acids of *Poecilia reticulata* embryos following high salinity adaptation of the viviparous female. *Biochem. System. Ecol.* **3,** 111–119.

Dépêche, J., Gilles, R., Daufresne, S., and Chiapello, H. (1979). Urea content and urea production via the ornithine–urea cycle pathway during the ontogenic development of two teleost fishes. *Comp. Biochem. Physiol.* **63A,** 51–56.

Deplano, M., Connes, R., Diaz, J. P., and Barnabe, G. (1991). Variation in the absorption of macromolecular proteins in larvae of the sea bass *Dicentrarchus labrax* during transition to the exotrophic phase. *Mar. Biol.* **110,** 29–36.

DeVlaming, V., Baltz, D., Anderson, S., Fitzgerald, R., Delahunty, G., and Barkley, M. (1983). Aspects of embryo nutrition and excretion among viviparous embiotocid teleosts: potential endocrine involvements. *Comp. Biochem. Physiol.* **74A,** 189–198.

DiMichele, L., and Powers, D. A. (1984). The relationship between oxygen consumption rate and hatching in *Fundulus heteroclitus. Physiol. Zool.* **57,** 46–51.

Dingemanse, M., and Lamers, W. H. (1995). Gene expression and development of hepatic nitrogen metabolic pathways. *In* "Nitrogen Metabolism and Excretion" (P. J. Walsh and P. A. Wright, eds.), pp. 229–242. CRC Press, Boca Raton, FL.

Evans, D. H. (1993). Osmotic and ionic regulation. *In* "The Physiology of Fishes" (D. H. Evans, ed.), pp. 315–341. CRC Press, Boca Raton, FL.

Evans, D. H., Oikari, A., Kormanik, G. A., and Mansberger, L. (1982). Osmoregulation by the prenatal spiny dogfish, *Squalus acanthias. J. Exp. Biol.* **101,** 295–305.

Felskie, A. K., Anderson, P. M., and Wright, P. A. (1998). Expression and activity of carbamoyl phosphate synthetase III and ornithine urea cycle enzymes in various tissues of four fish species. *Comp. Biochem. Physiol.* **119B,** 355–364.

Fenwick, J. C., Wendelaar Bonga, S. E., and Flik, G. (1999). *In vivo* bafilomycin-sensitive Na^+ uptake in young freshwater fish. *J. Exp. Biol.* **202,** 3659–3666.

Finn, R. N. (1994). Physiological energetics of developing marine fish embryos and larvae. Dr. scient. thesis, University of Bergen, Norway.

Finn, R. N., Fyhn, H. J., and Evjen, M. S. (1991) Respiration and nitrogen metabolism of Atlantic halibut eggs (*Hippoglossus hippoglossus* L.). *Mar. Biol.* **108,** 11–19.

Finn, R. N., Fyhn, H. J., and Evjen, M. S. (1995a). Physiological energetics of developing embryos and yolk-sac larvae of Atlantic cod *Gadus morhua*. I. Respiration and nitrogen metabolism. *Mar. Biol.* **124,** 355–369.

Finn, R. N., Rønnestad, I., and Fyhn, H. J. (1995b). Respiration, nitrogen and energy metabolism of developing yolk-sac larvae of Atlantic halibut (*Hippoglossus hippoglossus* L.). *Comp. Biochem. Physiol.* **111A,** 647–671.

Finn, R. N., Widdows, J., and Fyhn, H. J. (1995c). Calorespirometry of developing embryos and yolk-sac larvae of turbot (*Scophthalmus maximus* L.). *Mar. Biol.* **122,** 157–163.

Finn, R. N., Henderson, R. J., and Fyhn, H. J. (1995d). Physiological energetics of developing embryos and larvae of Atlantic cod (*Gadus morhua* L.) II. Lipid metabolism and enthalpy balance. *Mar. Biol.* **124,** 371–379.

Finn, R. N., Fyhn, H. J., Henderson, R. J., and Evjen, M. S. (1996). The sequence of catabolic substrate oxidation and enthalpy balance of developing embryos and yolk-sac larvae of turbot (*Scophthalmus maximus* L.). *Comp. Biochem. Physiol.* **115A,** 133–151.

Finn, R. N., Fyhn, H. J., Norberg, B., Munholland, J., and Reith, M. (2000). Oocyte hydration as a key feature in the adaptive evolution of teleost fishes to seawater. *In* "Proc. 6th Int. Symp. Reprod. Physiol. Fish." (B. Norberg, O. S. Kjesbu, G. L. Taranger, E. Anderson, S. O. Stefansson, eds.), pp. 281–291. Inst. Mar. Res. and Univ. Bergen, Bergen.

Fulton, T. W. (1891). The comparative fecundity of sea-fishes. 9. *Ann. Rep. Fish. Bd. Scotland,* **1890, Part I,** 243–268.

Fulton, T. W. (1898). On the growth and maturation of the ovarian eggs of teleost fishes. 16. *Ann. Rep. Fish. Bd. Scotland,* **1897, Part III,** 88–134.

Fyhn, H. J. (1989). First-feeding in marine fish larvae: are free amino acids the source of energy? *Aquaculture* **80,** 111–120.

Fyhn, H. J. (1990). Energy production in marine fish larvae with emphasis on free amino acids as a potential fuel. *In* "Comparative Physiology. Nutrition in Wild and Domestic Animals" (J. Mellinger, ed.), pp. 176–192. Karger, Basel.

Fyhn, H. J. (1993). Multiple functions of free amino acids during embryogenesis in marine fishes. *In* "Physiological and Biochemical Aspects of Fish Development" (B. T. Walther and H. J. Fyhn, eds.), pp. 285–289. University of Bergen, Norway.

Fyhn, H. J., and Govoni, J. J. (1995). Endogenous nutrient mobilization during embryogenesis in two marine fish species, Atlantic menhaden (*Brevoortia tyrannus*) and spot (*Leiostomus xanthusus*). *In* "Early Life History of Fish" (H. Rosenthal, J. Verreth, and I. Huse, eds.), Vol. III, pp. 64–69. ICES Mar. Sci. Symp. **201.**

Fyhn, H. J., and Serigstad, B. (1987). Free amino acids as energy substrate in developing eggs and larvae of the cod *Gadus morhua*. *Mar. Biol.* **96,** 335–341.

Fyhn, H. J., Serigstad, B., and Mangor-Jenson, A. (1987). Free amino acids in developing eggs and yolk-sac larvae of the cod, *Gadus morhua* L. *Sarsia* **72,** 363–365.

Fyhn, H. J., Finn, R. N., Helland, S., Rønnestad, I., and Lømsland, E. R. (1993). Nutritional value of phytoplankton and zooplankton as live food for marine fish larvae at start-feeding. *In* "Fish Farming Technology" (H. Reinertsen, L. A. Dahle, L. Jørgensen, and K. Tvinnereim, eds.), pp. 121–126. Balkema Publ., Rotterdam, the Netherlands.

Fyhn, H. J., Rønnestad, I., and Berg, L. (1995). Variation in protein and free amino acid content of marine copepods during the spring bloom. *Spec. Publ. Eur. Aquacult. Soc.* **24,** 321–324.

Fyhn, H. J., Finn, R. N., Reith, M., and Norberg, B. (1999). Yolk protein hydrolysis and oocyte free amino acids as key features in the adaptive evolution of teleost fishes to seawater. *Sarsia* **84,** 451–456.

Gallucci, E., Micelli, S., and Lippe, C. (1971). Non-electrolyte permeability across thin lipid membranes. *Arch. Int. Physiol. Biochim.* **79,** 881–887.

Gatesoupe, F. J. (1986). The effect of starvation and feeding on the free amino acid composition of sea bass larvae (*Dicentrachus labrax*). *Ocèanis* **12,** 207–222.

Grasdalen, H., and Jørgensen, L. (1987). In vivo NMR-studies of fish eggs. Monitoring of metabolite levels, intracellular pH, and the freezing and permeability of water in developing eggs of plaice, *Pleuronectes platessa* L. *Sarsia* **72,** 359–361.

Greeley, M. S., Calder, D. R., and Wallace, R. A. (1986). Changes in teleost yolk proteins during oocyte maturation: correlation of yolk proteolysis with oocyte hydration. *Comp. Biochem. Physiol.* **84B,** 1–9.

Griffith, R. W. (1991). Guppies, toadfish, lungfish, coelacanths and frogs: a scenario for the evolution of urea retention in fishes. *Environ. Biol. Fishes* **32,** 199–218.

Groot, C., and Margolis, L. (1991). "Pacific Salmon Life Histories." University of British Columbia Press, Vancouver, Canada.

Guggino, W. B. (1980). Salt balance in embryos of *Fundulus herteoclitus* and *F. bermudae* adapted to seawater. *Am. J. Physiol.* **238,** R42–R49.

Gunasekera, R. M., Shim, K. F., and Lam, T. J. (1996a). Influence of protein content of brood stock diets on larval quality and performance in Nile tilapia, *Oreochromis niloticus*. (L.). *Aquaculture* **146,** 121–134.

Gunasekera, R. M., Shim, K. F., and Lam, T. J. (1996b). Effects of dietary protein level on spawning performance and amino acid composition of eggs of Nile tilapia, *Oreochromis niloticus*. *Aquaculture* **146,** 245–259.

Gunasekera, R. M., de Silva, S. S., and Ingram, B. A. (1999). The amino acid profiles in developing eggs and larvae of the freshwater percichthyid fishes, trout cod, *Maccullochella macquariensis* and Murray cod, *Maccullochella peelii peelii*. *Aquat. Living Resour.* **12,** 255–261.

Guraya, S. S. (1986). "The Cell and Molecular Biology of Fish. Oogenesis." Karger, Basel.

Hagenmaier, H. E. (1974). The hatching process in fish embryos—IV. The enzymological properties of a highly purified enzyme (chorionase) from the hatching fluid of the rainbow trout, *Salmo gairdneri* Rich. *Comp. Biochem. Physiol.* **49B,** 313–324.

Hagenmaier, H. E., Smitz, I., and Føhles, J. (1976). Zum Vorkommen von Isopeptidbindungen in der Eihlle der Regenbogenforelle (*Salmo gairdneri* Rich.). *Hoppe-Seyler's Z. Physiol. Chem.* **357,** 1436–1438.

Halstead, L. B. (1985). The vertebrate invasion of fresh water. *Phil. Trans. R. Soc. Lond.* **309B,** 243–258.

Harper, A. E. (1974). Nonessential amino acids. *J. Nutr.* **104,** 965–967.

Harper, A. E. (1983). Dispensable and indispensable amino acid interrelationships. In "Amino Acids—Metabolism and Medical Applications" (G. L. Blackburn., J. P. Grant, and V. R. Young, eds.), pp. 105–121. John Wright & PSG Inc., Boston, MA.

Hartling, R. C., and Kunkel, J. G. (1999). Developmental fate of the yolk protein lipovitellin in embryos and larvae of winter flounder, *Pleuronectes americanus*. *J. Exp. Zool.* **284,** 686–695.

Haug, T. (1990). Biology of the Atlantic halibut *Hippoglossus hippoglossus* (L. 1758). *Adv. Mar. Biol.* **26,** 1–70.

Hayashi, S., Fujiwara, S., and Noguchi, T. (1989). Degradation of uric acid in fish liver peroxisomes. *J. Biol. Chem.* **264,** 3211–3215.

Helbing, C., Gergely, G., and Atkinson, B. G. (1992). Sequential up-regulation of thyroid hormone β-receptor, ornithine transcarbamylase, and carbamyl phosphate synthetase mRNAs in the liver of *Rana catesbeiana* tadpoles during spontaneous and thyroid hormone-induced metamorphosis. *Dev. Gen.* **13,** 289–301.

Helfman, G. S., Collette, B. B., and Facey, D. E. (1997). "The Diversity of Fishes." Blackwell Science, Malden, MA.

Helland, S., Triantaphyllidis, G. V., Fyhn, H. J., Evjen, M. S., Lavens, P., and Sorgeloos, P. (2000). Modulation of the free amino acid pool and protein content in populations of the brine shrimp *Artemia*. *Mar. Biol.* **137,** 1005–1016.

Helvik, J. V., Oppen-Berntsen, D. O., and Walther, B. T. (1991). The hatching mechanisms in Atlantic halibut (*Hippoglossus hippoglossus*). *Int. J. Dev. Biol.* **35,** 9–16.

Heming, T. A., and Buddington, R. K. (1988). Yolk absorption in embryos and larval fishes. *In* "Fish Physiology" (W. S. Hoar and D. J. Randall, eds.), Vol. 11, pp. 407–446. Academic Press, San Diego.

Hjort, J. (1914). Fluctuations in the great fisheries of Northern Europe viewed in the light of biological research. *Rapp. P.V. Reun. Cons. Int. Explor. Mèr* **20**, 1–228.

Hølleland, T., and Fyhn, H. J. (1986). Osmotic properties of eggs of the herring *Clupea harengus*. *Mar. Biol.* **91**, 377–383.

Hornsey, D. J. (1978). Permeability coefficients of the egg-case membrane of *Scyliorhinus canicula* L. *Experientia* **34**, 1596–1597.

Houlihan, D. F., McCarthy, I. D., Carter, C. G., and Martin, F. (1995). Protein turnover and amino acid flux in fish larvae. *ICES Mar. Sci. Symp.* **201**, 87–99.

Huxtable, R. J. (1992). Physiological actions of taurine. *Physiol. Rev.* **72**, 101–163.

Hwang, P. P. (1989). Distribution of chloride cells in teleost larvae. *J. Morphol.* **200**, 1–8.

Hwang, P. P. (1990). Salinity effects on development of chloride cells in the larvae of ayu (*Plecoglossus altivelis*). *Mar. Biol.* **107**, 1–7.

Inui, Y., and Miwa, S. (1985). Thyroid hormone induces metamorphosis of flounder larvae. *Gen. Comp. Endocrinol.* **60**, 450–454.

Izquierdo, M. S. (1996). Review article: essential fatty acid requirements of cultured marine fish larvae. *Aquacult. Nutr.* **2**, 183–191.

Jørgensen, L., and Grasdalen, H. (1990). Phosphate metabolites in larvae of halibut (*Hippoglossus hippoglossus*) studied by *in vivo* NMR. *In* "Program. Development and Aquaculture of Marine Larvae" (B. T. Walther, H. J. Fyhn, and A. Mangor-Jensen, eds.), p. 26. University of Bergen, Norway.

Kamler, E. (1992). "Early Life History of Fish: An Energetics Approach." Chapman & Hall, London.

Kaushik, S. J., Dabrowski, K., and Luquet, P. (1982). Patterns of nitrogen excretion and oxygen consumption during ontogenesis of common carp (*Cyprinus carpio*). *Can. J. Fish. Aquat. Sci.* **39**, 1095–1105.

Kendall, A. W., Ahlstrom, E. H., and Moser, H. G. (1984). Early life history stages of fishes and their characters. *In* "Ontogeny and Systematics of Fishes" (H. G. Moser, W. J. Richards, D. M. Cohen, M. P. Fahay, A. W. Kendall, and S. L. Richardson, eds.), Spec. Publ. 1, pp. 11–22. *Am. Soc. Ichthyol. Herpetol.* Allen Press, Lawrence, KS.

Kjesbu, O. S., and Kryvi, H. (1989). Oogenesis in cod, *Gadus morhua* L., studied by light and electron microscopy. *J. Fish Biol.* **34**, 735–746.

Kjesbu, O. S., and Kryvi, H. (1993). A histological examination of oocyte final maturation in cod (*Gadus morhua* L.). *In* "Physiological and Biochemical Aspects of Fish Development" (B. T. Walther and H. J. Fyhn, eds.), pp. 86–93. University of Bergen, Norway.

Kjesbu, O. S., Klungsøyr, J., Kryvi, H., Witthames, P. R., and Greer Walker, M. (1991). Fecundity, atresia, and egg size of captive Atlantic cod (*Gadus morhua*) in relation to proximate body composition. *Can. J. Fish. Aquat. Sci.* **48**, 2333–2343.

Kjørsvik, E., and Reiersen, A. L. (1992). Histomorphology of the early yolk-sac larvae of the Atlantic halibut (*Hippoglossus hippoglossus* L.)—an indication of the timing of functionality. *J. Fish Biol.* **41**, 1–19.

Kjørsvik, E., Mangor-Jensen, A., and Holmefjord, I. (1990). Egg quality in fishes. *Adv. Mar. Biol.* **26**, 71–113.

Klumpp, D. W., and von Westernhagen, H. (1986). Nitrogen balance in marine fish larva: influence of developmental stage and prey density. *Mar. Biol.* **93**, 189–199.

Kolkovski, S., Tandler, A., Kissil, G. W., and Gertler, A. (1993). The effect of dietary exogenous digestive enzymes on ingestion, assimilation, growth and survival of gilthead sea bream (*Sparus aurata*, Sparidae, Linneus) larvae. *Fish Physiol. Biochem.* **12**, 203–209.

Kolkovski, S., Tandler, A., and Izquierdo, M. S. (1997). Effects of live food and dietary digestive enzymes on the efficiency of microdiets for seabass (*Dicentrarchus labrax*) larvae. *Aquaculture* **148,** 313–322.

Kong, H., Edberg, D. D., Korte, J. J., Salo, W. L., Wright, P. A., and Anderson, P. M. (1998). Nitrogen excretion and expression of carbamoyl-phosphate synthetase III activity and mRNA in extrahepatic tissues of largemouth bass (*Micropterus salmoides*). *Arch. Biochem. Biophys.* **350,** 157–168.

Kormanik, G. A. (1989). The egg case of *Raja erinacea* plays only a minimal role as an ionic/osmotic barrier. *Bull. Mt. Desert Isl. Biol. Lab.* **28,** 12–13.

Kormanik, G. A. (1992). Ion and osmoregulation in prenatal elasmobranchs: evolutionary implications. *Am. Zool.* **32,** 294–302.

Kormanik, G. A. (1993). Ionic and osmotic environmental of developing elasmobranch embryos. *Environ. Biol. Fish.* **38,** 233–240.

Kormanik, G. A. (1995). Maternal-fetal transfer of nitrogen in chondrichthyans. *In* "Nitrogen Metabolism and Excretion" (P. J. Walsh and P. A. Wright, eds.), pp. 243–258. CRC Press, Boca Raton, FL.

Kormanik, G. A., and Evans, D. H. (1986). The acid–base status of prenatal pups of the dogfish, *Squalus acanthias*. *J. Exp. Biol.* **125,** 173–179.

Kormanik, G. A., Lofton, A. J., and O'Leary-Liu, N. (1992). Egg case permeability to ammonia and urea in two species of skates (*Raja* sp.). *Bull. Mt. Desert Isl. Biol. Lab.* **31,** 27–28.

Kormanik, G. A., Billings, C., and Chadwick, Y. (1998). Glutamine synthetase and glutaminase in several tissues from late-term embryos of the dogfish, *Squalus acanthias*. *Bull. Mt. Desert Isl. Biol. Lab.* **37,** 99–100.

Korsgaard, B. (1994). Nitrogen distribution and excretion during embryonic postyolk sac development in *Zoarces viviparus*. *J. Comp. Physiol.* **164B,** 42–46.

Korsgaard, B. (1997). Ammonia and urea in the maternal–fetal trophic relationship of the viviparous blenny (eelpout) *Zoarces viviparus*. *Physiol. Zool.* **70,** 712–717.

Korsgaard, B., Mommsen, T. P., and Wright, P. A. (1995). Nitrogen excretion in teleostean fish: adaptive relationships to environment, ontogenesis, and viviparity. *In* "Nitrogen Metabolism and Excretion" (P. J. Walsh and P. A. Wright, eds.), pp. 259–287. CRC Press, Boca Raton, FL.

Korte, J. J., Salo, W. L., Cabrera, V. M., Wright, P. A., Felskie, A. K., and Anderson, P. M. (1997). Expression of carbamoyl-phosphate synthetase III mRNA during the early stages of development and in muscle of adult rainbow trout (*Oncorhynchus mykiss*). *J. Biol. Chem.* **272,** 6270–6277.

Koven, W. M., Tandler, A., Kissil, G. W., and Sklan, D. (1992). The importance of n-3 highly unsaturated fatty acids for growth in larval *Sparus aurata* and their effect on survival, lipid composition and size distribution. *Aquaculture* **104,** 91–104.

Kurokawa, T., Tanaka, H., Kagawa, H., and Otha, H. (1996). Absorption of protein molecules by the rectal cells in eel larvae *Anguilla japonica*. *Fish. Sci.* **62,** 832–833.

Laale, H. W. (1980). The perivitelline space and egg envelopes of bony fishes: a review. *Copeia* **1980,** 210–226.

Lande, M. B., Donovan, J. M., and Zeidel, M. L. (1995). The relationship between membrane fluidity and permeabilities to water, solutes, ammonia, and protons. *J. Gen. Physiol.* **106,** 67–84.

Langeland, J. A., and Kimmel, C. B. (1997). Fishes. *In* "Embryology" (S. F. Gilbert and A. M. Raunio, eds.), pp. 383–407. Sinauer Associates, Sunderland, MA.

Lazo, J. P., Dinis, M. T., Holt, J. G., Faulk, C., and Arnold, C. R. (2000). Co-feeding microparticulate diets with algae: toward eliminating the need of zooplankton at first-feeding red drum (*Sciaenops ocellatus*). *Aquaculture* **188,** 339–351.

Li, J., Eygensteyn, J., Lock, R. A. C., Verbost, P. M., van der Heijden, A. J. H., Wendelaar Bonga, S. E., and Flik, G. (1995). Branchial chloride cells in larvae and juveniles of freshwater tilapia *Oreochromis mossambicus*. *J. Exp. Biol.* **198,** 2177–2184.

Li, X., Jenssen, E., and Fyhn, H. J. (1989). Effects of salinity on egg swelling in Atlantic salmon (*Salmo salar*). *Aquaculture* **76,** 317–334.

Lindley, T. E., Scheiderer, C. L., Walsh, P. J., Wood, C. M., Bergman, H. L., Bergman, A. L., Laurent, P., Wilson, P., and Anderson, P. M. (1999). Muscle as the primary site of urea cycle enzyme activity in an alkaline lake-adapted tilapia, *Oreochromic alcalicus grahami. J. Biol. Chem.* **274**, 29858–29861.

Lombardi, J., and Files, T. (1993). Egg capsule structure and permeability in the viviparous shark, *Mustelus canis. J. Exp. Zool.* **267**, 76–85.

Long, J. A. (1995). "The Rise of Fishes. 500 Million Years of Evolution." John Hopkins University Press, Baltimore, MD.

Lønning, S., and Davenport, J. (1980). The swelling egg of the long rough dab, *Hippoglossoides platessoides limandoides* (Bloch). *J. Fish Biol.* **17**, 359–378.

Lopez-Alvarado, J., and Kanazawa, A. (1994). Effect of dietary arginine levels on growth of red sea bream larvae fed diets supplemented with crystalline amino acids. *Fish. Sci.* **60**, 435–439.

Lopez-Alvarado J., Langdon, C. J., Teshima, S. I., and Kanazawa, A. (1994). Effects of coating and encapsulation of crystalline amino acids on leaching in larval feeds. *Aquaculture* **122**, 335–346.

Lyndon, A. R., Davidson, T., and Houlihan, D. F. (1993). Changes in tissue and plasma free amino acid concentrations after feeding in Atlantic cod. *Fish Physiol. Biochem.* **10**, 365–375.

Mangor-Jensen, A. (1987). Water balance in developing eggs of the cod, *Gadus morhua* L. *Fish Physiol. Biochem.* **3**, 17–24.

Mangor-Jensen, A., and Waiwood, K. G. (1995). The effect of light exposure on buoyancy of halibut eggs. *J. Fish Biol.* **47**, 18–25.

Mangor-Jensen, A., Waiwood, K. G., and Peterson, R. H. (1993). Water-balance in eggs of striped bass (*Morone saxatilis*). *J. Fish Biol.* **43**, 345–353.

Mangor-Jensen, A., Holm, J. C., Rosenlund, G., Lie, Ø., and Sandnes, K. (1994). Effect of dietary vitamin C on maturation and egg quality of cod (*Gadus morhua* L.). *J. World Aquacult. Soc.* **25**, 30–40.

Mani-Ponset, L., Guyot, E., Diaz, J. P., and Connes, R. (1996). Utilization of yolk reserves during post-embryonic development in three teleostean species: the sea bream *Sparus aurata,* the sea bass *Dicentrarchus labrax,* and the pike-perch *Stizostedion lucioperca. Mar. Biol.* **126**, 539–547.

Mårstøl, M., Kjesbu, O. S., Solemdal, P., and Fyhn, H. J. (1993). Free amino acid content as a potential selection criterion for egg quality in marine fishes. *In* "Physiological and Biochemical Aspects of Fish Development" (B. T. Walther, and H. J. Fyhn, eds.), pp. 99–103. University of Bergen, Norway.

Matsubara T., and Koya, Y. (1997). Course of proteolytic cleavage in three classes of yolk proteins during oocyte maturation in barfin flounder *Verasper moseri,* a marine teleost spawning pelagic eggs. *J. Exp. Zool.* **278**, 189–200.

Matsubara, T., and Sawano, K. (1995). Proteolytic cleavage in three classes of yolk protein during vitellogenin uptake and oocyte maturation in barfin flounder (*Verasper moseri*). *J. Exp. Zool.* **272**, 34–45.

Matsubara, T., Ohkubo, N., Andoh, T., Sullivan, C. V., and Hara, A. (1999). Two forms of vitellogenin, yielding two distinct lipovitellins, play different roles during oocyte maturation and early development of barfin flounder, *Verasper moseri,* a marine teleost that spawns pelagic eggs. *Dev. Biol.* **213**, 18–32.

McCormick, J. H., Broderius, S. J., and Fiandt, J. T. (1984). Toxicity of ammonia to early life stages of the green sunfish *Lepomis cyanellus. Environ. Poll.* **36**, 147–163.

Meadows, J., Smith, R. C., and Reeves, J. (1986). Uric acid protects membranes and linolenic acid from ozone-induced oxidation. *Biochem. Biophys. Res. Commun.* **137**, 536–541.

Mellinger, J. (1994). La flottabilité des oefs de téléostéens. *L'Année Biologique* **33**, 117–1138.

Mellinger, J. (1995). L'utilisation des lipides au cours du développement des poissons. *L'Année Biologique* **34**, 137–168.

Merchie, G. (1995). Nutritional effect of vitamin C on the growth and physiological condition of the larvae of aquaculture organisms. PhD thesis, University of Gent, Belgium.

Milroy, T. H. (1898). The physical and chemical changes taking place in the ova of certain marine teleosteans during maturation. 16. *Ann. Rep. Fish. Bd. Scotland* **1897, Part III,** 135–152.

Misiaszek, C. M. (1996). The development of ion regulation in larval rainbow trout, *Oncorhynchus mykiss.* M.Sc. thesis, Department of Biology, McMaster University, Hamilton, Ontario, Canada.

Miwa, S., and Inui, Y. (1987). Effects of various doses of thyroxine and triiodothyronine on the matamorphosis of flounder (*Paralichthys olivaceus*). *Gen. Comp. Endocrinol.* **67,** 356–363.

Mommsen, T. P., and Hochachka, P. W. (1988). The purine nucleotide cycle as two temporally separated metabolic units: a study on trout muscle. *Metabolism* **37,** 552–556.

Mommsen, T. P., and Walsh, P. J. (1988). Vitellogenesis and oocyte assembly. *In* "Fish Physiology" (W. S. Hoar, and D. J. Randall, eds.), Vol. 11A, pp. 347–406. Academic Press, London.

Mommsen, T. P., and P. W. Walsh. (1989). Evolution of urea synthesis in vertebrates: the piscine connection. *Science* **243,** 72–75.

Mommsen, T. P., and Walsh, P. J. (1991). Urea synthesis in fishes: evolutionary and biochemical perspectives. *In* "Biochemistry and Molecular Biology of Fishes" (P. W. Hochachka and T. P. Mommsen, eds.), Vol. 1, pp. 137–163. Elsevier, Amsterdam.

Montgomergy, D. R., Buffington, J. M., Peterson, N. P., Schuett-Hames, D., and Quinn, T. P. (1996). Stream-bed scour, egg burial depths, and the influence of salmonid spawning on bed surface mobility and embryo survival. *Can. J. Fish. Aquat. Sci.* **53,** 1061–1070.

Moore, S., and Stein, W. (1954). A modified ninhydrin reagent for the photometric determination of amino acids and related compounds. *J. Biol. Chem.* **211,** 907–913.

Moyle, P. B. (1993). "Fish. An Enthusiast's Guide." University of California Press, Berkeley.

Murakami, M., Iuchi, I., and Yamagami, K. (1991). Partial characterization and subunit analysis of major phosphoproteins of egg yolk in the fish, *Oryzias latipes. Comp. Biochem. Physiol.* **100B,** 587–593.

Naas, K. E., Næss, T., and Harboe, T. (1992). Enhanced feeding of halibut larvae (*Hippoglossus hippoglossus* L.) in green water. *Aquaculture* **105,** 143–156.

Næss, T., Germain-Henry, M., and Naas, K. E. (1995). First feeding of Atlantic halibut (*Hippoglossus hippoglossus*) using different combinations of *Artemia* and wild plankton. *Aquaculture* **130,** 235–250.

Needham, J. (1931). "Chemical Embryology." Cambridge University Press, London.

Needham, J. (1942). "Biochemistry and Morphogenesis." Cambridge University Press, London.

Noguchi, T., Takada, Y., and Fujiwara, S. (1979). Degradation of uric acid to urea and glyoxylate in peroxisomes. *J. Biol. Chem.* **254,** 5272–5275.

Okumura, H., Kayaba, T., Kazeto, Y., Hara, A., Adachi, S., and Yamauchi, K. (1995). Changes in the electrophoretic pattern of lipovitellin during oocyte development in the Japanese eel *Anguilla japonica. Fish. Sci.* **61,** 529–530.

Oppen-Bernsten, D. O., Bogsnes, A., and Walther, B. T. (1990a). The effects of hypoxia, alkalinity, and neurochemicals on hatching of Atlantic salmon (*Salmo salar*) eggs. *Aquaculture* **86,** 417–430.

Oppen-Bernsten, D. O., Helvik, J. V., and Walther, B. T. (1990b). The major structural proteins of cod (*Gadus morhua*) eggshell and protein crosslinking during teleost egg hardening. *Dev. Biol.* **137,** 258–265.

Østby, G., Finn, R. N., and Fyhn, H. J. (2000). Osmotic aspects of final oocyte maturation in Atlantic halibut. *In* "Proc. 6th Int. Symp. Reprod. Physiol. Fish." (B. Norberg, O. S. Kjesbu, G. L. Taranger, E. Anderson, and S. O. Stefansson, eds.), p. 185. Inst. Mar. Res. and Univ. Bergen, Norway.

Packard, G. C., and Packard, M. J. (1986). Nitrogen excretion by embryos of a gallinaceous bird and a reconsideration of the evolutionary origin of uricotely. *Can. J. Zool.* **64,** 691–693.

Paley, R. K., Twitchen, I. D., and Eddy, F. B. (1993). Ammonia, Na^+, K^+, and Cl-levels in rainbow trout yolk-sac fry in response to external ammonia. *J. Exp. Biol.* **180,** 273–284.

Parra, G., Rønnestad, I., and Yúfera, M. (1999). Energy metabolism in eggs and larvae of the senegal sole. *J. Fish Biol.* **55**, 205–214.
Pilley, C. M., and Wright, P. A. (2000). The mechanisms of urea transport in early life stages of rainbow trout (*Oncorhynchus mykiss*). *J. Exp. Biol.* **203**, 3199–3207.
Pittman, K. A. (1991). Aspects of the early life history of the Atlantic halibut (*Hippoglossus hippoglossus* L.): embryonic and larval development and the effects of temperature. Dr. scient. thesis, University of Bergen, Norway.
Potts, W. T. W., and Rudy, P. P. (1969). Water balance in the eggs of the Atlantic salmon *Salmo salar*. *J. Exp. Biol.* **50**, 223–237.
Poupard, G., André, M., Durliat, M., Ballagny, C. Boef, G., and Babin, P. J. (2000). Apolipoprotein E gene expression correlates with endogenous lipid nutrition and yolk syncytial layer lipoprotein synthesis during fish development. *Cell Tissue Res.* **300**, 251–261.
Rahaman-Noronha, E. (1995). Nitrogen partitioning and excretion in embryonic rainbow trout (*Oncorhychus mykiss*). M.Sc. thesis, Department of Zoology, University of Guelph, Ontario, Canada.
Rahaman-Noronha, E., O'Donnell, M. J., Pilley, C. M., and Wright, P. A. (1996). Excretion and distribution of ammonia and the influence of boundary layer acidification in embryonic rainbow trout (*Oncorhynchus mykiss*). *J. Exp. Biol.* **199**, 2713–2723.
Rainuzzo, J. R., Farestveit, R., and Jørgensen, L. (1993). Fatty acid and amino acid composition during embryonic and larval development in plaice (*Pleuronectes platessa*). In "Physiology and Biochemistry of Marine Fish Larvae" (B. T. Walther and H. J. Fyhn, eds.), pp. 290–295. University of Bergen, Norway.
Randall, D. J., Wood, C. M., Perry, S. F., Bergman, H., Maloiy, G. M. O., Mommsen, T. P., and Wright, P. A. (1989). Urea excretion as a strategy for survival in a fish living in a very alkaline environment. *Nature* **337**, 165–166.
Read, L.J. (1968a). Ornithine–urea cycle enzymes in early embryos of the dogfish *Squalus suckleyi* and the skate *Raja binoculata*. *Comp. Biochem. Physiol.* **24**, 669–674.
Read, L.J. (1968b). A study of ammonia and urea production and excretion in the fresh-water-adapted form of the Pacific lamprey, *Entosphenus tridentatus*. *Comp. Biochem. Physiol.* **26**, 455–466.
Reitan, K. I., Rainuzzo, J. R., Øie, G., and Olsen, Y., (1993). Nutritional effects of algal addition in first feeding turbot (*Scophthalmus maximus* L.) larvae. *Aquaculture* **118**, 257–275.
Reitan, K. I., Bolla, S., and Olsen, Y. (1994). A study of the mechanism of algal uptake in yolk-sac larvae of Atlantic halibut (*Hippoglossus hippoglossus* L.). *J. Fish Biol.* **44**, 303–310.
Reith, M., Munholland, J., Kelly, J., Finn, R. N., and Fyhn, H. J. (2001). Lipovitellins derived from two forms of vitellogenin are differentially processed during oocyte maturation in haddock (*Melanogrammus aeglefinus*). *J. Exp. Zool.* **291**, 58–67.
Rice, S. D., and Bailey, J. E. (1980). Survival, size and emergence of pink salmon, *Onchorhynchus gorbuscha*, alevins after short- and long-term exposures to ammonia. *Fish. Bull.* **78**, 641–648.
Rice, S. D., and Stokes, R. M. (1974). Metabolism of nitrogenous wastes in the eggs and alevins of rainbow trout, *Salmo gairdneri*, Richardson. In "The Early Life History of Fish" (J. H. S. Blaxter, ed.), pp. 325–337. Springer-Verlag, New York.
Rice, S. D., and Stokes, R. M. (1975). Acute toxicity of ammonia to several developmental stages of rainbow trout, *Salmo gairdneri*. *Fish. Bull.* **73**, 207–210.
Riis-Vestergaard, J. (1987). Physiology of teleost embryos related to environmental challenges. *Sarsia* **72**, 351–358.
Rombough, P. J. (1988). Respiratory gas exchange, aerobic metabolism, and effects of hypoxia during early life. In "Fish Physiology" (W. S. Hoar and D. J. Randall, eds.), Vol. 11A, pp. 59–161. Academic Press, San Diego.
Rombough, P. J. (1998). Partitioning of oxygen uptake between the gills and skin in fish larvae: a novel method for estimating cutaneous oxygen uptake. *J. Exp. Biol.* **201**, 1763–1769.
Rombough, P. J. (1999). The gill of fish larvae. Is it primarily a respiratory or an ionregulatory structure? *J. Fish Biol.* **55** (Supplement A).

Rombough, P. J., and Moroz, B. M. (1990). The scaling and potential importance of cutaneous and branchial surfaces in respiratory gas exchange in young chinook salmon (*Oncorhynchus tshawytscha*). *J. Exp. Biol.* **154**, 1–12.

Rombough, P. J., and Moroz, B. M. (1997). The scaling and potential importance of cutaneous and branchial surfaces in respiratory gas exchange in larval and juvenile walleye *Stizostedion vitreum*. *J. Exp. Biol.* **200**, 2459–2468.

Rombough, P. J., and Ure, D. (1991). Partitioning of oxygen uptake between cutaneous and branchial surfaces in larval and juvenile chinook salmon *Oncorhynchus tshawytscha*. *Physiol. Zool.* **64**, 717–727.

Rønnestad, I. (1992). Utilization of free amino acids in marine fish and larvae. Dr. thesis, University of Bergen, Norway.

Rønnestad, I. (1993). No efflux of free amino acids from yolk-sac larvae of Atlantic halibut (*Hippoglossus hippoglossus* L.). *J. Exp. Mar. Biol. Ecol.* **167**, 39–45.

Rønnestad, I. (1995). Interpretation of ontogenetic changes in composition studies of fish eggs and larvae: presenting relative data can lead to erroneous conclusions. *Aquacult. Nutr.* **1**, 199.

Rønnestad, I., and Fyhn, H. J. (1993). Metabolic aspects of free amino acids in developing marine fish eggs and larvae. *Rev. Fish. Sci.* **1**, 239–259.

Rønnestad, I., and Naas, K. E. (1993). Oxygen consumption and ammonia excretion in larval Atlantic halibut (*Hippoglossus hippoglossus* L.) at first feeding: a first step towards an energetic model. *In* "Fish Larval Physiology and Biochemistry" (B. T. Walther and H. J. Fyhn, eds.), pp. 279–284. University of Bergen, Norway.

Rønnestad, I., Finn, R. D., Groot, E. P., and Fyhn, H. J. (1992a). Utilization of free amino acids related to energy metabolism of devloping eggs and larvae of lemon sole *Microstomus kitt* reared in the laboratory. *Mar. Ecol. Progr. Ser.* **88**, 195–205.

Rønnestad, I., Fyhn, H. J., and Gravningen, K. (1992b). The importance of free amino acids to the energy metabolism of eggs and larvae of turbot (*Scophthalmus maximus*). *Mar. Biol.* **114**, 517–525.

Rønnestad, I., Groot, E. P., and Fyhn, H. J. (1993). Compartmental distribution of free amino acids and protein in developing yolk-sac larvae of Atlantic halibut (*Hippoglossus hippoglossus*). *Mar. Biol.* **116**, 349–354.

Rønnestad, I., Koven, W. M., Tandler, A., Harel, M., and Fyhn, H. J. (1994). Energy metabolism during development of eggs and larvae of gilthead sea bream (*Sparus aurata*). *Mar. Biol.* **120**, 187–196.

Rønnestad, I., Finn, R. D., Lein, I., and Fyhn, H. J. (1995). Compartmental changes in the contents of total lipid, lipid classes and their associated fatty acids in developing yolk-sac larvae of Atlantic halibut, *Hippoglossus hippoglossus* (L.). *Aquacult. Nutr.* **1**, 119–130.

Rønnestad, I., Robertson, and Fyhn, H. J. (1996). Free amino acids and protein content in pelagic and demersal eggs of tropical marine fishes. *In* "The Fish Egg" (D. D. MacKinlay and M. Eldridge, eds.), pp. 81–84. Am Fish. Soc. San Francisco.

Rønnestad, I., Lie, Ø., and Waagbø, R. (1997). Vitamin B$_6$ in Atlantic halibut, *Hippoglossus hippoglossus* L.—endogenous utilization and retention in fish larvea fed natural zooplankton. *Aquaculture* **157**, 337–345.

Rønnestad, I., Koven, W. M., Tandler, A., Harel, M., and Fyhn, H. J. (1998a). Utilisation of yolk fuels in developing eggs and larvae of European sea bass (*Dicentrarchus labrax*). *Aquaculture* **162**, 157–170.

Rønnestad, I., Hemre, G. I. Finn R. N., and Lie, Ø. (1998b). Alternate sources and dynamics of vitamin A and its incorporation into the eyes during the early endothropic and exothropic larval stages of Atlantic halibut (*Hippoglossus hippoglossus* L). *Comp. Biochem. Physiol.* **119A**, 787–793.

Rønnestad, I., Helland, S., and Lie, Ø. (1998c). Feeding *Artemia* to larvae of Atlantic halibut (*Hippoglossus hippoglossus* L.) results in lower larval vitamin A content compared with feeding copepods. *Aquaculture* **164**, 159–164.

5. NITROGEN METABOLISM AND EXCRETION

Rønnestad, I., Thorsen, A., and Finn, R. N. (1999a). Fish larval nutrition: a review of recent advances in the roles of amino acids. *Aquaculture* **177,** 201–216.
Rønnestad, I., Hamre, K., Lie, Ø., and Waagbø, R. (1999b). Ascorbic acid and α-tocopherol levels in larvae of Atlantic halibut before and after exogenous feeding. *J. Fish Biol.* **55,** 720–731.
Rønnestad, I., Perez Dominguez, R., and Tanaka, M. (2000a). Ontogeny of digestive tract functionality in Japanese flounder, *Paralichthys olivaceus* studied by *in vivo* microinjection: pH and assimilation of free amino acids. *Fish Physiol. Biochem.* **22,** 225–235.
Rønnestad, I., Conceição, L. E. C., Aragão, C., and Dinis, M. T. (2000b). Free amino acids are absorbed faster and assimilated more efficiently than protein in postlarval Senegal sole (*Solea senegalensis*). *J. Nutr.* **130,** 2809–2812.
Rust, B. R. (1995). Quantitative aspects of nutrient assimilation in six species of fish larvae. PhD thesis, School of Fisheries, University of Washington, Seattle, WA.
Rust, B. R., Hardy, R. W., and Stickney, R. R. (1993). A new method for force-feeding larval fish. *Aquaculture* **116,** 341–352.
Saha, N., and Ratha, B. K. (1989). Comparative study of ureogenesis in freshwater, air-breathing teleosts. *J. Exp. Zool.* **252,** 1–8.
Segner, H., and Verreth, J. (1995). Metabolic enzyme activities in larvae of the African catfish, *Clarias gariepinus:* changes in relation to age and nutrition. *Fish Physiol. Biochem.* **14,** 385–398.
Segner, H., Storch, V., Reinecke, M., Kloas, W., and Hanke, W. (1994). The development of functional organ systems in turbot (*Scophthalmus maximus* L.). *Mar. Biol.* **119,** 471–486.
Selman, K., and Wallace, R. A. (1989). Cellular aspects of oocyte growth in teleosts. *Zool. Sci.* **6,** 211–231.
Selman, K., Petrino, T. R., and Wallace, R. A. (1994). Experimental conditions for oocyte maturation in the zebra fish, *Brachydanio rerio*. *J. Exp. Zool.* **269,** 538–550.
Seoka, M., Takii, K., Takaoka, O., Nakamura, M., and Kumai, H. (1997). Biochemical phases in embryonic red sea bream development. *Fish. Sci.* **63,** 122–127.
Shahsavarani, A. (2000). Effects of temperature on embryonic physiology of Arctic char (*Salvelinus alpinus*). M.Sc. thesis, Department of Zoology, University of Guelph, Ontario, Canada.
Shearer, K. D. (2000). Experimental design, statistical analysis and modelling of dietary nutrient requirement studies for fish: a critical review. *Aquacult. Nutr.* **6,** 91–102.
Shelbourne, J. E. (1965). The abnormal development of plaice embryos and larvae in marine aquaria. *J. Mar. Biol. Ass. U.K.* **35,** 177–192.
Shen, A. C. Y., and Leatherland, J. F. (1978). Structure of the yolk sac epithelium and gills in the early developmental stages of rainbow trout (*Salmo gairdneri*) maintained in different ambient salinities. *Env. Biol. Fish.* **3,** 345–354.
Shields, R. J., Bell, J. G., Luizi, F. S., Gara, B., Bromage, N. R., and Sargent, J. R. (1999). Natural copepods are superior to enriched Artemia nauplii as feed for halibut larvae (*Hippoglossus hippoglossus*) in terms of survival, pigmentation and retinal morphology: relation to dietary essential fatty acids. *J. Nutr.* **129,** 1186–1194.
Simic, M. G., and Jovanovic, S. V. (1989). Antioxidation mechanisms of uric acid. *J. Am. Chem. Soc.* **111,** 5778–5782.
Sire, M.-F., Babin, P. J., and Vernier, J.-M. (1994). Involvement of the lysosomal system in yolk protein deposit and degradation during vitellogenesis and embryonic development in trout. *J. Exp. Zool.* **269,** 69–83
Sivaloganathan, B., Walford, J., Ip, Y. K., and Lam, T. J. (1998). Free amino acids and energy metabolism in eggs and larvae of seabass, *Lates calcarifer*. *Mar. Biol.* **131,** 695–702.
Smith, C. P., and Wright, P. A. (1999). Molecular characterization of an elasmobranch urea transporter. *Am. J. Physiol.* **276,** R622–R626.
Smith, H. W. (1932). Water regulation and its evolution in the fishes. *Q. Rev. Biol.* **7,** 1–26.
Smith, H. W. (1953). "From Fish to Philosopher." Little, Brown and Co., Boston.

Smith, S. (1947). Studies in the development of the rainbow trout (*Salmo irideus*). *J. Exp. Biol.* **23**, 357–378.
Srivastava, R. K., and Brown, J. A. (1992). Assessment of egg quality in Atlantic salmon, *Salmo salar*, treated with testosterone—II. Amino acids. *Comp. Biochem. Physiol.* **103A**, 397–402.
Srivastava, R. K., Brown, J. A., and Shahidi, F. (1995). Changes in the amino acid pool during embryonic development of cultured and wild Atlantic salmon (*Salmo salar*). *Aquaculture* **131**, 115–124.
Steele, S. L., Chadwick, T. D., and Wright, P. A. (2001). Ammonia detoxification and localization of urea cycle enzymes activity in embryos of the rainbow trout (*Oncorhynchus mykiss*) in relation to early tolerance to high environmental ammonia levels. *J. Exp. Biol.* (in press).
Sturman, J. A. (1993). Taurine in development. *Physiol. Rev.* **73**, 119–147.
Suzuki, T., and Suyama, M. (1983). Free amino acids and phosphopeptides in the extracts of fish eggs. *Bull. Japan Soc. Sci. Fish.* **49**, 1747–1753.
Suzuki, T., Shirai, T., and Hirano, T. (1991). Anserine and free amino acid contents in the egg and fry stages of developing rainbow trout (*Oncorhynchus mykiss*). *Comp. Biochem. Physiol.* **100A**, 105–108.
Takada, Y., and Noguchi, T. (1986). Ureidoglycollate lyase, a new metalloenzyme of peroxisomal urate degradation in marine fish liver. *Biochem. J.* **235**, 391–397.
Takeuchi, T., Dedi, J., Ebisawa, C., Watanabe, T., Seikai, T., Hosoya, K., and Nakazoe, J.-I. (1995). The effect of β-carotene and vitamin A enriched *Artemia nauplii* on the malformation and color abnormality of larval Japanese flounder. *Fish. Sci.* **61**, 141–148.
Terjesen, B. (1995). Nitrogen metabolism in developing eggs and larvae of the African catfish *Clarias gariepinus* (Burcell 1822). Cand. scient. thesis, University of Bergen, Norway.
Terjesen, B. F., Chadwick, T. D., Verreth, J. A., Rønnestad, I., and Wright, P. A. (2001). Pathways for urea production during early life of an air-breathing teleost, the African catfish (*Clarias gariepinus* Burchell 1822). *J. Exp. Biol.* (in press).
Terjesen, B. F., Verreth, J., and Fyhn, H. J. (1997). Urea and ammonia excretion by embryos and larvae of the African catfish *Clarias gariepinus* (Burchell 1822). *Fish Physiol. Biochem.* **16**, 311–321.
Terjesen, B. F., Mangor-Jensen, A., and Fyhn, H. J. (1998). Ammonia dynamics in relation to Atlantic halibut (*Hippoglossus hippoglossus* L.). *Fish Physiol. Biochem.* **18**, 189–201.
Terjesen, B. F., Rønnestad, I., Norberg, B., and Anderson, P.M. (2000). Detection and basic properties of carbamoyl phosphate synthetase III during teleost ontogeny: a case study in the Atlantic halibut (*Hippoglossus hippoglossus*). *Comp. Biochem. Physiol.* **126**, 521–535.
Thompsen, T. (2000). Free amino acids and protein content in developing embryos and larvae of five salmonid species. Cand. scient. thesis, University of Bergen, Norway.
Thoroed, S. M., and Fugelli, K. (1994). Free amino compounds and cell volume regulation in erythrocytes from different marine fish species under hypoosmotic conditions: the role of a taurine channel. *J. Comp. Physiol. B* **164**, 1–10.
Thorsen, A. (1995). Oogenesis in marine bony fishes; physiological mechanisms of oocyte hydration and egg buoyancy. Dr. thesis, University of Bergen, Norway.
Thorsen, A., and Fyhn, H. J. (1991) Osmotic effectors during preovulatory swelling of marine fish eggs. *In* "Reproductive Physiology of Fish. Fish Symp. 91" (A. P. Scott *et al.*, eds.), pp. 312–314. Sheffield University, England.
Thorsen, A., and Fyhn, H. J. (1996). Final oocyte maturation *in vivo* and *in vitro* in marine fishes with pelagic eggs; yolk protein hydrolysis and free amino acid content. *J. Fish Biol.* **48**, 1195–1209.
Thorsen, A., Fyhn, H. J., and Wallace, R. A. (1993). Free amino acids as osmo effectors for oocyte hydration in marine fishes. *In* "Physiological and Biochemical Aspects of Fish Development" (B. T. Walther and H. J. Fyhn, eds.), pp. 94–98. University of Bergen, Norway.
Thorsen, A., Kjesbu, O. S., Fyhn, H. J., and Solemdal, P. (1996). Physiological mechanisms of buoyancy in eggs from brackish water cod. *J. Fish Biol.* **48**, 457–477.

Tonheim, S., Wm. Koven, W., and Rønnestad, I. (2000). Enrichment of *Artemia* with free methionine. *Aquaculture* **190**, 223–235.
Torres, J. J., Brightman, R. I., Donnelly, J., and Harvey, J. (1996). Energetics of larval red drum, *Sciaenops ocellatus*. Part I: Oxygen consumption, specific dynmaic action, and nitrogen excretion. *Fish. Bull.* **94**, 756–765.
Trinkhaus, J. P. (1992). The midblastula transition, the YSL transition and the onset of gastrulation in *Fundulus*. *Development* **94**, 756–765.
Trinkhaus, J. P. (1993). The yolk syncytial layer of *Fundulus:* its origin and history and its significance for early embryogenesis. *J. Exp. Zool.* **265**, 258–284.
Trinkhaus, J. P. (1996). Ingression during early gastrulation of *Fundulus*. *Dev. Biol.* **177**, 356–370.
Tyler, C. R., and Sumpter, J. P. (1996). Oocyte growth and development in teleosts. *Rev. Fish Biol. Fish.* **6**, 287–318.
Tytler, P., Bell, M. V., and Robinson, J. (1993). The ontogeny of osmoregulation in marine fish: effects of changes in salinity and temperature. *In* "Physiology and Biochemistry of Marine Fish Larvae" (B. T. Walther and H. J. Fyhn, eds.), pp. 249–258. University of Bergen, Norway.
Van der Meeren, T. (1991). Algae as first food for cod larvae, *Gadus morhua* L.: filter feeding or ingestion by accident? *J. Fish Biol.* **39**, 225–237.
Van Waarde, A. (1988). Biochemistry of non-protein nitrogenous compounds in fish including the use of amino acids for anaerobic energy production. *Comp. Biochem. Physiol.* **91B**, 207–228.
Walford, J., Lim, T. M., and Lam, T. J. (1991). Replacing live foods with microencapsulated diets in the rearing of seabass (*Lates calcarifer*) larvae: do the larvae ingest and digest protein-membrane microcapsules? *Aquaculture* **92**, 225–235.
Wallace, R. A., and Selman, K. (1978). Oogenesis in *Fundulus heteroclitus* I. Preliminary observations on oocyte maturation *in vivo* and *in vitro*. *Dev. Biol.* **62**, 354–369.
Wallace, R. A., and Selman, K. (1981). Cellular and dynamic aspects of oocyte growth in teleosts. *Am. Zool.* **21**, 325–343.
Wallace, R. A., and Selman, K. (1985). Major protein changes during vitellogenesis and maturation of *Fundulus* oocytes. *Dev. Biol.* **110**, 492–498.
Wallace, R. A., and Selman, K. (1990). Ultrastructural aspects of oogenesis and oocyte growth in fish and amphibians. *J. Electr. Micros. Tech.* **16**, 175–201.
Wallace, R. A., Boyle, S. M., Grier, H. J., Selman, K., and Petrino, T. R. (1993). Preliminary observations on oocyte maturation and other aspects of reproductive biology in captive female snook, *Centropomus undecimalis*. *Aquaculture* **116**, 257–273.
Walsh, P. J. (1997). Evolution and regulation of ureagenesis and ureotely in fishes. *Ann. Rev. Physiol.* **59**, 299–323.
Walsh, P. J., Heitz, M., Campbell, C. E., Cooper, G. J., Medina, M., Wang, Y. S., Goss, G. G., Vincek, V., Wood, C. M., and Smith, C. P. (2000). Molecular identification of a urea transporter in gill of the ureotelic gulf toadfish (*Opsanus beta*). *J. Exp. Biol.* **203**, 2357–2364.
Watanabe, Y. (1984). An ultrastructural study of intracellular digestion of horseradish peroxidase by the rectal epithelium cells in larvae of a freshwater cottid fish *Cottus nozawae*. *Bull. Japan Soc. Sci. Fish.* **50**, 409–416.
Watanabe, T. (1993). Importance of docosahexaenoic acid in marine larval fish. *J. World Aquacult. Soc.* **24**, 152–161.
Watanabe, T., and Kiron, V. (1994). Prospects in larval fish dietetics. *Aquaculture* **124**, 223–251.
Wells, P. R., and Pinder, A. W. (1996). The respiratory development of Atlantic salmon. II. Partitioning of oxygen uptake among gills, yolk sac and body surfaces. *J. Exp. Biol.* **199**, 2737–2744.
Wilkie, M. P., Wang, Y., Walsh, P. J., and Youson, H. J. (1999). Nitrogenous waste excretion by the larvae of phylogenetically ancient vertebrate: the sea lamprey (*Petromyzon marinus*). *Can. J. Zool.* **77**, 707–715.

Wood, C. M. (1993). Ammonia and urea metabolism and excretion. *In* "The Physiology of Fishes" (D. H. Evans, ed), pp. 379–425. CRC Press, Boca Raton, FL.
Wourms, J. P., Grove, B. D., and Lombardi, J. (1988). The maternal–embryonic relationship in viviparous fishes. *In* "Fish Physiology" (W. S. Hoar and D. J. Randall, eds.), Vol. 11B, pp. 1–134. Academic Press, San Diego.
Wright, P. A. (1993). Nitrogen excretion and enzyme pathways for ureagenesis in freshwater tilapia (*Oreochromis niloticus*). *Physiol. Zool.* **66**, 881–901.
Wright, P. A., and Land, M. D. (1998). Urea production and transport in teleost fishes. *Comp. Biochem. Physiol.* **119A**, 47–54.
Wright, P. A., Heming, T., and Randall, D. (1986). Downstream pH changes in water flowing over the gills of rainbow trout. *J. Exp. Biol.* **126**, 499–512.
Wright, P. A., Randall, D. J., and Perry, S. F. (1989). Fish gill water boundary layer: a site of linkage between carbon dioxide and ammonia excretion. *J. Comp. Physiol. B* **158**, 627–635.
Wright, P. A., Felskie, A. K., and Anderson, P. M. (1995). Induction of ornithine–urea cycle enzymes and nitrogen metabolism and excretion in rainbow trout (*Oncorhynchus mykiss*) during early life stages. *J. Exp. Biol.* **198**, 127–135.
Yamagami, K. (1988). Mechanisms of hatching in fish. *In* "Fish Physiology" (W. S. Hoar and D. J. Randall, eds.), Vol. 11A, pp. 447–499. Academic Press, San Diego.
Yamamoto, T. S. (1984). Ammonia release by chum salmon eggs at the initiation of their embryonic development. *Dev. Growth Differ.* **26**, 95–104.
Yamamoto, T., Unuma, T., and Akiyama, T. (2000). The influence of dietary protein and fat levels on tissue free amino acid levels of fingerling rainbow trout (*Oncorhynchus mykiss*). *Aquaculture* **182**, 353–372.
Yancey P. H., Clark, M. E., Hand, S. C., Bowlus, R. D., and Somero, G. N. (1982). Living with water stress: evolution of osmolyte systems. *Science* **217**, 1214–1222.
Yarzhombek, A. A., and Maslennikova, N. V. (1971). Nitrogenous metabolites of the eggs and larvae of various fishes. *J Ichthyol.* **11**, 276–281.
Yasumasu, S., Yamada, K., Akasaka, K, Mitsunaga, K., Iuchi, I., Shimada, H., and Yamagami, K. (1992). Isolation of cDNAs for LCE and HCE, two constituent proteases of the hatching enzyme of *Oryzias latipes,* and concurrent expression of their mRNAs during development. *Dev. Biol.* **153**, 250–258.
Yasumasu, S., Shimada, I. I., Inohaya, K., Iuchi, I., Yasumasu I., and Yamagami, K. (1996). Different exon–intron organizations of the genes for two astacin-like protease high choriolytic enzyme (choriolysin H) and low choriolytic enzyme (chorioly L), the constituents of the fish hatching enzyme. *Eur. J. Biochem.* **237**, 752–758.
Yokoyama, M., and Nakazoe, J. (1996). Intraperitoneal injection of sulfur amino acids enhance the hepatic cysteine dioxygenase activity in rainbow trout (*Oncohynchus mykiss*). *Fish Physiol. Biochem.* **15**, 143–148.
Yokoyama, M., and Nakazoe, J. (1998). Effect of oral administartion of L-cystine on hypotaurine level in rainbow trout. *Fish. Sci.* **64**, 144–147.
Yùfera, M., Pascual, E., and Fernández-Diaz, C. (1999). A highly efficient microencapsulated food for rearing early larvae of marine fish. *Aquaculture* **177**, 249–256.

6

INFLUENCE OF FEEDING, EXERCISE, AND TEMPERATURE ON NITROGEN METABOLISM AND EXCRETION

CHRIS M. WOOD

I. Introduction
II. Feeding and Nitrogen Metabolism
III. Metabolic Fuel Usage and Nitrogen Metabolism
IV. Exercise and Nitrogen Metabolism
   A. Sustainable Aerobic Swimming
   B. Anaerobic Burst Swimming
V. Temperature and Nitrogen Metabolism
   A. Nonfed Fish
   B. Actively Feeding Fish
VI. Other Nitrogen Products?
   A. Urea versus Ammonia
   B. Other Compounds
VII. Excretion Mechanisms
VIII. Elasmobranchs
IX. Concluding Remarks
   References

I. INTRODUCTION

The literature on nitrogen metabolism is unbalanced. The vast majority of fish swim actively, eat regularly, grow continually, live in nonextreme environments, and, at least if they are teleosts, excrete mainly ammonia by way of the gills. Their major environmental variables are temperature and food availability. However, as illustrated by other chapters in this volume, and a host of recent reviews (Mommsen and Walsh, 1992; Wood, 1993, 1995; Jobling, 1994; Korsgaard et al., 1995; Wright, 1995; Walsh, 1997a; Wilkie, 1997), most experimental studies have concentrated on fish that are confined in small chambers and therefore not swimming, feeding, or growing; the fish are often subjected to severe environmental challenges, but the temperature is usually held constant. The species studied are often selected because they are suspected to have unusual end products or pathways of nitrogenous waste excretion, rather than because they are representative of the great majority of teleosts. The primary aim of the present review is to redress this imbalance by examining the separate and interactive effects of feeding, exercise, and temperature on nitrogen metabolism in representative teleost fish. Two additional areas highlighted are the possibility of additional nitrogen products that may be excreted, and the mechanism(s) of ammonia excretion, both in actively feeding fish. Finally, the meager information available on these topics for elasmobranchs is summarized, in the hope of stimulating further research on this most interesting but poorly studied group.

II. FEEDING AND NITROGEN METABOLISM

When a fish eats another fish, or a commercial diet made from fish meal, it is eating about 40% protein on a dry weight basis. Assuming a daily ration of 4% for an intensively feeding fish and 0.16 gram of nitrogen per gram of protein (Cho, 1990), then the fish is eating 2.56 g/kg or 183,000 μmol of nitrogen per day. If this were all excreted, then the average excretion rate would be 7625 μmol·kg^{-1}·h^{-1}. Typically, measured nitrogen excretion rates for nonfed fish confined in boxes are less than 500 μmol·kg^{-1}·h^{-1}. Of course, because fish grow all their lives, a substantial portion of this ingested nitrogen is not excreted but rather converted to protein growth. Nevertheless, the comparison serves to illustrate the widespread misconception that the examination of nitrogen metabolism in nonfed fish (which by definition cannot be growing) provides a picture of their normal physiology. Instead, it provides a picture of their physiology during starvation, a special circumstance that can be considered a component of normal physiology in only a few circumstances (e.g., anorexia during migration, overwintering at cold temperature, or nest guarding). The critical influence of feeding on physiologic function has been routinely ignored, often for the sake of experimental convenience.

When experiments are performed on quiescent fish that are not being fed (a common procedure in many laboratories), nitrogen excretion rates tend to drop markedly, stabilizing after about a week of starvation (Fromm, 1963; Iwata, 1970; Smith and Thorpe, 1976; Rychly and Marina, 1977; Kaushik and Teles, 1985). These abnormally low rates of nitrogen excretion represent the so-called *endogenous* fraction, the minimum unavoidable loss that results from the normal protein, nucleic acid, purine, and pyrimidine base turnover required for body maintenance. The difference between this "baseline" rate seen in a starved fish and the much higher rates occurring in fed fish approximates the *exogenous* fraction, the portion not retained as growth from the diet.

Because confinement in small volumes is necessary to measure nitrogen waste excretion in individual fish, and it is often inconvenient or impossible to have the fish feed or grow under such conditions, a viable alternative is to work with fish in groups, living in their normal feeding and holding tanks. Theoretically, there is absolutely nothing wrong with this approach, because both naturally and in aquaculture, fish tend to live and feed in groups. However, statisticians have an unfortunate habit of demanding that the number of tanks of fish be replicated to the same degree as the number of fish in individual experiments, a factor that has discouraged this approach. Fortunately, a number of researchers have disregarded this criticism and, thereby, produced some very useful data.

A landmark review is that of Handy and Poxton (1993), who gathered together a large amount of data derived from this "bulk in tank" approach, from the aquaculture of mainly marine species, much of it from literature that is difficult to access for most researchers. This synthesis demonstrated that nitrogen excretion rates depend critically on feeding, and may often rise above 3000 μmol·kg^{-1}·h^{-1}. Wood (1995) similarly reviewed nitrogen waste excretion data for salmonids in aquaculture, and concluded that rates in actively feeding fish ranged as high as 2500 μmol·kg^{-1}·h^{-1}, and probably even higher in salmon feeding in the wild. Subsequent studies have confirmed these very high rates in actively feeding salmonids and the great sensitivity of absolute rates to ration (e.g., Dockray et al., 1996, 1998; Linton et al., 1997, 1998a,b). Leung et al. (1999) have summarized similar recent data for a variety of nonsalmonid marine fishes. Figure 1, from the data of Alsop and Wood (1997), illustrates the marked effects on nitrogen waste excretion and metabolic rate of switching rainbow trout from a maintenance ration (1% per day) to either starvation or twice daily satiation feeding (3% per day); a difference of over 6-fold in nitrogen excretion rate resulted, but only 1.7-fold in metabolic rate (O_2 consumption).

Three classic studies from the earlier literature are particularly noteworthy. Despite the difficulties of motivating individually confined fish to eat, a heroic study by Beamish and Thomas (1984) on adult rainbow trout accomplished just this feat. Not only were individual fish trained to take a fixed ration from forceps inserted through a feeding port, they were also fitted with a chronic bladder catheter to collect urinary nitrogen excretion and housed in a chamber that allowed collection of their fecal nitrogen excretion. Rations tested were low (0.5 and 1.0% per day), but, nevertheless, both total nitrogen excretion and nitrogen retention increased in response to increased ration, and in response to increased nitrogen content in the ration. Fecal nitrogen losses were small, reflecting high nitrogen absorption efficiency (>93%) through the digestive tract. Overall, nitrogen retention was about 50%, and the great majority of nitrogen excretion occurred by way of the gills in the form of ammonia-N, with urea-N as a minor but significant component; urinary nitrogen excretion was less than 4% of the total. There have been no more thorough studies on the nitrogen budgets of individual fish since the work of Beamish and Thomas (1984), and their conclusions remain valid today.

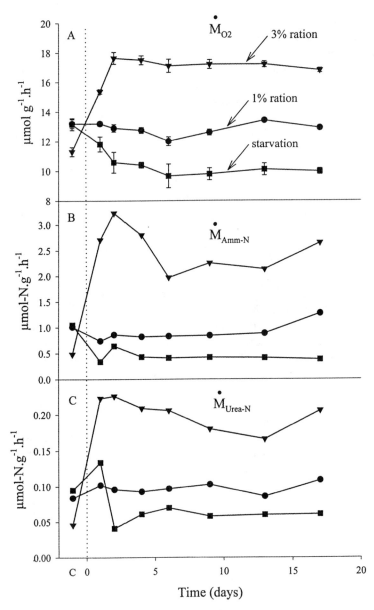

Fig. 1. Changes in the rates of (A) O_2 consumption (M_{O2}); (B) ammonia-N excretion (M_{Amm-N}); and (C) urea-N excretion (M_{Urea-N}) at 15°C in groups of about 120 juvenile rainbow trout (~8 g) over a 17-day period following the switch to daily rations of 3%, 1%, or starvation at day 0. Prior to day 0, all fish were fed a 1% ration. Note the much larger relative changes in M_{Amm-N} and M_{Urea-N} than in M_{O2}. Mean data for duplicate tanks; standard error bars in (A) are for the measurements, not the tanks. (Recalculated from Alsop and Wood, 1997. Reproduced with permission of the Company of Biologists Ltd.)

In another landmark study, Brett and Zala (1975) used an "in-tank" monitoring system with high temporal resolution to monitor rates of ammonia-N and urea-N excretion and O_2 consumption in a group of 30 juvenile sockeye salmon held together in a single tank. Following a once-daily feeding of 3% ration, ammonia-N excretion increased 3.5-fold to reach a peak 5 h later, and thereafter declined (Fig. 2). Notably, the O_2 consumption rate showed a very different pattern, rising *prior* to feeding, peaking coincident with feeding, and declining prior to the ammonia-N peak, presumably reflecting an entrainment of the fishes' activity level ("excitement") to the time when food was provided. The importance

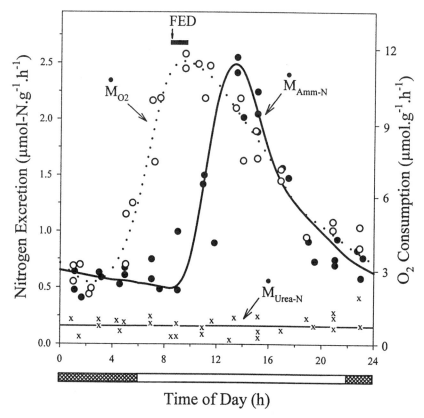

Fig. 2. Diurnal variation in the rates of O_2 consumption (M_{O2}, broken line, open circles), ammonia-N excretion (M_{Amm-N}, closed circles, thick solid line) and urea-N excretion (M_{Urea-N}, crosses, thin solid line) at 15°C in a group of about 30 freshwater-adapted juvenile sockeye salmon (~30 g) fed a 3% ration once daily between 8:30 and 9:30 A.M. Ammonia-N excretion and O_2 consumption rates have been scaled to the same peak value. Note that the O_2 consumption rates (right axis) rise prior to feeding and are about 5 times greater than the nitrogen excretion rates (left axis), which peak 4 h after feeding. (Recalculated from Brett and Zala, 1975. Reproduced with the permission of Her Majesty the Queen in Right of Canada, 2000, Fisheries and Oceans Canada.)

of this work is that it was the first to illustrate the marked and rapid variation in nitrogen excretion that occurs with feeding, and to dissociate it from changes in metabolic rate. The work was also important because urea-N excretion remained unchanged over the daily cycle, at about the same level seen in fish subjected to prolonged starvation, suggesting that urea-N was entirely part of the *endogenous* fraction. The latter was later replicated in Atlantic salmon (Wiggs et al., 1989; Fivelstad et al., 1990), Arctic charr (Fivelstad et al., 1990) and sturgeon (Gershanovich and Pototskij, 1992) but disputed by experiments on seabass (Guerin-Ancey, 1976), carp (Kaushik, 1980), rainbow trout (Kaushik, 1980; Beamish and Thomas, 1984; Alsop and Wood, 1997; see Fig. 1C), and Nile tilapia (Wright, 1993) where urea-N excretion was responsive to alterations in ration, dietary composition, and starvation. The discrepancy likely reflects differences in diets and/ or different metabolic pathways among species. For example, if excreted urea-N is produced from dietary arginine it may contribute to the *exogenous* fraction; if produced from uricolysis of nucleic acid and purine breakdown products associated with normal turnover, it will be mainly *endogenous*. However if uricolysis is activated in the face of environmental or metabolic disturbance (pH challenge, for example; see Section VI.A), this distinction may no longer be valid.

An ingenious study by Brown and Cameron (1991a,b) clarified the relationship between nitrogen metabolism and postprandial metabolic rate in fish. It is well known that in fish, as in other vertebrates, feeding is associated with an elevation of O_2 consumption—that is, metabolic rate—commonly called specific dynamic action (SDA) (Jobling, 1981a; Beamish and Trippel, 1990). Part of this is due to an increase in activity and general excitement associated with the presentation of food, as seen by Brett and Zala (1975) (see Fig. 2). However, even when this is corrected for by forcing the fish to swim continually while feeding (LeGrow and Beamish, 1986), the stimulation remains substantial—typically up to a doubling of metabolic rate. The cost of mechanical processing of the food appears to be relatively small (Tandler and Beamish, 1979; Jobling and Spencer Davies, 1980). In mammals, a protein meal is the most potent stimulator of SDA, reflecting the oxidative costs of deamination, transamination, and urea synthesis. However, it is the excretion of ammonia-N, rather than of urea-N, which is mostly stimulated by feeding in fish. Nevertheless, by analogy, it was originally assumed that the situation was similar in fish (e.g., Beamish, 1974; Knights, 1985); that is, SDA reflected the cost of ammonia production, and that was why SDA increased with the protein content of the meal (Jobling and Spencer Davies, 1980; LeGrow and Beamish, 1986). The innovative investigation by Brown and Cameron (1991a,b) on confined channel catfish put to rest any causative relationship between elevated ammonia-N excretion after feeding and elevated metabolic costs. Instead, their results provided confirmation of the original suggestion of Jobling (1981a)—that the cost of stimulated protein synthesis (i.e., growth) is the major cause of SDA. In brief, using direct intravascular infusions of essential amino acids through indwelling catheters (to avoid mechanical costs), these workers demonstrated that the net elevation in ammonia-N excretion accounted for only 21% of the amino-N load, whereas the amino acids were completely cleared from the bloodstream in a temporal pattern that paralleled SDA. By using the [^{3}H]-phenylalanine flooding dose technique, they demonstrated that the protein synthesis rate was massively stimulated in both muscle and liver at this time, explaining the disappearance of the amino acids, and that the elevated rate could explain at least 80% of the SDA. Furthermore, when fish were pretreated with the protein synthesis inhibitor cyclohexamide, ammonia-N excretion was unaffected, but SDA was completely eliminated.

These findings are in accord with others showing that protein synthesis rates are greatly stimulated by feeding in fish (Lied *et al.*, 1983; McMillan and Houlihan, 1988, 1989; Fauconneau *et al.*, 1989; Lyndon *et al.*, 1992; Houlihan *et al.*, 1995c). Plasma and intracellular free amino acid levels rise following feeding (Kaushik and Teles, 1985; Ash *et al.*, 1989; Brown and Cameron, 1991a; Espe *et al.*, 1993; Walsh and Milligan, 1995). This rise, together with accompanying hormonal signals, is a likely driver of the synthetic machinery. Estimates of exact costs vary, but there is general agreement that protein synthesis is a very expensive process—at least 4 ATP hydrolyzed per peptide bond from basic theory, more if amino acid transport processes and simultaneous breakdown ("futile cycling") are taken into account. This being the case, and considering the fact that fish normally grow continually throughout their lives, it is likely that the protein synthetic cost of growth is the largest component of their energy budget. This topic has been reviewed in detail by Houlihan *et al.* (1995a,b) and Carter and Houlihan (Chapter 2). Suffice it to note here that since structural protein is so costly to manufacture, it would seem to make sense to minimize its use as a metabolic fuel.

III. METABOLIC FUEL USAGE AND NITROGEN METABOLISM

Despite the preceding argument, historically it has been believed that protein is a major metabolic fuel in fish (e.g., Driedzic and Hochachka, 1978; van Waarde, 1983; Jobling, 1994), especially during exercise (Krueger *et al.*, 1968; van den Thillart, 1986; Davison, 1989; Weber and Haman, 1996). The exact origin of this idea is unclear—certainly the marked depletion of protein that occurs during the spawning migration of anorexic salmon (Idler and Clemens, 1959; Mommsen *et al.*, 1980) is one cogent piece of evidence, but this is an extreme situation, and indeed lipid is used up before protein. The dominance of protein as a fuel has been backed up by some *compositional* studies (e.g., Krueger *et al.*, 1968; Beamish *et al.*, 1989), but not others (e.g., Brett, 1973; Christiansen *et al.*, 1989). Brett (1995) has pointed out the many potential errors to the compositional approach;

additional ones are that it assumes that fuels depleted in fish sacrificed at various times during an exercise regime represent fuels burned, and not fuels interconverted or excreted (Bever *et al.*, 1981; Lauff and Wood, 1996a). Other lines of evidence come from measurements of a relatively high respiratory quotient (RQ) (e.g., Kutty, 1968; van den Thillart, 1986), combined with an often-stated but poorly founded assumption that since fish burn very little carbohydrate, a high RQ must indicate substantial protein oxidation (e.g., van den Thillart, 1986).

Is this true? Lauff and Wood (1996a,b, 1997) have developed an extension of the respirometric approach, termed the *instantaneous* method, which casts light on this question. The instantaneous approach adapts standard mammalian metabolic theory (Kleiber, 1992) to fish metabolism in order to monitor fuel oxidation nondestructively from simultaneous measurements of O_2 consumption, CO_2 production, and most importantly nitrogen waste excretion in the external water. The method assumes that metabolism is completely aerobic, and that steady-state conditions apply (i.e., waste products are not held back), assumptions that appear reasonable for fish at rest or swimming at submaximal velocities.

In brief, the ratio of measured nitrogen waste excretion (M_N) to O_2 consumption (M_{O2}) yields the nitrogen quotient (NQ) (M_N/M_{O2}); from knowledge of the typical metabolism of fish protein, an NQ = 0.27 represents the condition when aerobic respiration is fueled 100% by protein (van den Thillart and Kesbeke, 1978). Therefore, under any steady-state condition, the percent of metabolism fueled by protein is calculated as 100% $\times$ NQ/0.27. The measured partitioning of M_N between ammonia-N and urea-N can then be used to predict the RQ of this protein component from standard theory (usually around 0.94–0.97 in teleosts, rather than the 0.83 of mammals; van den Thillart and Kesbeke, 1978; Kleiber, 1992). Once the relative contribution of protein to total fuel use is known, the contribution of the protein RQ to the overall RQ can be subtracted. The known RQs for carbohydrate (1.00) and lipid (0.71) can then be used to factor out the remaining contributions of these two fuel sources, respectively. Note that since the true RQ of protein in fish is quite close to that of carbohydrate, measurements of NQ are essential if RQ is to be correctly interpreted (c.f. van den Thillart, 1986).

The results of the instantaneous analysis have been informative. First, in resting nonfed fish, lipid is the dominant fuel (35–68%) and the contribution of protein oxidation to overall aerobic metabolic rate is low: 14–30% in four studies on rainbow trout (Lauff and Wood, 1996a,b; Alsop and Wood, 1997; Kieffer *et al.*, 1998), and 16–30% in the Nile tilapia (Alsop *et al.*, 1999). Recalculation of data from four earlier studies indicates similarly low protein oxidation rates in sockeye salmon (19–36%; Brett and Zala, 1975), Atlantic salmon (26%; Wiggs *et al.*, 1989), plaice (27%; Jobling, 1980), and freshwater mullet (34%; Kutty and Peer Mohamed, 1975). The low protein contribution in salmonids tends to increase during progressive starvation, but protein never becomes the dominant fuel (Brett and Zala, 1975; Lauff and Wood, 1996a). Surprisingly, and contrary to popular belief, carbohydrate is the second most important fuel after lipid, and its contribution increases to a much greater extent during starvation (Lauff and Wood, 1996a). Note that reanalysis of the data of Kutty (1972) indicating >85% protein usage in the Mozambique tilapia, and Sukumaran and Kutty (1977) indicating >44% protein usage in the Madurai catfish portrays a very different pattern. Potential reasons for this discrepancy may be because only a short period of food deprivation (24–36 h) was used, so that the fish were still burning amino acids directly from the diet (see below) and/or because metabolism was partially anaerobic (c.f. van den Thillart and Kesbeke, 1978) under the conditions of the experiments as concluded by Kutty (1972).

Second, when fish are fed, the oxidation of protein clearly increases. For example, Kutty (1978) reanalyzed the data (Fig. 2) of Brett and Zala (1975) to illustrate the marked variation in protein oxidation over the feeding cycle. Translating to the assumptions of the instantaneous approach of Lauff and Wood (1996a,b), protein oxidation ranged from a low of 19% during the period of excitement prior to feeding to more than 90% during the peak of ammonia-N excretion 5 h after feeding, with an overnight plateau around 36%. Applying the same analysis to the data of Alsop and Wood (1997) on rainbow trout (Fig. 1), protein fueled 50–70% of metabolism in the satiation-fed fish, 25% in the fish on maintenance ration, and only 15% in the starved fish. Long-term feeding studies with rainbow trout similarly yielded values in the 50–85% range for satiation feeding regimes (Dockray *et al.,* 1996; Linton *et al.,* 1997, 1998a,b; D'Cruz *et al.,* 1998) and 25–50% for restricted ration regimes (Dockray *et al.,* 1996; D'Cruz *et al.,* 1998; Linton *et al.,* 1999). When fish eat a protein-rich meal, amino acids flood into the bloodstream, reaching a peak after a few hours (Brown and Cameron, 1991a; Espe *et al.,* 1993). Whatever is not needed for protein synthesis can be deaminated and oxidized in the citric acid cycle for the immediate provision of energy.

IV. EXERCISE AND NITROGEN METABOLISM

Does it make sense for an exercising fish to oxidize the very machinery (muscle structural protein) that is powering the swimming? In addressing this question, two types of exercise must be considered.

A. Sustainable Aerobic Swimming

This is the type of exercise that fish *normally* perform most of the time during their everyday lives. The instantaneous approach shows that when fish are made to swim at sustainable, submaximal velocities, the contribution of protein oxidation to overall fuel use stays the same, or in most trials actually *decreases* with increasing velocity. Measurements of M_N either remain unchanged or increase to a lesser extent than M_{O_2} in both trout (Lauff and Wood, 1996b, 1997; Kieffer et al., 1998) and tilapia (Alsop et al., 1999) (Fig. 3). Similarly, reanalysis of published data suggests a fall in the contribution of protein oxidation during spontaneous swimming in both freshwater mullet (Kutty and Peer Mohamed, 1975) and Atlantic salmon (Wiggs et al., 1989). As velocity is increased, lipid remains the dominant fuel, but the relative contribution of carbohydrate increases, just as during starvation (Lauff and Wood, 1996b; Kieffer et al., 1998). In accord with these observations, swimming at low speeds has been shown to stimulate protein synthesis rates to a greater extent than protein degradation rates in muscle tissues of trout, so that net protein accretion rates are improved (Houlihan and Laurent, 1987). In turn, these observations agree with a large literature demonstrating that continuous low-speed exercise improves growth, and particularly growth in protein content, in most though not all studies (reviewed by Davison, 1989, 1997). In fasted Arctic char, short-term low-speed swimming elevated the plasma concentrations of selected amino acids (Barton et al., 1995), a signal that could help activate protein synthesis. However, it is not known whether this also occurs in fed fish, such as the Arctic char, which grew faster when continuously exercised (Christiansen et al., 1989; Christiansen and Jobling (1990).

An interesting feature of the response to aerobic exercise is the fact that urea-N excretion increases to a greater relative extent than ammonia-N excretion as velocity rises (Fig. 3) (Lauff and Wood, 1996b; Alsop and Wood, 1997; Alsop et al., 1999). The cause of this phenomenon is unknown. Lauff and Wood (1996b) suggested that it could reflect greater recruitment of white muscle fibers at higher swimming speeds and accompanying adenylate turnover, a by-product of which is urea produced by uricolysis. An alternative proposal by Anderson (Chapter 7) is that increased urea-N is produced via an inter-organ ornithine–urea cycle that serves to detoxify ammonia-N resulting from adenylate breakdown in white muscle. Regardless, if either explanation is correct, then this portion of urea-N excretion should have been excluded from the fuel use calculations because it was derived from adenylates, not amino acids, meaning that true protein utilization during exercise is even lower.

Another consistent finding from the instantaneous approach is that the relative contribution of protein oxidation appears to increase with the duration of sustained swimming (Fig. 3D), manifested as either an absolute increase of ammonia-N excretion (Kutty, 1972; Sukumaran and Kutty, 1977; Alsop et al., 1999) or its constancy in the face of declining M_{O_2} with time (Lauff and Wood, 1996b, 1997). At least in part, this may reflect the effects of starvation during the exercise test (see above) as much as it does prolonged exercise. Thus, when trout were trained for 2 weeks (during which time they were fed daily), and subsequently swim-tested, they exhibited substantially lower values of M_N, much lower rates of protein oxidation, and higher rates of lipid utilization during exercise

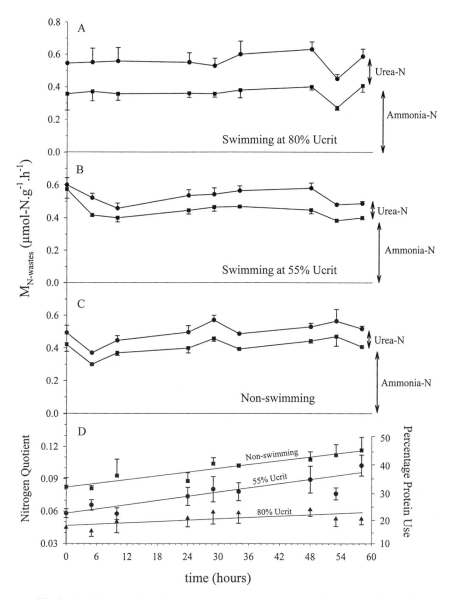

Fig. 3. The influence of aerobic, sustainable exercise on ammonia-N and urea-N excretion at 15°C in juvenile rainbow trout swimming in individual respirometers over a 58-h period. Fish (~ 17 g, 11 cm) were swum at speeds corresponding to (A) 80% or (B) 55% of their maximum sustainable velocity (U_{crit} = 3.84 body lengths·s^{-1}), or (C) were left in a mild current that would not induce swimming. Feeding was suspended 48 h prior to test. Panel (D) shows the percentage use of protein in oxidative metabolism, as calculated from the nitrogen quotient. Note the general lack of change of ammonia-N excretion with speed or time, but the higher urea-N excretion at the highest speed. Means ± SEM. (Data from Lauff and Wood, 1996b. Reproduced with the permission of Springer-Verlag GmBH & Co. Ltd.)

(Lauff and Wood, 1997). It would appear that training actually reorganizes metabolism so as to spare the muscle, with the result that protein growth is promoted over lipid growth in fish that are both fed and continually swum (Christiansen *et al.*, 1989; Lauff and Wood, 1997).

This raises the question of key importance: What happens to nitrogen waste excretion when fish are feeding *during exercise,* which is of course the normal behavior that fish perform naturally? In other words, is the *exogenous* fraction of nitrogen excretion, the portion not retained as growth from the diet, larger, lower, or the same in fish that are swimming during feeding than in fish that are not swimming but are fed the same ration? To my knowledge, the definitive experiment has not been done, for the simple reason that fish tend to become spontaneously more active anyway during feeding, whereas fish that are forced to swim against a current appear to decrease their spontaneous or "nonspecific" activity (e.g., LeGrow and Beamish, 1986; Christiansen and Jobling, 1990), that is, the actual degree of exercise cannot be perfectly controlled. However, in view of the convincing literature that net protein accretion rates, protein conversion efficiency from the diet, and overall growth rates are all improved if fish are continually swimming (Houlihan and Laurent, 1987; Christiansen *et al.,* 1989; Davison, 1989, 1997), it seems likely that the *exogenous* fraction of nitrogen excretion will be lower if fish swim aerobically while feeding; that is, amino acids from the free pool will be funneled preferentially toward protein synthesis, rather than toward deamination and oxidation. Because feeding seems to preferentially elevate the concentrations of essential amino acids in the blood plasma (Brown and Cameron, 1991a; Espe *et al.,* 1993), whereas swimming preferentially elevates nonessential amino acids (Barton *et al.,* 1995), it may be that the combination is most effective in stimulating protein synthesis. Interestingly, the SDA effect of feeding continues unabated (Beamish, 1974; Alsop and Wood, 1997) or may even increase (Muir and Niimi, 1972; Blaikie and Kerr, 1996) during submaximal exercise. Because SDA mainly represents the cost of elevated protein synthesis (see above), this indicates that carbohydrate or lipid are used to a greater extent, not only to power exercise itself, but to power protein synthesis during exercise.

B. Anaerobic Burst Swimming

This type of exercise, while occasionally critical to survival, probably represents only a small portion of the daily energy budget of most fish. However, exhaustive exercise (usually by chasing) has become a favorite model for metabolic and acid-base studies; Wood and Wang (1999) note that more than 200 papers have been written on exhaustively exercised trout alone, and their review provides a guide to recent studies and earlier reviews. For nitrogen metabolism, the key event is a massive and rapid generation of ammonia-N by the white muscle. Almost all of this results from the deamination of adenylates through one arm of the purine nucleotide cycle; less than 2% is attributable to the depletion of aspartate (Mommsen and Hochachka, 1988). ATP stores are degraded through ADP and the adenylate kinase reaction to AMP, which in turn is deaminated via AMP deaminase, resulting in stoichiometrically equivalent increases in inosine monophosphate (IMP) and ammonia in the intracellular compartment of the muscle (Fig. 4A). Because the intracellular compartment of the white muscle accounts for about 35% of the fish's body mass, and typical increases in intracellular concentrations of ammonia are in the range of 5–10 mmol·liter^{-1} (Wood, 1988; Wright et al., 1988; Mommsen and Hochachka, 1988; Wang et al., 1994, 1996), the load is considerable, around 2500 μmol·kg^{-1} of total body mass. Plasma ammonia levels do increase by as much as fivefold, but on an absolute basis this is a negligible elevation (<200 μmol·liter^{-1}) relative to intracellular levels (Fig. 4A). At least in adult trout, only a modest increase occurs in ammonia-N excretion to the water through the gills amounting to about an extra 200 μmol·kg^{-1}·h^{-1} during the next 4 h (Fig. 4B), and this probably helps clear excess H$^+$ ions from the blood (Wood, 1988). The time of elevated branchial excretion corresponds to the period during which most of the ammonia is cleared from the muscle, but quantitatively there is a large discrepancy (Fig. 4). Experiments with a perfused tail–trunk preparation have shown that ammonia release rates from muscle to plasma really are low (Wang et al., 1996, 1998) and correspond to the measured excretion rates to the water. The obvious conclusion is that the majority of the ammonia generated is held back inside the muscle cells and removed *in situ,* rather than excreted.

This retention has several obvious advantages: It provides an effective base (NH$_3$) to buffer H$^+$ ions produced (with lactate and pyruvate) by glycolysis (Dobson and Hochachka, 1987), it serves to activate phosphofructokinase (Su and Storey, 1994), thereby maintaining glycolytic flux (Mommsen and Hochachka, 1988), and, most importantly, it provides a ready source of ammonia-N for resynthesis of ATP. During recovery, AMP deaminase is shut down, while the other arm of the purine nucleotide cycle (adenylsuccinate synthetase and adenylsuccinate lyase) effectively removes IMP and ammonia, and replenishes the adenylate pool (Mommsen and Hochachka, 1988). Note the inverse symmetry between the removal of IMP and ammonia, and the regeneration of ATP during recovery (Wang et al., 1994; Fig. 4A). Ammonia is actually removed by resynthesis of glutamate from α-ketoglutarate. Glutamate in turn reacts with oxaloacetate (from malate and fumarate) to furnish the required aspartate as the nitrogen donor for adenylate resynthesis, thereby regenerating α-ketoglutarate. It would be interesting to know if there was also excess urea-N production at this time of increased adenylate turnover in white muscle (perhaps via uricolysis, or an interorgan ornithine–urea cycle; see above), but the measurements do not appear to have been made.

Given that ammonia is often considered a highly diffusive molecule, how is it retained in muscle tissue? The answer is discussed in detail by Wood and Wang

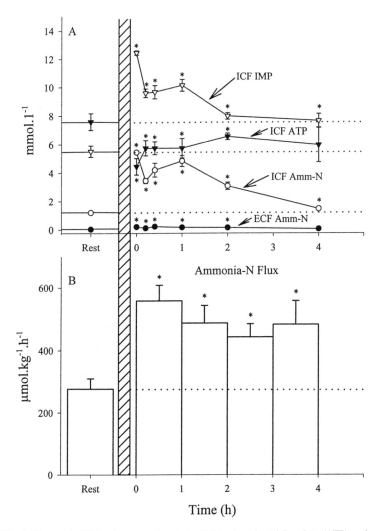

Fig. 4. Changes in (A) inosine monophosphate (IMP), adenosine triphosphate (ATP), and ammonia-N in the intracellular fluid (ICF) of white muscle and ammonia-N in extracellular fluid (ECF = blood plasma); and (B) branchial ammonia-N excretion (M_{Amm-N}) at 15°C in adult rainbow trout (~250 g) following 6 min of exhaustive anaerobic burst exercise. Note the reciprocal changes in ICF ammonia-N and IMP versus ATP, and the much larger changes in ICF ammonia-N than ECF ammonia-N. Note also the small absolute rates of branchial ammonia-N excretion relative to ICF ammonia-N. Means ± SEM. Asterisks indicate significant difference from value at rest. [Data from (A) Wang et al. 1994, and (B) Wood, 1988. Both reproduced with the permission of the Company of Biologists Ltd.]

(1999). Very simply, the retention can be explained by passive phenomena, as originally suggested by Wright et al. (1988) and Wright and Wood (1988), a theory that was initially quite controversial (c.f. Heisler, 1990). There is no evidence for active retention. The theory derives from physicochemical principles first articulated by Boron and Roos (1976) for the distribution of weak bases where there is significant permeability to the charged form. Ammonia (as NH_4^+) thereby distributes as an ion via the Nernst relationship according to the muscle membrane potential (-85 to -102 mV), rather than as a weak base via the Jacobs-Stewart relationship (Jacobs and Stewart, 1936) according to the pH gradient (0.4–0.8 units), because the NH_4^+ to NH_3 permeability ratio (pNH_4^+/pNH_3) of the muscle cell membrane is relatively high (Wood and Wang, 1999). This means that at passive equilibrium, intracellular total ammonia concentration will be about 35 times higher than extracellular ammonia, rather than about 5 times higher if only the pH gradient were involved. The theory has been confirmed by direct measurements of ammonia distribution ratios and pH gradients *in vivo* in exercised fish (Wright et al., 1988; Wright and Wood, 1988; Tang et al., 1992; Wang et al., 1994), together with determinations of muscle membrane potential *in situ* (Beaumont et al., 2000), and experimental manipulations of both it and transmembrane pH gradients in a perfused tail–trunk preparation (Wang et al., 1996). Indeed, pNH_4^+/pNH_3 appears to actually increase after exhaustive exercise, thereby favoring passive retention (Wang et al., 1994, 1996). This ability to store and tolerate high concentrations of ammonia in the muscle mass may also be of survival value in situations where branchial ammonia excretion is inhibited such as high external pH (Wilkie and Wood, 1995) and copper intoxication (Beaumont et al., 2000), though contractility and swimming performance appear to be negatively impacted (Randall and Brauner, 1991; Beaumont et al., 2000).

Although the above scenario attributes the retained ammonia plus the small excess ammonia-N excretion after exhaustive exercise to adenylate deamination, note that in at least one study on very small trout, excess ammonia-N excretion was much greater, and exceeded ATP depletion by several-fold (Scarabello et al., 1992). In that experiment, electrical stimulation was employed, which may have elevated the stress component and, therefore, cortisol mobilization, as well as possibly depolarizing the membrane potential of muscle so as to favor ammonia washout. Milligan (1997) has shown that even in adult trout exercised to exhaustion without shocks, there is a substantial (up to threefold) but selective mobilization of free amino acids (only 6 of 22 individual amino acids increased) in the blood plasma, white muscle, and liver for up to 4 h after exercise. In the muscle, part of the elevation appeared to occur actually during the exercise itself (see also Storey, 1991). Pharmacological blockade of cortisol elevation completely eliminated this amino acid surge in all compartments, which Milligan (1997) attributed to stimulation of proteolysis in the liver by cortisol. It is conceivable that some of these mobilized amino acids may be deaminated and oxidized to help fuel glycogen resynthesis (and thereby contribute to excess ammonia-N excretion after exercise), while others are directed toward protein synthesis for muscle repair.

V. TEMPERATURE AND NITROGEN METABOLISM

Apart from food availability, daily and seasonal variations in temperature are probably the most important *normal* influence of the environment on nitrogen metabolism in fish. However, in nature it is likely that feeding and temperature are dependent rather than independent variables. Nevertheless, we will start with the simpler situation, those few studies where nitrogen waste excretion has been measured at different temperatures in the absence of feeding.

A. Nonfed Fish

In general, such studies have demonstrated that nitrogen waste excretion is highly sensitive to temperature, much more so than is O_2 consumption, which generally has a Q_{10} less than 2.0. Maetz (1972) reported a Q_{10} of 4.0 for ammonia-N excretion in goldfish abruptly transferred from 16°C to 6°C, but the Q_{10} dropped to 1.9 when the fish were loaded with a great excess of ammonia. Maetz interpreted this to mean that the majority of the acute temperature sensitivity was in the metabolic production mechanism, rather than in the branchial excretion mechanism. In long-term acclimated rainbow trout (Kieffer *et al.*, 1998), the Q_{10} values for ammonia-N excretion between 5°C and 15°C were lower (1.4 to 2.9 in trout at different sustainable exercise levels), but still above those for M_{O2} (1.3–1.5). The Q_{10} remained high even after exhaustive exercise (Kieffer and Tufts, 1996). Similarly, in long-term acclimated Nile tilapia between 15 and 30°C, the Q_{10} for ammonia-N excretion was 2.8 versus 1.7 for O_2 consumption (Alsop *et al.*, 1999). Q_{10} values for urea-N excretion were comparable to those for ammonia-N excretion. Comparably high sensitivities of ammonia-N excretion to temperature have been reported in larval carp (Kaushik *et al.*, 1982), various sturgeon species (Gershanovich and Pototskij, 1996), the plaice (Jobling, 1981b), the surf steenbras (Cockcroft and Du Preez, 1989), and in the mangrove snapper and areolated grouper (Leung *et al.*, 1999). The overall conclusion is that as temperature increases, an increasing percentage of aerobic metabolism in nonfed fish is fueled by the oxidation of protein.

B. Actively Feeding Fish

Here, the situation is much more complex because net protein accretion and nitrogen storage are also occurring. Protein synthesis increases with temperature (e.g., Fauconneau and Arnal, 1985; Loughna and Goldspink, 1985; Watt et al., 1988). A recent meta-analysis of the literature by McCarthy and Houlihan (1997) indicated that white muscle and whole-body protein synthesis rates actually rise in an exponential fashion as temperature increases, and this conclusion has now been reinforced by an experimental study on a single species, the marine wolffish, fed to satiation at four different acclimation temperatures (McCarthy et al., 1999). Nevertheless, it is likely that this overall relationship disguises some important variation. For example, protein synthesis rates exhibit considerable thermal acclimation, such that the acute temperature-dependent relationship is displaced upward by cold acclimation and vice versa (Loughna and Goldspink, 1985; Watt et al., 1988). Furthermore when food is restricted or withheld entirely, such as in studies by Foster et al. (1992) on adult cod and Mathers et al. (1993) on rainbow trout fry, the increase in protein synthesis with temperature is either greatly attenuated or does not occur.

The interaction of temperature and ration is well illustrated by our recent work on the long-term responses of rainbow trout to global warming, where climate change has been simulated by adding 2°C to the natural annual thermal cycle for inshore Lake Ontario, Canada. Protein turnover rates proved to be surprisingly sensitive to this small (2°C) temperature difference, increasing on average by 10–30% in gill and liver in four different 3-month seasonal exposures during which trout were fed to satiation (Reid et al., 1995, 1997, 1998; Morgan et al., 1998). Absolute rates were generally lower in the cold winter months (4–10°C) than in the hot summer months (13–24°C), but the stimulation caused by +2°C was much greater in winter. Overall, these increases in protein synthesis, degradation, and net accretion correlated well with similar increases in food consumption. When ration was restricted in one summer study, protein turnover and net accretion rates decreased, rather than increased, in response to a chronic 2°C elevation (Morgan et al., 1999). An additional complexity is that even in the presence of unlimited ration, once a certain critical optimum temperature for protein synthesis is exceeded, then protein accretion rates drop precipitously, largely through an increase in protein degradation rates (Reid et al., 1995, 1997, 1998). The optimal temperature for net protein accretion may be higher than the optimal temperature for net growth because of differing optima for food consumption, protein synthesis, and protein degradation rates (McCarthy et al., 1999).

These global warming simulation studies with rainbow trout appear to be the only temperature investigations in which M_N and M_{O2} measurements have been performed in parallel to protein turnover and feeding measurements. In general, they indicate that nitrogenous waste excretion increases in concert with increases in feeding, protein synthesis, and overall metabolic rate as temperature rizes. The majority of these changes occur in the dominant ammonia-N fraction, rather than in the urea-N fraction. The stimulatory effect of slow seasonal temperature rise on all of these metabolic functions is much greater in the cold winter months than the

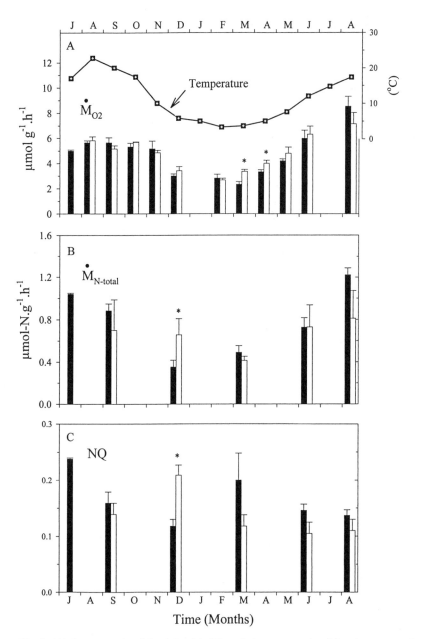

Fig. 5. (A) O₂ consumption (M_{O2}; left axis); (B) total nitrogen excretion ($M_{N\text{-total}}$) as measured with a nitrogen oxidizer; and (C) Nitrogen quotient (NQ) in rainbow trout over a 14-month period during which the fish were exposed to the natural thermal regime of inshore Lake Ontario (solid bars)

warm summer months (Dockray *et al.*, 1996; Linton *et al.*, 1997, 1998a,b). Similarly the stimulatory effect of +2°C on metabolic functions is much greater in the winter, and in both winter and summer it is eliminated when ration is restricted (Dockray *et al.*, 1998; D'Cruz *et al.*, 1998; Linton *et al.*, 1999). In winter, increases in M_N keep pace with or exceed increases in M_{O2}, so the percent of aerobic metabolism fueled by the oxidation of protein stays constant or increases as temperature rises (Linton *et al.*, 1998b; D'Cruz *et al.*, 1998), similar to the pattern discussed earlier for nonfed fish (Kieffer *et al.*, 1998; Alsop *et al.*, 1999). However, in summer, the contribution of protein to aerobic respiration tends to decrease as temperature rises in actively feeding trout (Linton *et al.*, 1997). When protein synthesis drops at critically high temperatures, both nitrogenous waste excretion and metabolic rate fall simultaneously, with the former effect being more marked, so protein oxidation decreases even further (Dockray *et al.*, 1996; Linton *et al.*, 1997). These patterns are well illustrated by periodic measurements of M_{O2} and M_N in the 14-month study of Linton *et al.* (1998a) with satiation-fed trout, a period during which the fish increased their original body mass by 35-fold (Fig. 5)! Note, however, that the "critical high temperature" was lower in the second year than the first, perhaps due to differences in age, size, or thermal history (Fig. 5).

VI. OTHER NITROGEN PRODUCTS?

A. Urea versus Ammonia

A handful of unusual teleosts, such as the Lake Magadi tilapia (Randall *et al.*, 1989; Wood *et al.*, 1989), the gulf toadfish (Mommsen and Walsh, 1989; Walsh, 1997b), and several species of catfish of the Indian subcontinent (Saha and Ratha, 1998), excrete urea as their major nitrogen waste either all the time or under certain environmental conditions. In these cases, urea-N production is attributable to a full expression of the enzymes of the ornithine–urea cycle (OUC) in their liver and sometimes other tissues; details are reviewed by Anderson (Chapter 7) and Ip *et al.* (Chapter 4). The Lahontan cutthroat trout and several other species endemic

---

or this natural regime + 2°C (open bars). The natural thermal regime is shown on the right axis in panel (A). The exposure started in early July (J) 1995 and ended in late August (A) 1996. The fish were fed to satiation once per day. Over the experimental period, the fish grew from from ~11 to ~365 g, so the data are scaled to 1 kg using the mass exponent 0.824 determined by Cho (1990). Note the marked changes in absolute rates with the temperature, the stimulatory effect of + 2°C on $M_{N-total}$ and NQ in winter (December) and the inhibitory effect in summer, especially at peak water temperatures (August). Means ± SEM of 4 replicate tanks for each treatment, each containing approximately 80 fish at the start and 40 at the end of the experiment (reduction due to periodic sampling). Asterisks indicate a significant difference due to + 2°C. (Data from Linton *et al.*, 1998a. Reproduced with permission of the Minister of Public Works and Government Services Canada, 2001.)

to alkaline (pH 9.4) Pyramid Lake, Nevada, USA (Wright et al., 1993; McGeer et al., 1994), and a cyprinid endemic to alkaline (pH 9.8) Lake Van, Turkey (Danulat and Kempe, 1992), excrete urea-N at higher rates than most teleosts,, with corresponding reductions in ammonia-N excretion, but ammonia-N still predominates. Available evidence suggests that these fish produce urea via uricolysis or arginolysis for which the enzymes are present; one or more key enzymes of the OUC are absent.

The focus here, however, is on more representative teleosts living in nonextreme environments. There is general accord that the vast majority of such fish excrete predominantly ammonia-N plus lesser amounts of urea-N, the latter again formed by uricolysis or arginolysis. Wood (1993), Handy and Poxton (1993), and Jobling (1994), in quantitatively reviewing a large number of studies, all noted that there appears to be a general tendency for greater urea-N excretion rates in marine teleosts (up to 40% of ammonia-N) than in freshwater teleosts (5–35%). The reason is unknown; perhaps it allows marine fish to use urea as a minor osmolyte, as suggested by the results of Wright et al. (1995b). When challenged with environments that tend to inhibit ammonia-N excretion across the gills (e.g., high external ammonia, high pH), and/or internal ammonia accumulation and alkalosis, many "standard" teleosts (goldfish, various salmonids, killifish, Nile tilapia) exhibit an increase in urea-N excretion, which has usually been attributed to an activation of the enzymes of uricolysis as a mechanism for detoxification of ammonia (Olson and Fromm, 1971; Wood et al., 1989; Wilkie and Wood, 1991, 1996; Wilkie et al., 1993, 1994; Wright, 1993; McKenzie et al., 1999; Patrick and Wood, 1999). However, such fish eventually reestablish a pattern where ammonia-N excretion predominates. Even in the adult largemouth bass, which is unusual in expressing a full complement of the OUC enzymes in its liver, ammonia-N excretion normally predominates, and exposure to high external ammonia evokes only a transient increase in urea-N excretion (Kong et al., 1998).

Conflicting with this general pattern of ammoniotelism, it is now well established that at least two "standard" teleosts (the rainbow trout and the Atlantic cod) living in circumneutral environments produce urea-N during their embryonic stages, apparently via a full expression of the OUC enzymes at this time (Wright et al., 1995a; Korte et al., 1997; Wright and Land, 1998; Chadwick and Wright, 1999). The capacity appears to be lost at some point after hatch, and some of the OUC enzymes, particularly hepatic carbomyl phosphate synthetase III, are no longer present in juveniles and adults (see Chapter 5). However, it is not inconceivable that under certain conditions (e.g., continuous dietary nitrogen loading by intensive feeding), the OUC could continue to function and promote ureotelism in juvenile and adult life in "standard" teleosts such as salmonids. In this regard, there have been several surprising reports from hatchery studies where the effluents from large outdoor ponds containing a known biomass of intensively fed salmonids were analyzed diurnally and seasonally. In one study on chinook salmon fingerlings (Burrows, 1964), the nitrogenous composition of the effluent changed from almost exclusively urea-N to largely ammonia-N as loading density and temperature increased over the summer. In another study on juvenile coho salmon (McLean and Fraser, 1974), urea-N contributed anywhere from 0 to 78% of the sum of ammonia-N plus urea-N excretion on different days, and at different times of individual days. Urea-N excretion predominated on days of high light intensity, whereas higher densities and temperatures were again associated with a predominance of ammonia-N excretion. These hatchery observations have never been replicated. They appear to have been very carefully performed, but controls for urea production by microbial activity were not carried out. Large-scale interconversion of ammonia to urea (or vice versa) by microbial activity in the ponds seems unlikely but not impossible.

Recently we had the opportunity to perform a similar study at a hatchery, monitoring the outflows from outdoor tanks that each contained ~10,000 juvenile lake char or lake char-brook char hybrids fed by "demand" feeders (M. P. Wilkie, Y. Wang, and C. M. Wood, unpublished results). Both ammonia-N and urea-N excretion underwent dramatic fluctuations during the day, but urea-N excretion was at maximum 35% of the total, and averaged less than 10% overall. However because of cool spring temperatures, daily ration was low ($\sim 0.7\% \cdot day^{-1}$) and nitrogen excretion rates in these char were less than 20% of those reported by Burrows (1964) and McLean and Fraser (1974). Because so few studies have yet been performed on intensively feeding fish under "real-world" conditions, it is important to keep an open mind on this question of possible ureotelism in juvenile and adult teleosts.

B. Other Compounds

An equally important issue is the possibility that nitrogen products *other than* ammonia-N and urea-N might be excreted in amounts that are quantitatively significant. Most workers would probably wish to avoid this issue entirely, because simple spectrophotometric assays exist for ammonia-N and urea-N that are sensitive, rapid, and economical; the same is not true for the tedious, more expensive procedures for measuring total nitrogen and the variety of possible other compounds that could contribute to the total. Nevertheless, there are two reasons to suspect that other nitrogen products may be excreted. The first is the marked discrepancy observed between protein oxidation measured respirometrically by the instantaneous approach from the excretion of ammonia-N and urea-N (see Section III) and the much larger amount estimated from the compositional approach by protein depletion from the carcass (Lauff and Wood, 1996a). This discrepancy applies not only to the loss of protein nitrogen, but also to the total loss of calories from the carcass, which may greatly exceed that estimated from O_2 consumption integrated over the test period (Krueger *et al.*, 1968; Brett, 1995; Lauff and Wood,

1996a). The second line of evidence comes from those few studies (all in the older literature!) that have actually measured total nitrogen excretion (by Kjeldahl digestion or a nitrogen oxidizer) and compared it directly to the sum of ammonia-N plus urea-N excretion. In the Pacific staghorn sculpin (Wood, 1958) and two species of anchovies (McCarthy and Whitledge, 1972), 13%–19% of total nitrogen excretion remained unaccounted for, even after small measured excretions of creatine/creatinine-N and trimethylamine/ trimethylamine oxide-N were subtracted. Other reported discrepancies were 23–31% in two carp specimens (Smith, 1929), 12% in the surf steenbras (Cockcroft and Du Preez, 1989), 23–32% in Atlantic salmon and Arctic char (Fivelstad et al., 1990), and 24% in the rainbow trout (Olson and Fromm, 1971). The latter study identified virtually all of this discrepancy (23%) as the excretion of protein nitrogen. Feeding was not controlled in these studies, but Beamish and Thomas (1984) reported data indicating that the discrepancy varied on different dietary regimes in rainbow trout.

Recently, we have started to measure total nitrogen excretion in our metabolic work with trout (G. DeBoeck, D. Alsop, and C. M. Wood, unpublished) and have invariably found that it exceeds the sum of ammonia-N plus urea-N excretion (Fig. 6). The discrepancy is highly variable, but it tends to be greater (sometimes

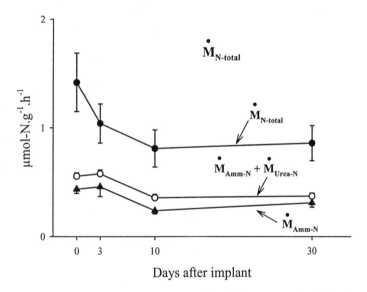

Fig. 6. A comparison of direct measurements of total nitrogen excretion ($M_{N\text{-total}}$) as determined with a nitrogen oxidizer versus the sum of ammonia-N excretion ($M_{Amm\text{-}N}$) and urea-N excretion ($M_{Urea\text{-}N}$) at 14°C in rainbow trout (10–20 g) over a 30-day treatment following placement of a coconut oil implant in the peritoneal cavity. This was a sham treatment in an endocrine study; the elevation in rates at days 0–3 was probably due to the disturbance of the initial surgery. The fish were fed a daily ration corresponding to approximately half of satiation. Means ± 1 SEM. (Unpublished data of G. DeBoeck, D. Alsop, and C. M. Wood.)

greater than twofold the sum of ammonia-N plus urea-N) in actively feeding fish, and declines with starvation. An identical pattern has also been seen in the Pacific midshipman (Walsh et al., 2001). This finding, of course, helps reconcile the discrepancies in the instantaneous versus the compositional approaches to protein oxidation, but raises numerous important questions. What is the unmeasured nitrogen product? How is it excreted? Should it be considered a nitrogenous product of protein oxidation? The latter is perhaps the most important, because if it really is a product of aerobic respiration, then most of what is written about the generally low reliance on protein as a metabolic fuel in Section III may well be wrong! For example, in the data set of Fig. 6 (where O_2 consumption was measured simultaneously), protein oxidation would account for only about 22% of metabolism based on ammonia-N plus urea-N excretion, but around 50% based on total nitrogen excretion.

Our present working hypothesis is that the original findings of Olson and Fromm (1971) are correct, and that most of the discrepancy represents protein nitrogen (or amino acid nitrogen) excretion—that is, nonoxidized moieties. If this is the case, then we are correct in not using this "extra nitrogen" in fuel use calculations, but it nevertheless represents an important vehicle for both calorie and nitrogen losses. Amino acids could be lost by "leakage" across the gills as plasma levels surge after feeding (Brown and Cameron, 1991a; Espe et al., 1993), together with metabolic rate, gill blood flow, and perhaps gill permeability to amino acids. Nevertheless, it makes little teleological sense to waste such a valuable energy source in this fashion. More likely, the major portion of the unidentified nitrogen reflects loss of mucoprotein, an unavoidable cost of an aquatic lifestyle. In the kelp bass, Bever et al. (1981) found that a significant portion of the [^{14}C]radioactivity from labeled amino acids injected into the bloodstream quickly appeared in body mucus. In rainbow trout, Miller and MacKay (1982) reported net mucous excretion rates to the water under control conditions averaging an incredible 4 $g \cdot kg^{-1} \cdot h^{-1}$, which increased by about 60% after exposure to pH 3.6–4.0. Given any reasonable estimate of nitrogen content for mucus, these mucoprotein excretion rates would dwarf the sum of normal ammonia-N plus urea-N excretion rates. Clearly, more work is needed in this potentially important area.

VII. EXCRETION MECHANISMS

The majority of ammonia-N and urea-N is excreted across the gills rather than through the kidney (see Wood, 1993, for a quantitative summary of many studies); the skin may also play some small role, especially in marine teleosts. The "unmeasured nitrogen" excretion (see Section VI) also appears to be via the gills and/or skin, at least based on the original bladder catheterization experiments of Smith (1929). On a relative basis, ammonia-N excretion through the kidney may increase greatly in response to metabolic acidosis, largely by tubular secretion (Wood *et al.*, 1999), but this remains a small percentage of the overall total. Renal excretion of urea-N appears to be minimized by active tubular reabsorption, at least in the freshwater rainbow trout (McDonald and Wood, 1998). Branchial excretion of urea-N has traditionally been considered to occur by simple diffusion. However, in light of recent findings of a facilitated diffusion transporter for urea-N in the gills of the ureotelic gulf toadfish, it is possible that such transporters might also occur in the gills of a wide variety of "standard" ammoniotelic teleosts (see Chapter 8 and Walsh *et al.*, 2001). Certainly, urea-N excretion across the gills of rainbow trout can be greatly accelerated in response to exogenous loading, thereby maintaining homeostasis of plasma urea-N concentrations (McDonald and Wood, 1998). Interestingly, there is one report of urea-N concentration in the chloride cells of gills of the European eel in seawater (Masoni and Garcia-Romeu, 1972).

The area of greatest research focus and controversy has been the mechanism(s) of ammonia-N excretion across the gills. The literature and arguments surrounding this issue have been reviewed in detail by Walsh (1997a) and Wilkie (1997), and will not be revisited here. In seawater, the situation remains unclear: apical Na^+/NH_4^+ exchange, passive transcellular NH_3 diffusion and paracellular NH_4^+ diffusion, and substitution of NH_4^+ for K^+ on basolateral $Na^+,K^+/ATPase$, and $Na^+,K^+/2Cl^-$ cotransporters may all be important. However, in freshwater, it now appears that a consensus may be emerging. Wilkie (1997) concludes: "In freshwater, ammonia excretion likely takes place via passive NH_3 diffusion down favourable blood-gill water P_{NH3} gradients—facilitated by acidification of the expired gill water—by carbonic anhydrase-catalyzed hydration of CO_2. The likely absence of apical electroneutral Na^+/H^+ exchange probably rules out a role for (apical) Na^+/NH_4^+ exchange—and the low cationic permeability of freshwater gills makes significant diffusion of NH_4^+ unlikely." In light of most available evidence, this conclusion appears very reasonable. However, a note of caution must be sounded. Virtually all the data on which this is based comes from starved, quiescent fish, confined in boxes. Plasma total ammonia levels in such fish are usually very low, often less than 100 $\mu mol \cdot liter^{-1}$ when obtained by chronic cannulation (see Wood, 1993, for a critique of methods). Unfortunately, there appear to be no chronic cannulation data for actively feeding fish, but plasma ammonia concentrations obtained by caudal puncture rise dramatically after a meal (Kaushik and Teles, 1985), and may be over 500 $\mu mol \cdot liter^{-1}$ in satiation-fed trout (Linton *et al.*, 1997). Even in starved trout (cannulated), plasma ammonia levels may reach 200–400 $\mu mol \cdot liter^{-1}$ after intensive exercise (Wood, 1988; Wright and Wood, 1988; Wang *et al.*, 1994). Heisler (1990) reported that the excellent relationship between the blood-to-water P_{NH3} gradient and net branchial ammonia-N flux in cannulated trout broke down when plasma total ammonia levels exceeded 200 $\mu mol \cdot liter^{-1}$, and suggested that a threshold (i.e., $>200 \mu mol \cdot liter^{-1}$) was passed for activation of an outward transport mechanism, additional to the role of simple diffusion.

There are numerous reports in the literature of "apparent" Na^+/NH_4^+ exchange; for example, in one study (Wright and Wood, 1985; further analyzed by Wood, 1989), Na^+ uptake and ammonia excretion were linearly related once the diffusive movement of NH_3 was subtracted. It is also known that raising blood total ammonia levels by infusion of ammonium salts markedly stimulates Na^+ uptake in approximate proportion to the increase in net ammonia excretion (e.g., Wilson et al., 1994). Although this was originally explained as a side effect of the accompanying acidosis (increased H^+-linked Na^+ uptake, and associated diffusion trapping of NH_3 in the gill water by elevated H^+ excretion; Wilson et al., 1994), recent experiments have shown that the same phenomenon occurs even when NH_4HCO_3 is infused, which induces alkalosis rather than acidosis (Salama et al., 1999). The Na^+ uptake/ammonia-N excretion linkage appears to be direct but loose, because pharmacological blockade of Na^+ uptake only slightly reduces ammonia-N excretion under these conditions (Wilson et al., 1994). Similarly, experimentally varying Na^+ uptake rate by rapidly changing water Na^+ levels has only a minor influence on net ammonia-N flux (Salama et al., 1999). In trout, Na^+ uptake is stimulated during sustained exercise (Postlethwaite and McDonald, 1995) and after exhaustive exercise (Wood, 1988). It would be interesting to know if the same phenomenon occurs after feeding. Some sort of mechanistic linkage between active Na^+ uptake and ammonia-N excretion across the gills remains a real possibility under natural conditions of ammonia loading from feeding and/or exercise.

VIII. ELASMOBRANCHS

The elasmobranchs are often viewed as a parallel evolutionary line to the teleosts, and there has been much theoretical interest in the comparison of their adaptive strategies relative to those of teleosts. For example, Kirschner (1993) concluded that the energetic efficiency of their evolutionary "choice" to retain urea for osmoregulation (Smith, 1936) so as to avoid drinking seawater was about equal to that of the teleost "choice" to drink seawater and excrete the excess salt at the gills. However, this theoretical interest has translated into remarkably few experimental studies, probably because elasmobranchs have little importance in aquaculture or recreational and commercial fishing. Earlier reviews have focused on their nitrogen metabolism (Perlman and Goldstein, 1988; Mommsen and Walsh, 1991; Wood, 1993; Wright, 1995; Goldstein and Perlman, 1995), and recently Ballantyne (1997) has summarized what is known about their overall metabolism in great detail. From these, it is clear that almost nothing is known about the effects on nitrogen metabolism of the three factors that form the focus of this chapter: feeding, exercise and temperature. The main purpose of this section is to draw attention to this deficit in the hope of stimulating new research.

Available information indicates that all marine elasmobranchs are strongly ureotelic, reflecting a full expression of the OUC in the liver (Chapter 7). The OUC is also present in the freshwater rays of the Amazon basin (Potomotrygonidae family), but these unusual fish are ammoniotelic, excreting large amounts of ammonia-N and small amounts of urea-N just like most teleosts (e.g., Goldstein and Forster, 1971; Barcellos et al., 1997). Urea is not retained in the blood fluids of freshwater rays. In contrast, marine elasmobranchs in full-strength seawater retain 300–500 mmol·liter^{-1} of urea (600–1000 mmol·liter^{-1} of urea-N) in their blood plasma, and excrete large amounts of urea-N and very small amounts of ammonia-N, almost exclusively at the gills (e.g., Wood et al., 1995; see also Chapter 8). Trimethylamine oxide excretion may also be important in nitrogen balance, amounting to 10–20% of urea-N excretion (Goldstein and Palatt, 1974). Renal nitrogen waste excretion is small because of apparent active reabsorption of urea-N at the kidney (Schmidt-Nielsen and Rabinowitz, 1964; Schmidt-Nielsen et al., 1972). Given the massive blood-to-water urea gradient across the gills, it is remarkable that urea-N excretion is not greater. Wood et al. (1995) calculated that branchial urea-N permeability in the dogfish shark was only about 7% of that in a typical teleost, while branchial ammonia-N permeability was only 4%! Based only on circumstantial evidence, branchial urea-N impermeability has been variously attributed to an unusual lipid composition of gill membranes (Boylan, 1967), an active "back-transport" mechanism in gill (Wood et al., 1995), or a combination (Pärt et al., 1998). New direct evidence for the presence of an ATP-dependent Na-urea antiporter and an unusually high cholesterol:phospholipid ratio in the basolateral membrane of the shark gill indicates that both factors are important (Fines et al., 2001). Branchial ammonia-N impermeability has been attributed to a "scavenging" by high-affinity glutamine synthetase in the gills (Wood et al., 1995). It seems likely, therefore, that the system is primarily designed to retain nitrogen, rather than to excrete it!

All the above work has been performed on nonfed fish confined in chambers. It would be very interesting to know what happens when a marine elasmobranch feeds. Haywood (1973) found that plasma urea concentrations progressively declined and osmoregulation was impaired when pyjama sharks were starved, but both recovered quickly after refeeding (Fig. 7). Plasma urea similarly fell, and both alanine and ammonia were released into the blood by the caudal musculature during starvation in the spiny dogfish (Leech et al., 1979). Armour et al. (1993) fed lesser spotted dogfish on high-protein and low-protein diets for 1 month. Plasma urea concentrations were unaffected in the low-protein diet animals, but [^{14}C]urea turnover studies indicated decreased metabolic production and clearance rates of urea-N, plus an inability to elevate plasma urea concentration in the face of an osmotic challenge. These data reinforce the idea that elasmobranchs may be nitrogen limited, with a resultant strategy aimed primarily at nitrogen retention, but unfortunately nitrogen excretion rates were not measured in any of

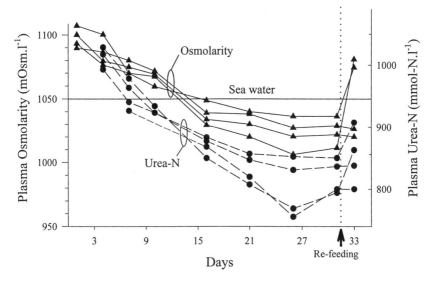

Fig. 7. The effect of progressive starvation for 30 days on plasma osmolarity (triangles; left axis) and plasma urea-N concentration (circles; right axis) at 13°C in four previously well-fed pyjama sharks (1–3 kg). On day 30, two of the sharks were fed (vertical dotted line), and their plasma osmolarity and urea-N levels immediately rose; the other two were not refed and exhibited no change. (Data recalculated from Haywood, 1973. Reproduced with the permission of Springer-Verlag GmBH & Co. Ltd.)

these studies. Mommsen and Walsh (1991) speculated that since urea-N is much more costly to make than ammonia-N, it would make sense for elasmobranchs to excrete extra *exogenous* nitrogen, over and above the needs of osmoregulation, in the form of ammonia-N rather than urea-N. The only information available to answer this speculation comes from a study in which dogfish shark were infused with ammonia-N (as neutralized NH_4Cl) at a rate of 1500 $\mu mol \cdot kg^{-1} \cdot h^{-1}$ for 6 h (Wood et al., 1995). Based on ration measurements in the wild for this species (Tanasichuk et al., 1991), the dietary nitrogen load is normally about 1000 $\mu mol \cdot kg^{-1} \cdot h^{-1}$, so this nitrogen loading rate appears to be quite reasonable. Both ammonia-N and urea-N excretion rose to similar extents during infusion, though the former more rapidly, and the entire ammonia-N load (actually 132%) was excreted within 18 h. When urea-N was infused at a fourfold higher rate (6000 $\mu mol \cdot kg^{-1} \cdot h^{-1}$ for 6 h), essentially none of it was excreted. Thus it appears likely that Mommsen and Walsh (1991) may have been at least partially correct, but NH_4Cl infusions are very different from natural feeding.

It would also be interesting to know what happens when a marine elasmobranch exercises, because glutamine appears to be a preferred oxidative fuel of muscle (Ballantyne, 1997), so nitrogen waste production might be expected to increase. However, there appear to be no measurements of nitrogen excretion in elasmobranchs performing sustainable aerobic exercise. Following exhaustive anaerobic exercise, Holeton and Heisler (1983) reported that ammonia-N excretion remained unchanged at the very low pre-exercise rate for up to 30 h postexercise in the larger spotted dogfish; urea-N excretion was not measured.

Very little is known about temperature effects. Acute temperature changes in the range of 1–15°C had no effect on branchial urea-N excretion in an externally irrigated, otherwise intact preparation of the gills of the spiny dogfish, but above 15°C, urea-N efflux increased steeply with a Q_{10} of about 6.0 up to 30°C (Boylan, 1967). Because only the temperature of the water irrigating the gills was apparently changed, and not the temperature of the rest of the dogfish, it is unlikely that these effects reflected alterations in the rate of metabolic production by the liver. In light of recent studies (Pärt *et al.*, 1998; Fines *et al.*, 2001), this breakpoint phenomenon can be interpreted as the result of a "phase change" in the gill lipids and/or a breakdown in the back-transport mechanism in the gill at higher temperatures. Ammonia-N flux was not measured in this study, but Heisler (1978) reported that it was extremely low and did not change when larger spotted dogfish were subjected to step temperature changes up or down between 10 and 20°C.

IX. CONCLUDING REMARKS

There are more than 22,000 species of teleosts and 600–700 species of elasmobranchs in the world, many of them now threatened by overfishing, climate change, and habitat destruction. Elasmobranchs are particularly vulnerable; this ancient group is now being decimated by collection for cancer therapies, high seas drift nets, souvenirs, and "live-finning." With teleosts, metabolism has been examined in perhaps several hundred species, enough to establish "standard" and unusual patterns, but few of these studies have been performed under natural conditions where fish eat, swim, and undergo temperature changes. With elasmobranchs, so few species have been studied that we do not even know what a "standard" pattern is, and the influence of feeding, exercise, and temperature remain virtually unknown. It is hoped that all these trends will be reversed in the new millennium.

ACKNOWLEDGMENTS

Supported by an NSERC research grant to C. M. W. I thank Angel Sing and Barb Reuter for excellent bibliographic assistance, and Mike Wilkie, Paul Anderson, and Pat Wright for helpful comments on the text.

REFERENCES

Alsop, D. H., and Wood, C. M. (1997). The interactive effects of feeding and exercise on oxygen consumption, swimming performance and protein usage in juvenile rainbow trout. *J. Exp. Biol.* **200**, 2337-2346.

Alsop, D. H., Kieffer, J. D., and Wood, C. M. (1999). The effects of temperature and swimming speed on instantaneous fuel use and nitrogenous waste excretion of the Nile tilapia. *Physiol. Biochem. Zool.* **72**, 474-483.

Armour, K. J., O'Toole, L. B., and Hazon, N. (1993). The effects of dietary protein restriction on the secretory dynamics of 1 α-hydroxycorticosterone and urea in the dogfish, *Scyliorhinus canicula*: a possible role for 1α-hydroxycorticosterone in sodium retention. *J. Endocrinol.* **138**, 275-282.

Ash, R., McLean, C., and Wescott, P. A. B. (1989). Arterio-portal differences and net appearance of amino acids in hepatic portal vein blood of the trout (*Salmo gairdneri*). In "Aquaculture—A Biotechnology in Progress" (N. DePauw, E. Japers, H. Ackefors, and N. Wilkens, eds.), pp. 801-806. European Aquaculture Society, Bredene, Belgium.

Ballantyne, J. S. (1997). Jaws: the inside story. The metabolism of elasmobranch fishes. *Comp. Biochem. Physiol.* **118B**, 703-742.

Barcellos, J. F. M., Wood, C. M., and Val, A. L. (1997). Ammonia and urea fluxes in *Potamotrygon* sp., a freshwater stingray of the Amazon. In "The Physiology of Tropical Fish, Symposium Proceedings" (A. L. Val, D. J. Randall, and D. MacKinlay, eds.), pp. 33-37. American Fisheries Society, San Francisco.

Barton, K. N., Gerrits, M. F., and Ballantyne, J. S. (1995). Effects of exercise on plasma nonesterified fatty acids and free amino acids in Arctic char (*Salvelinus alpinus*). *J. Exp. Zool.* **271**, 183-189.

Beamish, F. W. H. (1974). Apparent specific dynamic action of largemouth bass, *Micropterus salmoides*. *J. Fish. Res. Bd. Canada* **31**, 1763-1769.

Beamish, F. W. H., and Thomas, E. (1984). Effects of dietary protein and lipid on nitrogen losses in rainbow trout (*Salmo gairdneri*). *Aquaculture* **41**, 359-371.

Beamish, F.W.H., and Trippel, E. A. (1990). Heat increment: a static or dynamic dimension in bioenergetic models. *Trans. Am. Fish. Soc.* **119**, 649-661.

Beamish, F. W. H., Howlett, J. C., and Medland, T. E. (1989). Impact of diet on metabolism and swimming perfornmance in juvenile lake trout, *Salvelinus namaycush*. *Can. J. Fish. Aquat. Sci.* **46**, 384-388.

Beaumont, M. W., Taylor, E. W., and Butler, P. J. (2000). The resting membrane potential of white muscle from brown trout (*Salmo trutta*) exposed to copper in soft acidic water. *J. Exp. Biol.* **203**, 2229-2236.

Bever, K., Chenoweth, M., and Dunn, A. (1981). Amino acid gluconeogenesis and glucose turnover in kelp bass (*Paralabrax* sp.). *Am. J. Physiol.* **240**, R246-R252.

Blaikie, H. B., and Kerr, S. R. (1996). Effect of activity level on apparent heat increment in Atlantic cod, *Gadus morhua*. *Can. J. Fish. Aquat. Sci.* **53**, 2093-2099.

Boron, W. F., and Roos, A. (1976). Comparison of microelectrode, DMO, and methylamine methods for measuring intracellular pH. *Am. J. Physiol.* **231**, 799-801.

Boylan, J. W. (1967). Gill permeability in *Squalus acanthias*. In "Sharks, Skates and Rays" (Gilbert, P. W., Mathewson, R. F., and Rall, D. P., eds.), pp. 197-206. Johns Hopkins University Press, Baltimore, MD.

Brett, J. R. (1973). Energy expenditure of sockeye salmon, *Oncorhynchus nerka*, during sustained performance. *J. Fish. Res. Bd. Canada* **30**, 1799-1809.

Brett, J. R. (1995). Energetics. In "Physiological Ecology of Pacific Salmon" (C. Groot, L. Margolis, and W. C. Clarke, eds.). pp. 1-68. University of British Columbia Press, Vancouver.

Brett, J. R., and Zala, C. A. (1975). Daily pattern of nitrogen excretion and oxygen consumption of sockeye salmon (*Oncorhynchus nerka*) under controlled conditions. *J. Fish. Res. Bd. Canada* **32**, 2479–2486.

Brown, C. R., and Cameron, J. N. (1991a). The induction of specific dynamic action in channel catfish by infusion of essential amino acids. *Physiol. Zool.* **64**, 276–297.

Brown, C. R., and Cameron, J. N. (1991b). The relationship between specific dynamic action (SDA) and protein synthesis rates in the channel catfish. *Physiol. Zool.* **64**, 298–309.

Burrows, R. E. (1964). Effects of accumulated excretory products on hatchery-reared salmonids. *U.S. Fish Wild. Serv. Bur. Sport Fish. Wild Res. Rep.* **66**, 1–12.

Chadwick, T. D., and Wright, P. A. (1999). Nitrogen excretion and expression of urea cycle enzymes in the Atlantic cod (*Gadus morhua* L.): a comparison of early life stages with adults. *J. Exp. Biol.* **202**, 2653–2662.

Cho, C. Y. (1990). Fish nutrition, feeds, and feeding, with special emphasis on salmonid aquaculture. *Food Rev. Int.* **6**, 333–357.

Christiansen, J. S., and Jobling, M. (1990). The behaviour and the relationship between food intake and growth of juvenile Arctic charr, *Salvelinus alpinus* L., subjected to sustained exercise. *Can. J. Zool.* **68**, 2185–2191.

Christiansen, J. S., Ringo, E., and Jobling, M. (1989). Effects of sustained exercise on growth and body composition of first-feeding fry of Arctic charr, *Salvelinus alpinus*. *Aquaculture* **79**, 329–335.

Cockcroft, A. C., and Du Preez, H. H. (1989). Nitrogen and energy loss via nonfecal and fecal excretion in the marine teleost *Lithognathus lithognathus*. *Mar. Biol.* **101**, 419–425.

Danulat, E., and Kempe, S. (1992). Nitrogenous waste excretion and accumulation of urea and ammonia in *Chalcalburnus tarichi (Cyprinidae)* endemic to the extremely alkaline Lake Van (Eastern Turkey). *Fish Physiol. Biochem.* **9**, 377–386.

Davison, W. (1989). Training and its effects on teleost fish. *Comp. Biochem. Physiol.* **94A**, 1–10.

Davison, W. (1997). The effects of exercise training on teleost fish, a review of recent literature. *Comp. Biochem. Physiol.* **117A**, 67–75.

D'Cruz, L. M., Dockray, J. J., Morgan, I. J., and Wood, C. M. (1998). Physiological effects of sublethal acid exposure in juvenile rainbow trout on a limited or unlimited ration during a simulated global warming scenario. *Physiol. Zool.* **71**, 359–376.

Dobson, G. P., and Hochachka, P. W. (1987). Role of glycolysis in adenylate depletion and repletion during work and recovery in teleost white muscle. *J. Exp. Biol.* **129**, 125–140.

Dockray, J. J., Reid, S. D., and Wood, C. M. (1996). Effects of elevated summer temperatures and reduced pH on metabolism and growth of juvenile rainbow trout on unlimited ration. *Can. J. Fish Aquat. Sci.* **25**, 2752–2763.

Dockray, J. J., Morgan, I. J., Reid, S. D., and Wood, C. M. (1998). Responses of juvenile rainbow trout, under food limitation, to chronic low pH and a summer global warming scenario, alone and in combination. *J. Fish Biol.* **52**, 62–82.

Driedzic, W. R., and Hochachka, P. W. (1978). Metabolism in fish during exercise. *In* "Fish Physiology" (W. S. Hoar and D. J. Randall, eds.), Vol. 8, pp. 503–543. Academic Press, New York.

Espe, M., Lied, E., and Torrissen, K. R. (1993). Changes in plasma and muscle free amino acids in Atlantic salmon (*Salmo salar*) during absorption of diets containing different amounts of hydrolyzed cod muscle protein. *Comp. Biochem. Physiol.* **105A**, 555–562.

Fauconneau, B., and Arnal, M. (1985). *In vivo* protein synthesis in different tissues and the whole body of rainbow trout (*Salmo gairdneri* R.). Influence of environmental temperature. *Comp. Biochem. Physiol.* **82A**, 179–187.

Fauconneau, B., Breque, J., and Bielle, C. (1989). Influence of feeding on protein metabolism of Atlantic salmon (*Salmo salar* L.). *Aquaculture* **79**, 29–36.

Fines, G. A., Ballantyne, J. S., and Wright, P. A. (2001). Active urea transport and an unusual basolateral membrane composition in the gills of a marine elasmobranch. *Am. J. Physiol.* **280**, R16–R24.

Fivelstad, S., Thomassen, J. M., Smith, M. J., Kjartansson, H., and Sando, A.-B. (1990). Metabolite production rates from Atlantic salmon (*Salmo salar* L.) and Arctic char (*Salvelinus alpinus* L.) reared in single-pass land-based brackish water and sea-water systems. *Aquacult. Engi.* **9**, 1–21.

Foster, A. R., Houlihan, D. F., Hall, S. J., and Burren, L. J. (1992). The effects of temperature acclimation on protein synthesis rates and nucleic acid content of juvenile cod (*Gadus morhua* L.). *Can. J. Zool.* **70**, 2095–2102.

Fromm, P. O. (1963). Studies on renal and extra-renal excretion in a freshwater teleost, *Salmo gairdneri*. *Comp. Biochem. Physiol.* **10**, 121–128.

Gershanovich, A. D., and Pototskij, I. V. (1992). The peculiarities of nitrogen excretion in sturgeons (*Acipenser ruthenus*) (Pisces, Acipenseridae)—the influence of ration size. *Comp. Biochem. Physiol.* **103A**, 609–612.

Gershanovich, A. D., and Pototskij, I. V. (1996). The peculiarities of nitrogen excretion in sturgeons (*Acipenser ruthenus*) (Pisces, Acipenseridae)—2. Effects of water temperature, salinity, and pH. *Comp. Biochem. Physiol.* **111A**, 313–317.

Goldstein, L., and Forster, R. P. (1971). Urea biosynthesis and excretion in freshwater and marine elasmobranchs. *Comp. Biochem. Physiol.* **39B**, 415–421.

Goldstein, L., and Palatt, P. J. (1974). Trimethylamine oxide excretion rates in elasmobranchs. *Am. J. Physiol.* **227**, 1268–1272.

Goldstein, L., and Perlman, D. F. (1995). Nitrogen metabolism, excretion, osmoregulation, and cell volume regulation in elasmobranchs. *In* "Nitrogen Metabolism and Excretion" (P. J. Walsh and P. A. Wright, eds.), pp. 91–104. CRC Press, Boca Raton, FL.

Guerin-Ancey, O. (1976). Etude expérimentale de l'excretion azotée du bar (*Dicentrarchus labrax*) en cours de croissance. II. Effets du jeune sur l'excretion d'ammoniac et d'urée. *Aquaculture*, **9**, 187–194.

Handy, R. D., and Poxton, M. G. (1993). Nitrogen pollution in mariculture: toxicity and excretion of nitrogenous compounds by marine fish. *Rev. Fish Biol. Fish.* **3**, 205–241.

Haywood, G. P. (1973). Hypo-osmotic regulation coupled with reduced metabolic urea in the dogfish *Poroderma africanum*: an analysis of serum osmolarity, chloride, and urea. *Mar. Biol.* **23**, 121–127.

Heisler, N. (1978). Bicarbonate exchange between body compartments after changes of temperature in the larger spotted dogfish (*Scyliorhinus stellaris*). *Respir. Physiol.* **33**, 145–160.

Heisler, N. (1990). Mechanisms of ammonia elimination in fishes. *In* "Animal Nutrition and Transport Processes, Volume 2, Transport, Respiration, and Excretion: Comparative and Environmental Aspects, Comp. Physiol." (J. P. Truchot and B. Lahlou, eds.), pp. 137–151. Karger, Basel.

Holeton, G. F., and Heisler, N. (1983). Contribution of net ion transfer mechanisms to acid–base regulation after exhausting activity in the larger spotted dogfish (*Scyliorhinus stellaris*). *J. Exp. Biol.* **103**, 31–46.

Houlihan, D. F., and Laurent, P. (1987). Effects of exercise training on the performance, growth, and protein turnover of rainbow trout (*Salmo gairdneri*). *Can. J. Fish. Aquat. Sci.* **44**, 1614–1621.

Houlihan, D. F., Carter, C. G., and McCarthy, I. D. (1995a). Protein turnover in animals. *In* "Nitrogen Metabolism and Excretion" (P. J. Walsh and P. A. Wright, eds.). pp. 1–32. CRC Press, Boca Raton, FL.

Houlihan, D. F., Carter, C. G., and McCarthy, I. D. (1995b). Protein synthesis in fish. *In* "Biochemistry and Molecular Biology of Fishes" (P. W. Hochachka and T. P. Mommsen, eds.), Volume 4, pp. 191–220. Elsevier, Amsterdam.

Houlihan, D. F., Pedersen, B. H., Steffensen, J. F., and Brechin, J. (1995c). Protein synthesis, growth, and energetics in larval herring (*Clupea harengus*) at different feeding regimes. *Fish Physiol. Biochem.* **14**, 195–208.

Idler, D. R., and Clemens, W. A. (1959). The energy expenditures of Fraser River sockeye salmon during the spawning migration to Chilko and Stuart Lakes. *Intl. Pacific Salmon Fish. Comm. Prog. Rept.* **6,** 80 pp.

Iwata, K. (1970). Relationship between food and growth in young crucian carps, *Carassius auratus cuvieri,* as determined by the nitrogen balance. *Japan. J. Limnol.* **31,** 129–151.

Jacobs, M. H., and Stewart, D. R. (1936). The distribution of penetrating ammonium salts between cells and their surroundings. *J. Cell. Comp. Physiol.* **7,** 351–365.

Jobling, M. (1980). Effects of starvation on proximate chemical composition and energy utilization of plaice, *Pleuronectes platessa. J. Fish Biol.* **17,** 325–334.

Jobling, M. (1981a). The influences of feeding on the metabolic rate of fishes: a short review. *J. Fish Biol.* **18,** 385–400.

Jobling, M. (1981b). Some effects of temperature, feeding and body weight on nitrogenous excretion in young plaice *Pleuronectes platessa* L. *J. Fish Biol.* **18,** 87–96.

Jobling, M. (1994). "Fish Bioenergetics." Chapman and Hall, London.

Jobling, M., and Spencer Davies, P. (1980). Effects of feeding on metabolic rate and the specific dynamic action in plaice, *Pleuronectes platessa* L. *J. Fish Biol.* **16,** 629–638.

Kaushik, S. J. (1980). Influence of nutritional status on the daily patterns of nitrogen excretion in the carp (*Cyprinus carpio* L.) and the rainbow trout (*Salmo gairdneri* L.). *Reprod. Nutr. Dev.* **20,** 1751–1756.

Kaushik, S. J., and Teles, A. O. (1985). Effect of digestible energy on nitrogen and energy balance in rainbow trout. *Aquaculture* **50,** 89–101.

Kaushik, S. J., Drabowski, K., and Luquet, P. (1982). Patterns of nitrogen excretion and oxygen consumption during ontogenesis of common carp. *Can. J. Fish. Aquat. Sci.* **39,** 1095–1105.

Kieffer, J. D., and Tufts, B. L. (1996). The influence of environmental temperature on the role of the rainbow trout gill in correcting the acid–base disturbance following exhaustive exercise. *Physiol. Zool.* **69,** 1301–1323.

Kieffer, J. D., Alsop, D., and Wood, C. M. (1998). A respirometric analysis of fuel use during aerobic swimming at different temperatures in rainbow trout (*Oncorhynchus mykiss*). *J. Exp. Biol.* **201,** 3123–3133.

Kirschner, L. B. (1993). The energetics of osmoregulation in ureotelic and hypo-osmotic fishes. *J. Exp. Zool.* **267,** 19–26.

Kleiber, M. (1992). Respiratory exchange and metabolic rate. *In* "Handbook of Physiology" (S. R. Geiser, ed.), pp. 927–938. American Physiological Society, Bethesda, MD.

Knights, B. (1985). Energetics and fish farming. *In* "Fish Energetics" (P. Tytler and P. Calow, eds.), pp. 309–340. Johns Hopkins University Press, Baltimore, MD.

Kong, H., Edberg, D. D., Korte, J. J., Salo, W. L., Wright, P. A., and Anderson, P. M. (1998). Nitrogen excretion and expression of carbamoyl-phosphate synthetase III activity and mRNA in extrahepatic tissues of largemouth bass (*Micropterus salmoides*). *Arch. Biochem. Biophys.* **350,** 157–168.

Korsgaard, B., Mommsen, T. P., and Wright, P. A. (1995). Nitrogen excretion in teleostean fish: adaptive relationships to environment, ontogenesis, and viviparity. *In* "Nitrogen Metabolism and Excretion" (P. J. Walsh and P. A. Wright, eds.), pp. 259–288. CRC Press, Boca Raton, FL.

Korte, J. J., Salo, W. L., Cabrera, V. M., Wright, P. A., Felskie, A. K., and Anderson, P. M. (1997). Expression of carbamoyl-phosphate synthetase III mRNA during the early life stages of development and in the muscle of adult rainbow trout (*Oncorhynchus mykiss*). *J. Biol. Chem.* **272,** 6270–6277.

Krueger, H. M., Saddler, J. B., Chapman, G. A., Tinsley, I. J., and Lowry, R. P. (1968). Bioenergetics, exercise, and fatty acids of fish. *Am. Zool.* **8,** 119–129.

Kutty, M. N. (1968). Respiratory quotients in goldfish and rainbow trout. *J. Fish. Res. Bd. Canada* **25,** 1689–1728.

Kutty, M. N. (1972). Respiratory quotient and ammonia excretion in *Tilapia mossambica. Mar. Biol.* **16,** 126–133.

Kutty, M. N. (1978). Ammonia quotient in sockeye salmon (*Oncorhynchus nerka*). *J. Fish. Res. Bd. Canada* **35**, 1003–1005.

Kutty, M. N., and Peer Mohamed, M. (1975). Metabolic adaptations of mullet *Rhinomugil corsula* (Hamilton) with special reference to energy utilization. *Aquaculture* **5**, 253–270.

Lauff, R. F., and Wood, C. M. (1996a). Respiratory gas exchange, nitrogenous waste excretion, and fuel usage during starvation in juvenile rainbow trout. *J. Comp. Physiol.* **165B**, 542–551.

Lauff, R. F., and Wood, C. M. (1996b). Respiratory gas exchange, nitrogenous waste excretion and fuel usage during aerobic swimming in juvenile rainbow trout. *J. Comp. Physiol.* **166B**, 501–509.

Lauff, R. F., and Wood, C. M. (1997). Effects of training on respiratory gas exchange, nitrogenous waste excretion, and fuel usage during aerobic swimming in juvenile rainbow trout (*Oncorhynchus mykiss*). *Can. J. Fish. Aquat. Sci.* **54**, 560–571.

Leech, A. R., Goldstein, L., Cha, C. J., and Goldstein, J. M. (1979). Alanine biosynthesis during starvation in skeletal muscle of the spiny dogfish, *Squalus acanthias*. *J. Exp. Zool.* **207**, 73–80.

LeGrow, S. M., and Beamish, F. W. H. (1986). Influence of dietary protein and lipid on apparent heat increment of rainbow trout, *Salmo gairdneri*. *Can. J. Fish. Aquat. Sci.* **43**, 19–25.

Leung, K. M. Y., Chu, J. C. W., and Wu, R. S. S. (1999). Effects of body weight, water temperature and ration size on ammonia excretion by the areolated grouper (*Epinephelus areolatus*) and mangrove snapper (*Lutjanus argentimaculatus*). *Aquaculture* **170**, 215–227.

Lied, E., Rosenlund, G., Lund, B., and von der Decken, A. (1983). Effect of starvation and refeeding on *in vitro* protein synthesis in white trunk muscle of Atlantic cod (*Gadus morhua*). *Comp. Biochem. Physiol.* **76B**, 777–781.

Linton, T. K., Reid, S. D., and Wood, C. M. (1997). The metabolic costs and physiological consequences to juvenile rainbow trout of a simulated summer warming scenario in the presence and absence of sublethal ammonia. *Trans. Am. Fish. Soc.* **126**, 259–272.

Linton, T. K., Morgan, I. J., Walsh, P. J., and Wood, C. M. (1998a). Chronic exposure of rainbow trout to simulated climate warming and sublethal ammonia: a year long study of their appetite, growth, and metabolism. *Can. J. Fish. Aquat. Sci.* **55**, 576–586.

Linton, T. K., Reid, S. D., and Wood, C. M. (1998b). The metabolic costs and physiological consequences to juvenile rainbow trout of a simulated winter warming scenario in the presence and absence of sublethal ammonia. *Trans. Am. Fish. Soc.* **127**, 611–619.

Linton, T. K., Reid, S. D., and Wood, C. M. (1999). Effects of restricted ration on the growth and energetics of juvenile rainbow trout exposed to a summer of simulated warming and sublethal ammonia. *Trans. Am. Fish. Soc.* **128**, 758–763.

Loughna, P. T., and Goldspink, G. (1985). Muscle protein synthesis rates during temperature acclimation in a eurythermal (*Cyprinus carpio*) and a stenothermal (*Salmo gairdneri*) species of teleost. *J. Exp. Biol.* **118**, 267–276.

Lyndon, A. R., Houlihan, D. F., and Hall, S. J. (1992). The effect of short-term fasting and a single meal on protein synthesis and oxygen consumption in cod, *Gadus morhua*. *J. Comp. Physiol.* **162B**, 209–215.

Maetz, J. (1972). Branchial sodium exchange and ammonia excretion in the goldfish *Carassius auratus*. Effects of ammonia loading and temperature changes. *J. Exp. Biol.* **56**, 601–620.

Masoni, A., and Garcia-Romeu, F. (1972). Accumulation et excrétion de substances organiques par les cellules à chlorure de la branchie d'*Anguilla anguilla* L adapteé à l'eau de mer. *Z. Zellforsch.* **133**, 389–398.

Mathers, E. M., Houlihan, D. F., McCarthy, I. D., and Burden, L. J. (1993). Rates of growth and protein synthesis correlated with nucleic acid content in fry of rainbow trout, *Oncorhynchus mykiss*: effects of age and temperature. *J. Fish. Biol.* **43**, 245–263.

McCarthy, I. D., and Houlihan, D. F. (1997). The effect of temperature on protein metabolism in fish: the possible consequences for wild Atlantic salmon (*Salmo salar* L.) stocks in Europe as a result of global warming. *In* "Global Warming: Implications for Freshwater and Marine Fish" (C. M. Wood and D. G. McDonald, eds.), pp. 51–77. Cambridge University Press, Cambridge.

McCarthy, I. D., Moksness, E., Pavlov, D. A., and Houlihan, D. F. (1999). Effects of water temperature on protein synthesis and protein growth in juvenile Atlantic wolffish (*Anarhichas lupus*). *Can. J. Fish. Aquat. Sci.* **56,** 231–241.

McCarthy, J. J., and Whitledge, T. E. (1972). Nitrogen excretion by anchovy (*Engraulis mordax*) and jack mackeral (*Trachurus symmetricus*). *Fish. Bull.* **70,** 395–401.

McDonald, M. D., and Wood, C. M. (1998). Reabsorption of urea by the kidney of the freshwater rainbow trout. *Fish Physiol. Biochem.* **18,** 375–386.

McGeer, J. C., Wright, P. A., Wood, C. M., Wilkie, M. P., Mazur, C. F., and Iwama, G. K. (1994). Nitrogen excretion in four species of fish from an alkaline/saline lake. *Trans. Am. Fish. Soc.* **123,** 824–829.

McKenzie, D. J., Piraccini, G., Felskie, A., Romano, P., Bronzi, P., and Bolis, C. L. (1999). Effects of plasma total ammonia content and pH on urea excretion in Nile tilapia. *Physiol. Biochem. Zool.* **72,** 116–125.

McLean, W. E., and Fraser, F. J. (1974). Ammonia and urea production of coho salmon under hatchery conditions, Surveillance Rep. EPS, 5-PR-74-5. Environmental Protection Service Pacific Region.

McMillan, D. N., and Houlihan, D. F. (1988). The effect of re-feeding on tissue protein synthesis in rainbow trout. *Physiol. Zool.* **61,** 429–441.

McMillan, D. N., and Houlihan, D. F. (1989). Short-term responses of protein synthesis to re-feeding in rainbow trout. *Aquaculture* **79,** 37–46.

Miller, T. G., and MacKay, W. C. (1982). Relationship of secreted mucus to copper and acid toxicity in rainbow trout. *Bull. Environ. Contam. Toxicol.* **28,** 68–74.

Milligan, C. L. (1997). The role of cortisol in amino acid mobilization and metabolism following exhaustive exercise in rainbow trout (*Oncorhynchus mykiss* Walbaum). *Fish Physiol. Biochem.* **16,** 119–128.

Mommsen, T. P., and Hochachka, P. W. (1988). The purine nucleotide cycle as two temporally separated metabolic units: a study on trout muscle. *Metabolism* **37,** 552–556.

Mommsen, T. P., and Walsh, P. J. (1989). Evolution of urea synthesis in vertebrates: the piscine connection. *Science* **243,** 72–75.

Mommsen, T. P., and Walsh, P. J. (1991). Urea synthesis in fishes: evolutionary and biochemical perspectives. *In* "Biochemistry and Molecular Biology of Fishes" (P. W. Hochachka and T. O. Mommsen, eds.), Volume 1, pp. 137–163. Elsevier, New York.

Mommsen, T. P., and Walsh, P. J. (1992). Biochemical and environmental perspectives on nitrogen metabolism in fishes. *Experientia* **48,** 583–592.

Mommsen, T. P., French, C. J., and Hochachka, P. W. (1980). Sites and patterns of protein and amino acid utilization during the spawning migration of salmon. *Can. J. Zool.* **58,** 1785–1799.

Morgan, I. J., D'Cruz, L. M., Dockray, J. J., Linton, T. K., McDonald, D. G., and Wood, C. M. (1998). The effects of elevated winter temperature and sub-lethal pollutants (low pH, elevated ammonia) on protein turnover in the gill and liver of rainbow trout (*Oncorhynchus mykiss*). *Fish Physiol. Biochem.* **19,** 377–389.

Morgan, I. J., D'Cruz, L. M., Dockray, J. J., Linton, T. K., and Wood, C. M. (1999). The effects of elevated summer temperature and sublethal pollutants (ammonia, low pH) on protein turnover in the gill and liver of rainbow trout (*Oncorhynchus mykiss*) on a limited food ration. *Comp. Biochem. Physiol.* **123A,** 43–53.

Muir, B. S., and Niimi, A. J. (1972). Oxygen consumption of the euryhaline fish aholehole (*Kuhlia sandvicensis*) with reference to salinity, swimming, and food consumption. *J. Fish. Res. Bd. Canada* **29,** 67–77.

Olson, K. R., and Fromm, P. O. (1971). Excretion of urea by two teleosts exposed to different concentrations of ambient ammonia. *Comp. Biochem. Physiol.* **40A,** 9991–1007.

Pärt, P., Wright, P. A., and Wood, C. M. (1998). Urea and water permeability in dogfish gills (*Squalus acanthias*). *Comp. Biochem. Physiol.* **119A,** 117–123.

Patrick, M. L., and Wood, C. M. (1999). Ion and acid–base regulation in the freshwater mummichog (*Fundulus heteroclitus*): a departure from the standard model for freshwater teleosts. *Comp. Biochem. Physiol.* **122A**, 445–456.

Perlman, D. F., and Goldstein, L. (1988). Nitrogen metabolism. *In* "Physiology of Elasmobranch Fishes" (T. J. Shuttleworth, ed.), pp. 253–275. Springer-Verlag, Berlin.

Postlethwaite, E. K., and McDonald, D. G. (1995). Mechanisms of Na^+ and Cl^- regulation in freshwater rainbow trout (*Oncorhynchus mykiss*) during exercise and stress. *J. Exp. Biol.* **198**, 295–304.

Randall, D. J., and Brauner, C. J. (1991). Effects of environmental factors on exercise in fish. *J. Exp. Biol.* **160**, 113–126.

Randall, D. J., Wood, C. M., Perry, S. F., Bergman, H. L., Maloiy, G. M. O., Mommsen, T. P., and Wright, P. A. (1989). Urea excretion as a strategy for survival in a fish living in a very alkaline environment. *Nature* **337**, 165–166.

Reid, S. D., Dockray, J. J., Linton, T. K., McDonald, D. G., and Wood, C. M. (1995). Effects of a summer temperature regime representative of a global warming scenario on growth and protein synthesis in hardwater and softwater-acclimated juvenile rainbow trout. *J. Thermal. Biol.* **20**, 231–244.

Reid, S. D., Dockray, J. J., Linton, T. K., McDonald, D. G., and Wood, C. M. (1997). Effects of chronic environmental acidification and a summer global warming scenario: protein synthesis in juvenile rainbow trout. *Can. J. Fish. Aquat. Sci.* **54**, 2014–2024.

Reid, S. D., Linton, T. K., Dockray, J. J, McDonald, D. G., and Wood, C. M. (1998). Effects of chronic sublethal ammonia and a simulated summer global warming scenario: protein synthesis in juvenile rainbow trout. *Can. J. Fish. Aquat. Sci.* **55**, 1534–1544.

Rychly, J., and Marina, B. M. (1977). The ammonia excretion of trout during a 24-hour period. *Aquaculture* **11**, 173–178.

Saha, N., and Ratha, B. K. (1998). Ureogenesis in Indian air-breathing teleosts; adaptation to environmental constraints. *Comp. Biochem. Physiol.* **A120**, 195–208.

Salama, A., Morgan, I. J., and Wood, C. M. (1999). The linkage between sodium uptake and ammonia excretion in rainbow trout—kinetic analysis, the effects of $(NH_4)_2 SO_4$ and $NH_4 HCO_3$ infusion, and the influence of gill boundary layer pH. *J. Exp. Biol.* **202**, 697–709.

Scarabello, M., Heigenhauser, G. J. F., and Wood, C. M. (1992). Gas exchange, metabolite status, and excess postexercise oxygen consumption after repetitive bouts of exhaustive exercise in juvenile rainbow trout. *J. Exp. Biol.* **167**, 155–169.

Schmidt-Nielsen, B., and Rabinowitz, L. (1964). Methylurea and acetamide: active reabsorption by elasmobranch renal tubules. *Science* **146**, 1587–1588.

Schmidt-Nielsen, B., Truniger, B., and Rabinowitz, L. (1972). Sodium-linked urea transport by the renal tubule of the spiny dogfish, *Squalus acanthias*. *Comp. Biochem. Physiol.* **42A**, 13–25.

Smith, H. W. (1929). The excretion of ammonia and urea by the gills of fishes. *J. Biol. Chem.* **81**, 727–742.

Smith, H. W. (1936). The retention and physiological role of urea in the elasmobranchii. *Biol. Rev.* **11**, 49–82.

Smith, M. A. K., and Thorpe, A. (1976). Nitrogen metabolism and trophic input in relation to growth in freshwater and saltwater *Salmo gairdneri*. *Biol. Bull.* **150**, 135–151.

Soofiani, N. M., and Hawkins, A. D. (1982). Energetic costs at different levels of feeding in juvenile cod, *Gadus morhua* L. *J. Fish Biol.* **21**, 577–592.

Storey, K. B. (1991). Metabolic consequences of exercise in organs of rainbow trout. *J. Exp. Zool.* **260**, 157–164.

Su, J. Y., and Storey, K. B. (1994). Regulation of rainbow trout white muscle phosphofructokinase during exercise. *Int. J. Biochem.* **26**, 519–528.

Sukumaran, N., and Kutty, M. N. (1977). Oxygen consumption and ammonia excretion in the catfish

Mystus armatus, with special reference to swimming speed and ambient oxygen. *Proc. Ind. Acad. Sci.* **86B,** 195–206.

Tanasichuk, R. W., Ware, D. M., Shaw, W., and McFarlane, G. A. (1991). Variations in diet, daily ration, and feeding periodicity of Pacific hake (*Merluccius productus*) and spiny dogfish (*Squalus acanthias*) off the lower coast of Vancouver Island. *Can. J. Fish. Aquat. Sci.* **48,** 2118–2128.

Tandler, A., and Beamish, F. W. H. (1979). Mechanical and biochemical components of apparent specific dynamic action in largemouth bass, *Micropterus salmoides* Lacepede. *J. Fish Biol.* **14,** 343–350.

Tang, Y., Lin, H., and Randall, D. J. (1992). Compartmental distributions of carbon dioxide and ammonia in rainbow trout at rest and following exercise, and the effect of bicarbonate infusion. *J. Exp. Biol.* **169,** 235–249.

Van den Thillart, G. (1986). Energy metabolism of swimming trout (*Salmo gairdneri*). Oxidation rates of palmitate, glucose, lactate, alanine, leucine, and glutamate. *J. Comp. Physiol.* **B156,** 511–520.

Van den Thillart, G., and Kesbeke, F. (1978). Anaerobic production of carbon dioxide and ammonia by goldfish *Carassius auratus* (L.). *Comp. Biochem. Physiol.* **59A,** 393–400.

Van Waarde, A. (1983). Aerobic and anaerobic ammonia production by fish. *Comp. Biochem. Physiol.* **74B,** 675–684.

Walsh, P. J. (1997a). Nitrogen metabolism and excretion. *In* "The Physiology of Fishes," 2nd ed. (D. H. Evans, ed.), pp. 199–214. CRC Press, Boca Raton, FL.

Walsh, P. J. (1997b). Evolution and regulation of ureogenesis and ureotely in (Batrachoidid) fishes. *Ann. Rev. Physiol.* **59,** 299–323.

Walsh, P. J., and Milligan, C. L. (1995). Effects of feeding and confinement on nitrogen metabolism and excretion in the gulf toadfish *Opsanus beta. J. Exp. Biol.* **198,** 1559–1566.

Walsh, P. J., Wang, Y., Campbell, C. E., De Boeck, G., and Wood, C. M. (2001). Patterns of nitrogenous waste, excretion and gill urea transporter mRNA expression in several species of marine fish. *Mar. Biol.* (In press).

Wang, Y., Heigenhauser, G. J. F., and Wood, C. M. (1994). Integrated responses to exhaustive exercise and recovery in rainbow trout white muscle: acid–base, phosphogen, carbohydrate, lipid, ammonia, fluid volume and electrolyte metabolism. *J. Exp. Biol.* **195,** 227–258.

Wang, Y., Heigenhauser, G. J. F., and Wood, C. M. (1996). Ammonia transport and distribution after exercise across white muscle cell membranes in rainbow trout: a perfusion study. *Am. J. Physiol.* **271,** R738–R750.

Wang, Y., Henry, R. P., Wright, P. M., Heigenhauser, G. J. F., and Wood, C. M. (1998). Respiratory and metabolic functions of carbonic anhydrase in exercised white muscle of trout. *Am. J. Physiol.* **275,** R1766–R1799.

Watt, P. W., Marshall, P. A., Heap, S. P., Loughna, P. T., and Goldspink, G. (1988). Protein synthesis in tissues of fed and starved carp, acclimated to different temperatures. *Fish Physiol. Biochem.* **4,** 165–173.

Weber, J.-M., and Haman, F. (1996). Pathways for metabolic fuels and oxygen in high performance fish. *Comp. Biochem. Physiol.* **113A,** 33–38.

Wiggs, A. J., Henderson, E. B., Saunders, R. L., and Kutty, M. N. (1989). Activity, respiration, and excretion of ammonia by Atlantic salmon (*Salmo salar*) smolt and postsmolt. *Can. J. Fish. Aquat. Sci.* **46,** 790–795.

Wilkie, M. P. (1997). Mechanisms of ammonia excretion across fish gills. *Comp. Biochem. Physiol.* **118A,** 39–50.

Wilkie, M. P., and Wood, C. M. (1991). Nitrogenous waste excretion, acid–base balance, and ionoregulation in rainbow trout (*Oncorhynchus mykiss*) exposed to extremely alkaline water. *Physiol. Zool.* **64,** 1069–1086.

Wilkie, M. P., and Wood, C. M. (1995). Recovery from high pH exposure in rainbow trout: ammonia washout and the restoration of blood chemistry. *Physiol. Zool. Physiol. Zool.* **68,** 379–401.

Wilkie, M. P., and Wood, C. M. (1996). The adaptations of fish to extremely alkaline environments. *Comp. Biochem. Physiol.* **113B**, 665–673.

Wilkie, M. P., Wright, P. A., Iwama, G. K., and Wood, C. M. (1993). The physiological responses of the Lahontan cutthroat trout (*Oncorhynchus clarki henshawi*), a resident of highly alkaline Pyramid Lake (pH 9.4), to challenge at pH 10. *J. Exp. Biol.* **175**, 173–194.

Wilkie, M. P., Wright, P. A., Iwama, G. K., and Wood, C. M. (1994). The physiological adaptations of the Lahontan cutthroat trout (*Oncorhynchus clarki henshawi*) following transfer from well water to the highly alkaline waters of Pyramid Lake, Nevada (pH 9.4). *Physiol. Zool.* **67**, 355–380.

Wilson, R. W., Wright, P. M., Munger, S., and Wood, C. M. (1994). Ammonia excretion in rainbow trout (*Oncorhynchus mykiss*) and the importance of gill boundary layer acidification: lack of evidence for Na^+/NH_4^+ exchange. *J. Exp. Biol.* **191**, 37–58.

Wood, C. M. (1988). Acid–base and ionic exchanges at gills and kidney after exhaustive exercise in the rainbow trout. *J. Exp. Biol.* **146**, 461–481.

Wood, C. M. (1989). The physiological problems of fish in acid waters. *In* "Acid Toxicity and Aquatic Animals, Society for Experimental Biology Seminar Series" (R. Morris, D. J. A. Brown, E. W. Taylor, and J. A. Brown, eds.), pp. 125–148. Cambridge University Press, Cambridge.

Wood, C. M. (1993). Ammonia and urea metabolism and excretion. *In* "The Physiology of Fishes" (D. Evans, ed.), pp. 379–425. CRC Press, Boca Raton, FL.

Wood, C. M. (1995). Excretion. *In* "Physiological Ecology of the Pacific Salmon" (C. Groot, L. Margolis, and W. C. Clarke, eds.), pp. 381–438. Government of Canada Special Publications Branch, University of British Columbia Press, Vancouver.

Wood, C. M., and Wang, Y. (1999). Lactate, H^+, and ammonia transport and distribution in rainbow trout white muscle after exhaustive exercise. *In* "SEB Seminar Series: Regulation of Acid–Base Status in Animals and Plants" (S. Egginton, E. W. Taylor, and J. A. Raven, eds.), pp. 99–124. Cambridge University Press, Cambridge.

Wood, C. M., Perry, S. F., Wright, P. A., Bergman, H. L., and Randall, D. J. (1989). Ammonia and urea dynamics in the Lake Magadi tilapia, a ureotelic teleost fish adapted to an extremely alkaline environment. *Resp. Physiol.* **77**, 1–20.

Wood, C. M., Pärt, P., and Wright, P. A. (1995). Ammonia and urea metabolism in relation to gill function and acid–base balance in a marine elasmobranch, the spiny dogfish (*Squalus acanthias*). *J. Exp. Biol.* **198**, 1545–1558.

Wood, C. M., Milligan, L. M. and Walsh, P. J. (1999). Renal responses of trout to chronic respiratory and metabolic acidosis, and metabolic alkalosis. *Am. J. Physiol.* **277**, R482–R492.

Wood, J. D. (1958). Nitrogen excretion in some marine teleosts. *Can. J. Biochem. Physiol.* **36**, 1237–1242.

Wright, P. A. (1993). Nitrogen excretion and enzyme pathways for ureagenesis in freshwater tilapia (*Oreochromis niloticus*). *Physiol. Zool.* **66**, 881–901.

Wright, P. A. (1995). Three end products, many physiological roles. *J. Exp. Biol.* **198**, 273–281.

Wright, P. A., and Land, M. D. (1998). Urea production and transport in teleost fishes. *Comp. Biochem. Physiol.* **119A**, 47–54.

Wright, P. A., and Wood, C. M. (1985). An analysis of branchial ammonia excretion in the freshwater rainbow trout: effects of environmental pH change and sodium uptake blockade. *J. Exp. Biol.* **114**, 329–353.

Wright, P. A., and Wood, C. M. (1988). Muscle ammonia stores are not determined by pH gradients. *Fish. Physiol. Biochem.* **5**, 159–162.

Wright, P. A., Randall, D. J., and Wood, C. M. (1988). The distribution of ammonia and H^+ ions between tissue compartments in lemon sole (*Parophrys vetulus*) at rest, during hypercapnia, and following exercise. *J. Exp. Biol.* **136**, 149–175.

Wright, P. A., Iwama, G. K., and Wood, C. M. (1993). Ammonia and urea excretion in Lahontan cutthroat trout (*Oncorhynchus clarki henshawi*) adapted to highly alkaline Pyramid Lake (pH 9.4). *J. Exp. Biol.* **175**, 153–172.

Wright, P. A., Felskie, A. K., and Anderson, P. M. (1995a). Induction of ornithine–urea cycle enzymes and nitrogen metabolism and excretion in rainbow trout (*Oncorhynchus mykiss*) during rainbow trout early life stages. *J. Exp. Biol.* **198,** 273–281.

Wright, P. A., Pärt, P., and Wood, C. M. (1995b) Ammonia and urea excretion in the tidepool sculpin *Oligocottus maculosus:* sites of excretion, effects of reduced salinity and mechanisms of urea transport. *Fish Physiol. Biochem.* **19,** 111–123.

7

UREA AND GLUTAMINE SYNTHESIS: ENVIRONMENTAL INFLUENCES ON NITROGEN EXCRETION

PAUL M. ANDERSON

I. Introduction
II. Carbamoyl-Phosphate Synthetase III and Nitrogen Excretion in Fish
   A. Structure, Properties, and Evolutionary Relationships
   B. Regulation by *N*-Acetyl-L-Glutamate
III. Glutamine Synthetase and Ammonia Detoxification
   A. Elasmobranchs
   B. Teleosts
IV. Urea Cycle in Elasmobranchs
V. Urea Cycle in Teleosts: Adaptation to Unique Environmental Circumstances
   A. Largemouth Bass (*Micropterus salmoides*), Rainbow Trout (*Oncorhynchus mykiss*), and Other Nonureotelic Teleosts: Expression of CPSase III and OCTase in Liver and Muscle
   B. Alkaline Lake-Adapted Tilapia (*Oreochromis alcalicus grahami*): Ureotelism in Muscle as an Adaptation to High pH
   C. Ureotelism and Other Responses to Long-Term Air Exposure
   D. Gulf Toadfish (*Opsanus beta*) and Related Species: A Switch from Ammonotelism to Ureotelism Induced by Stress
   E. Embryogenesis and the Urea Cycle
VI. Summary
References

I. INTRODUCTION

The major end product of nitrogen metabolism in animals is ammonia, which is highly toxic and must be detoxified or excreted. Mammalian and other terrestrial and amphibian vertebrate species and the lungfishes are ureotelic,* that is, they maintain blood levels of ammonia below ≈ 0.03 mM by converting ammonia

---

***Ureotelic:* possess full complement of urea cycle enzymes and excrete most nitrogen as urea; *ureogenic:* potential for urea synthesis exists because full complement of urea cycle enzymes is present; *ureoosmotic:* possess full complement of urea cycle enzymes and urea synthesis and retention is adjusted to maintain osmolarity (Huggins *et al.,* 1969).

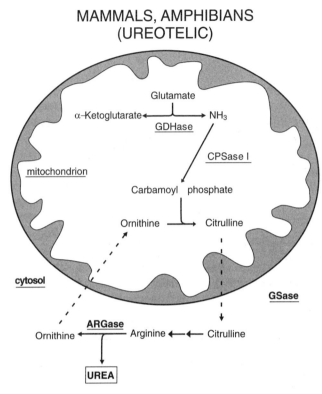

Fig. 1. Illustration of the ammonia-dependent urea cycle in liver of ureotelic terrestrial vertebrates. GDHase, glutamate dehydrogenase; ARGase, arginase; other abbreviations noted in the text. (Adapted from Anderson, 1991, with permission.)

to urea via the classical urea cycle in the liver (Fig. 1). The initial "fixation" of free ammonia is accomplished by conversion of ammonia to carbamoyl phosphate, catalyzed by mitochondrial carbamoyl-phosphate synthetase I (CPSase I) (Meijer et al., 1990; Campbell, 1991). Birds are uricotelic, converting ammonia into uric acid (Fig. 2); in this case the initial enzyme that catalyzes the "fixation" of free ammonia is mitochondrial glutamine synthetase (GSase). Campbell (1991) has noted that the mitochondrial formation of glutamine from ammonia by GSase in uricotelic species is analogous to the mitochondrial formation of citrulline from ammonia (via intermediate formation of carbamoyl phosphate catalyzed by CPSase I) by ureotelic species; both are neutral compounds and both exit the mitochondria, thus providing a mechanism for converting ammonia generated intramitochondrially to a form that would not carry a proton out of the mitochondria and disrupt oxidative phosphorylation (see also Campbell and Anderson, 1991).

As noted in other chapters in this book, the vast majority of teleost fish are

7. UREA AND GLUTAMINE SYNTHESIS

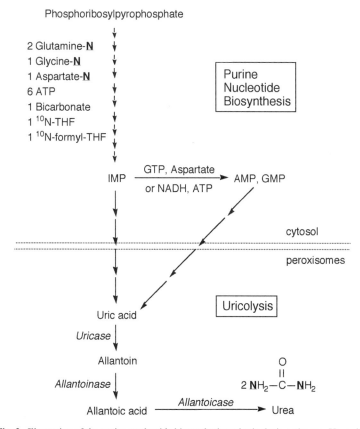

Fig. 2. Illustration of the purine nucleotide biosynthetic and uricolysis pathways. Urea nitrogens and source of urea nitrogens are indicated in bold. Names of uricolytic enzymes are italicized. THF, tetrahydrofolate.

ammonotelic, that is, ammonia generated in the liver and other tissues is not first detoxified by "fixation," but is simply excreted directly across the gills where it is diluted by the surrounding aqueous environment. Fish are more tolerant than other vertebrates to ammonia and many maintain plasma ammonia levels in the range of 0.2 mM or higher (Mommsen and Walsh, 1992; Wood, 1993).

In contrast to most teleost fishes, marine elasmobranchs (sharks, skates, and rays) are ureoosmotic and have an active urea cycle, synthesizing and retaining urea at high concentrations (0.3–0.6 M) primarily for the purpose of osmoregulation (Perlman and Goldstein, 1988; Anderson, 1991, 1995a; Ballantyne, 1997). The biochemical properties of the urea cycle in elasmobranch liver have been studied in considerable detail, revealing several features that are uniquely different from the urea cycle in ureotelic terrestrial vertebrate liver (Anderson, 1991, 1995a;

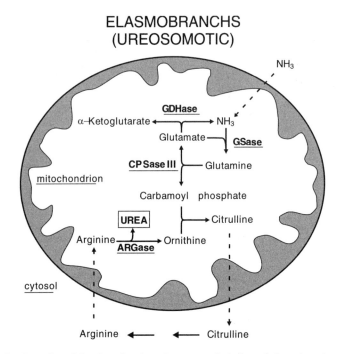

Fig. 3. Illustration of the glutamine-dependent urea cycle in liver of elasmobranchs. See Fig. 1 for abbreviations. (Adapted from Anderson, 1991, with permission.)

Campbell and Anderson, 1991) (Fig. 3). Most importantly, the first step of ammonia "fixation" in liver is catalyzed by GSase, which is localized exclusively in the mitochondria (Casey and Anderson, 1982); glutamine, rather than ammonia, then serves as the substrate for carbamoyl phosphate formation catalyzed by mitochondrial CPSase III (Anderson and Casey, 1984). Thus, the first step of ammonia "fixation" is analogous to that in uricotelic species, that is, formation of glutamine. However, in contrast to uricotelic species, glutamine generated in the mitochondria is not transported out, but is utilized in the mitochondria for carbamoyl phosphate and, ultimately, urea synthesis (Fig. 3). The well-established properties of the urea cycle in elasmobranchs have served as a comparative starting point for investigating the urea cycle pathway in teleost fish (for reviews, see Anderson, 1991, 1995a; Campbell and Anderson, 1991).

Although primarily and generally ammonotelic, most teleost fish do excrete a significant proportion of their total excreted nitrogen as urea [5–20%, but higher in some species, e.g., 30% in largemouth bass (*Micropterus salmoides*), and 50% in some air-breathing fishes; for summaries of the levels of excretion of ammonia and urea, see Campbell and Anderson, 1991; Wood, 1993; Graham, 1997; Saha and Ratha, 1998]. Excreted urea in teleost fish is generally considered to originate via one or more of three possible pathways: uricolysis, catabolism of dietary arginine by arginase, and/or the urea cycle (Wood, 1993; Mommsen and Walsh, 1992; Korsgaard *et al.*, 1995; Anderson, 1995a). As noted by Mommsen and Walsh (1992), it is often generally assumed that urea released by ammonotelic teleost species is derived through uricolysis. Most studies aimed at assessing the significance of the uricolytic pathway as the source of urea have focused on measuring tissue activities of the enzymes uricase, allantoinase, and allantoicase (Brown *et al.*, 1966; Wright, 1993; Wright *et al.*, 1993; McGeer *et al.*, 1994; see additional references in Anderson, 1995a) (Fig. 2). To serve as a major pathway for ammonia detoxification, the uricolytic pathway requires purine nucleotide biosynthesis and subsequent catabolism of purine nucleotides to xanthine and then to uric acid catalyzed by xanthine oxidase (Fig. 2). This pathway is characterized by (1) a flux rate that appears to be carefully regulated for the purpose of providing a balanced supply of purine nucleotides, not for the purpose of ammonia detoxification (Zalkin and Dixon, 1992); (2) subcellular localization in different cell compartments (purine biosynthesis in the cytosol, uricolysis in the peroxisomes; Hayashi *et al.*, 1989); and (3) a requirement for glutamine synthesis. As noted by Anderson (1995a), if a function of the uricolytic pathway in teleost fishes is ammonia detoxification, one might expect to find significant levels of GSase activity in liver localized in the mitochondria, as noted above for uricotelic species (see Campbell, 1991); this seems to be observed, however, only in those few teleost species that are clearly ureotelic. Thus, although uricolysis is likely the source of some or, in some cases perhaps, much of the urea excreted in teleost fishes, this may simply reflect normal catabolism of purine nucleotides, regulated by the demands of the cell for maintaining appropriate levels of adenine nucleotides for energy metabolism and purine nucleotides for DNA and RNA synthesis, as opposed to serving as a functional pathway for ammonia detoxification. Measurement of uricolytic enzymes may have as little relationship to the flux rate of ammonia through the purine nucleotide pathway as measurements of arginase have to the rate of flux of ammonia through the urea cycle pathway, as noted by Brown *et al.* (1966). Studies are needed to assess the significance of uricolysis in the context of the flux rate of ammonia through the purine nucleotide biosynthetic pathway and what percentage of urea excreted is accounted for by this pathway in a diversity of representative ammonotelic teleost species. Note, however, that with respect to the purine biosynthesis/degradation/uricolysis pathway, the urea cycle in fish also requires synthesis of one glutamine for each urea molecule excreted and is just as energetically expensive (see Fig. 3).

The action of arginase on dietary arginine, like uricolysis, no doubt accounts for some of the urea excreted by fish as a normal route for catabolism of arginine arising from dietary sources or protein degradation, but there is little information on how much this accounts for the low levels of excreted urea observed in most teleosts (Mommsen and Walsh, 1992; Korsgaard *et al.*, 1995).

Interest in the urea cycle and regulation of the expression of the urea cycle in teleost fish has increased in recent years due to documentation of the presence of a functional urea cycle in a few adult species of fish where detoxification of ammonia via ureotelism predominates, apparently as an adaptation to unusual environmental circumstances (e.g., stress, air exposure, high pH, exposure to high concentrations of ambient ammonia). In addition, the first two enzymes of the urea cycle, CPSase III and ornithine carbamoyltransferase (OCTase), respectively, have been demonstrated to be transiently expressed during early development in several teleost species (Wright *et al.*, 1995; Chadwick and Wright, 1999; Terjesen *et al.*, 2000). An additional recent observation unique to fish is the reported presence of these two enzymes in skeletal muscle tissue of several teleost species (Korte *et al.*, 1997; Felskie *et al.*, 1998; Kong *et al.*, 1998, 2000; Terjesen *et al.*, 2000) and of high levels of all urea cycle enzymes in muscle of one species (Lindley *et al.*, 1999). Some adult species are ureogenic, that is, low levels of all of the enzymes required for urea synthesis via the urea cycle are present in liver, but the physiological function of the potential urea cycle pathway and of the CPSase is not known (Anderson and Walsh, 1995; Kong *et al.*, 1998; Jow *et al.*, 1999). Finally, several studies have shown that synthesis and temporary storage of glutamine and/or other free amino acids, rather than urea synthesis, represent ammonia detoxification strategies resulting from exposure to increased levels of ambient ammonia or long term air exposure (Ip *et al.*, 1993; Jow *et al.*, 1999; see also Chapter 4). Thus, in contrast to ureoosmotic elasmobranchs or ureotelic terrestrial vertebrates, the expression of the urea cycle and the role of CPSase III and GSase in nitrogen excretion in adult teleosts appear to be highly variable, likely reflecting adaptation to unique environmental circumstances.

The primary focus of this chapter is on recent developments in our understanding of the expression of the urea cycle and the roles of GSase and CPSase III in nitrogen excretion and ammonia detoxification in fish, with emphasis on teleosts; it may be considered an extension of the review by Anderson (1995a). More comprehensive or specific related reviews include those by Perlman and Goldstein (1988), Anderson (1991, 1995a,b), Campbell and Anderson (1991), Griffith (1991), Mommsen and Walsh (1991, 1992), Atkinson (1992), Wood (1993), Wright (1995), Korsgaard *et al.* (1995), Ballantyne (1997), Graham (1997), and Walsh (1997).

II. CARBAMOYL-PHOSPHATE SYNTHETASE III AND NITROGEN EXCRETION IN FISH

A. Structure, Properties, and Evolutionary Relationships

Three different types of CPSases have been identified in higher eukaryotes based on the physiologically significant nitrogen-donating substrate utilized (ammonia or glutamine) and the requirement for the positive allosteric effector N-

acetyl-L-glutamate (AGA) for activity (Anderson, 1995b). CPSase I utilizes only ammonia as the nitrogen-donating substrate and requires the presence of AGA for activity (Ratner, 1973; Marshall, 1976; Meijer *et al.*, 1990):

$$NH_3 + 2\ ATP^{4-} + HCO_3^{1-} \xrightarrow{AGA,\ Mg^{2+}} 2\ ADP^{3-} + Pi^{2-} + NH_2CO_2PO_3^{2-} + H^{1+}. \quad (1)$$

CPSase I is localized in the matrix of liver mitochondria of ureotelic amphibians, terrestrial vertebrates, and lungfishes and catalyzes incorporation of ammonia into carbamoyl phosphate as the first step of the urea cycle. Mammalian CPSase I is also present in mitochondria of the small intestine, but is absent in all other tissues. CPSase I is the only CPSase that cannot utilize glutamine as the nitrogen-donating substrate (Anderson, 1995b).

CPSase II utilizes glutamine instead of ammonia as the physiologically significant nitrogen-donating substrate and does not require AGA for activity (Ratner, 1973; Jones, 1980; Evans, 1986):

$$Glutamine + H_2O + 2\ ATP^{4-} + HCO_3^{1-} \xrightarrow{Mg^{2+}} 2\ ADP^{3-} + Pi^{2-} + NH_2CO_2PO_3^{2-} + glutamate^{1-} + 2\ H^{1+}. \quad (2)$$

CPSase II is present in the cytosol of most cells of higher eukaryotes as part of a multifunctional complex (single polypeptide chain) that includes the next two enzymes of the pyrimidine pathway, aspartate transcarbamoylase and dihydroorotase. It is subject to feedback inhibition by UTP and to allosteric activation by 5-phosphoribosyl-1-pyrophosphate and its function is related to pyrimidine nucleotide biosynthesis. This multifunctional complex is commonly referred to as CAD (Jones, 1980; Evans, 1986).

The third major group of CPSases, CPSase III, found in invertebrates and fish, utilizes the amide group of glutamine as the nitrogen-donating substrate (like CPSase II) but requires AGA for activity (like CPSase I) (Trammel and Campbell, 1970, 1971; Anderson, 1976, 1980, 1981; Casey and Anderson, 1983):

$$Glutamine + H_2O + 2\ ATP^{4-} + HCO_3^{1-} \xrightarrow{AGA,\ Mg^{2+}} 2\ ADP^{3-} + Pi^{2-} + NH_2CO_2PO_3^{2-} + glutamate^{1-} + 2\ H^{1+}. \quad (3)$$

Except for the capability to use glutamine as nitrogen-donating substrate, the structure, properties, subcellular localization, and function of CPSase III are very similar to CPSase I. Like other amidotransferases (Zalkin, 1993), CPSase III can utilize ammonia as substrate, but the V_{max} is generally much lower than that with glutamine and the K_m for ammonia is much higher than the K_m for glutamine. The function of CPSase III in fish is clearly related to the urea cycle, and demonstration of the presence of CPSase III activity, as a likely rate-limiting step in the urea cycle, in fish tissue extracts is a logical first approach for assessing whether or not a functional urea cycle is present. This requires, as a minimum, demonstrating glutamine-dependent CPSase activity and showing that the activity is activated by AGA and is not affected by UTP. As has been noted previously (Anderson, 1995a), however, this is often difficult if CPSase II activity is also present and the level of CPSase III activity is low, as is often the case. Under these circumstances fractionation into mitochondrial and cytosolic fractions and/or gel filtration chromatography to separate CPSase III from CPSase II is required to provide definitive identification of CPSase III. The necessity of such separation is illustrated by several recent studies (Cao *et al.*, 1991; Anderson and Walsh, 1995; Felskie *et al.*, 1998; Terjesen *et al.*, 2000). It is important to note, however, that since the function of OCTase is unique to the urea cycle, establishing the presence of relatively high levels of OCTase activity is also strongly suggestive of the presence of urea cycle activity (see Tables I and II).

The nucleotide sequences of the cDNA for a number of CPSase IIIs have been determined and the deduced amino acid sequences compared with each other and with CPSase Is and CPSase IIs (Hong *et al.*, 1994; Korte *et al.*, 1997; Kong *et al.*, 1998, 2000; Lindley *et al.*, 1999). The high degree of sequence identity among all the CPSases is striking (Simmer *et al.*, 1990; Hong *et al.*, 1994). Nevertheless, the three classes of CPSase can be clearly distinguished when the amino acid identities are compared (Table III). Comparative analysis of the many CPSase sequences now available (particularly CPSase III and CPSase I) will certainly provide clues about (1) the molecular basis for the adaptations represented by the different CPSase IIIs and (2) the defining molecular evolutionary difference(s) between CPSase III and CPSase I whereby CPSase I has lost the capability for utilizing glutamine as substrate. For example, position #6 in Fig. 4 corresponds to the cysteine involved in formation of the γ-glutamyl cysteinyl intermediate required for all glutamine-dependent CPSases II and III (Zalkin, 1993), but which is a serine in CPSase I from human, rat, and *Xenopus laevis*. However, the CPSase I from the frog *Rana catesbeiana* has a cysteine in this position, yet it is a type I CPSase and cannot utilize glutamine as substrate (Helbing and Atkinson, 1994). This indicates that contrary to earlier suggestions (Nyunoya *et al.*, 1985), modifications other than a mutation that results in a serine instead of a cysteine are involved in the inability of a CPSase I to utilize glutamine as substrate. Modeling studies based on the recently determined X-ray structure (Thoden *et al.*, 1997) of *Escherichia coli* CPSase II indicate that the glutamine in position #10, which is conserved in all glutamine-dependent CPSases except the tilapia CPSase III, hydrogen bonds to the γ-glutamyl intermediate and may be essential along with other differences for utilization of glutamine as substrate (J. Thompson and P. M. Anderson, unpublished observations). In *R. catesbeiana* a negatively charged glutamate residue is present in this position, which may preclude utilization of glutamine even though *R. catesbeiana* has a cysteine in position #6. The tilapia CPSase III has a hydrophobic residue (leucine) rather than glutamine in position

Table I
Comparison of the Approximate Levels of CPSase Activities in Liver and Muscle of Various Fish Species

Species	Liver				Muscle			References
	Glutamine (nmol/min/g)	Glutamine + AGA (nmol/min/g)	% Inhibition by UTP	Glutamine (nmol/min/g)	Glutamine + AGA (nmol/min/g)	% Inhibition by UTP		
Common carp	1.6	1.6	95	0.16	0.25	64		Felski et al. (1998)
Channel catfish	2.0	2.0	87	0.17	0.18	39		Felski et al. (1998)
Goldfish	5.2	5.5	93	0.4	0.4	97		Felski et al. (1998)
Rainbow trout	12.6	12.6	92	0.19	0.32	25		Korte et al. (1997)
Bowfin	7.9	9.5	64	0.13	0.34	15		Felski et al. (1998)
Largemouth bass	4.7	16.0	13	1.0	1.9	5		Cao et al. (1991); Kong et al. (1998)
Marble goby	1.2	6.8	20	—	0.4	—		Jow et al. (1999)
Midshipman	2.4	26.1	7	0.1	1.3	0		Anderson and Walsh (1995); Julsrud et al. (1998)
Indian catfish	7	37	0	—	8.0	—		Saha et al. (1997); Saha and Ratha (1987)
Gulf toadfish	36	500	12	—	6.0	—		Anderson and Walsh (1995); Julsrud et al. (1998)
Tilapia (alkaline-lake adapted)	—	40	—	—	170	0		Lindley et al. (1999)
Spiny dogfish shark	33	240	0	—	—	—		Anderson (1980, 1981)
Rat (ammonia)	—	9,000	—	—	0	—		Cohen (1976)

Table II
Comparison of the Approximate Levels of OCTase Activities in Liver and Muscle of Various Fish Species

Species[a]	Liver (nmol/min/g)	Muscle (nmol/min/g)
Common carp	4	3
Channel catfish	350	1,000
Goldfish	170	140
Rainbow trout	2	230
Bowfin	520	280
Largemouth bass	2,100	430
Marble goby	4,370	3,000
Midshipman	13,000	700
Indian catfish	4,200	617
Gulf toadfish	72,000	820
Tilapia (alkaline lake-adapted)	4,370	3,000
Spiny dogfish shark	3,300	250
Rat	300,000	0

[a] Same references as Table I.

Table III
Comparison of Alkaline Lake-Adapted Tilapia CPSase III Sequence to Other CPSases[a]

CPSase	Species	% Identity
III	Rainbow trout	83
III	Largemouth bass	86
III	Gulf toadfish	81
III	Spiny dogfish shark	74
I	Rat	71
I	Human	69
I	Bullfrog (R. catesbeiana)	71
I	Frog (X. laevis)	71
II	Human (pyrimidine CAD)	49
II	Spiny dogfish shark (pyrimidine CAD) (multifunctional)	50
II	Escherichia coli (glutaminase and synthetase subunits)	38

[a] The percent identity of the amino acid sequence of the alkaline lake-adapted tilapia CPSase III to the other CPSases was obtained by alignment using the program ClustalW. Amino acid sequences can be found from the Data Bank under accession numbers 5499724 (alkaline lake-adapted tilapia III), 1518088 (rainbow trout III), 2245664 (largemouth bass III), 6538784 (gulf toadfish III), 530209 (spiny dogfish shark III), 117492 (rat I), 4033707 (human I), 2118284 (bullfrog I), T. E. Lindley and P. M. Anderson, unpublished observations (X. laevis I), 1709955 (human II), 3024509 (spiny dogfish shark II), 115627 and 115621 (glutaminase and synthetase subunits of E. coli II, respectively).

7. UREA AND GLUTAMINE SYNTHESIS

		1 2 3 4 5 6 7 8 9 10 11 12 13 14 15
Human I	289	P L F G I S T G N L I T G L A
Rat I	289	P L F G I S T G N I I T G L A
Xenopus I	282	P V F G V S M G N E I A A L A
Rana I	285	P I F G I C K G N E I A A L A
Tilapia III	288	P V F G I C M G N L I T A L A
Bass III	289	P V F G I C M G N Q I T A L A
Toadfish III	288	P V F G I C M G N Q I T A L A
Trout III	286	P V F G I C M G N Q I T A L A
Shark III	289	P V F G I C M G N Q L T A L A
Hamster II	247	P V F G I C L G H Q L L A L A
Human II	247	P V F G I C L G H Q L L A L A
Shark II	247	P L F G I C L G H Q I L S L A
E. coli II	264	P V F G I C L G H Q L L A L A

Fig. 4. Alignment of the amino acid sequences of several CPSase Is, IIs, and IIIs surrounding the glutamine-binding site for glutamine-dependent CPSases. Alignment was performed using the program ClustalW and displayed using the program SeqVu 1.1. For comparative reference in the text the positions are identified by numbering 1–15, left to right. The numbers in the column are the actual residue numbers of position #1 for the corresponding translated CPSase.

#10, which may explain the observed poor binding of glutamine and low turnover rate with glutamine as substrate compared to ammonia as substrate. It does not seem likely that the difference between CPSase III and CPSase I, which are unique to, and are conserved in, quite different phylogenetic groups, can be defined by a single amino acid residue. Nevertheless, this kind of comparative analysis together with modeling studies suggests that the differences are not great, and identification of these differences and their effects on structure and function are readily amenable to experimental verification.

CPSase III has been highly purified and characterized from both an elasmobranch (*Squalus acanthias*, spiny dogfish shark) (Anderson, 1981; Hong *et al.*, 1994) and a teleost (largemouth bass) (Casey and Anderson, 1983; Kong *et al.*, 1998). Although these two CPSase IIIs have quite similar properties and amino acid sequences, it should not be assumed that all CPSase IIIs in fish have similar properties. For example, as described below, the partially purified CPSase III from the alkaline lake-adapted tilapia *Oreochromis alcalicus grahami* has a lower V_{max} with glutamine as substrate than with ammonia as substrate and AGA has little effect on glutamine binding (Lindley *et al.*, 1999). Another example is the mitochondrial and AGA-dependent CPSase from the Indian catfish *Heteropneustes fossilis*, which is also just as active with ammonia as substrate and activities with ammonia and glutamine appear to be additive (Saha *et al.*, 1997). Although the latter is likely a reflection of a unique CPSase III, an alternative explanation suggested by the authors is that these fish have genes for both a CPSase III and a

CPSase I. This would be of considerable interest, given the current view that CPSase I of terrestrial ureotelic species evolved from CPSase III in fish by adaptation, rather than by gene duplication of CPSase III followed by diversification (Campbell and Anderson, 1991; Hong et al., 1994; Anderson, 1995b).

B. Regulation by N-Acetyl-L-Glutamate

CPSase I is essentially inactive in the absence of AGA and the regulatory role of AGA as a positive allosteric effector for CPSase I and urea synthesis is well established (Meijer et al., 1990; Meijer, 1995). This regulatory role in urea synthesis cannot necessarily be assumed in those species of fish where ureotelism has been documented, because CPSase III is active in the absence of AGA at high glutamine concentrations if excess Mg^{2+} is present, with a level of activity as high as 30% of the activity observed in the presence of AGA (Anderson, 1981; Casey and Anderson, 1983). On the other hand, a characteristic of CPSase III that may be of physiological importance is that the binding of the substrate glutamine and the allosteric effector AGA are synergistic, that is, the K_m for each decreases as the concentration of the other increases. Thus, small changes in the concentrations of both AGA and glutamine at lower and more physiological concentrations of glutamine could have a significant effect on the rate of carbamoyl phosphate synthesis catalyzed by CPSase III. Julsrud et al. (1998) have recently shown that AGA is present in liver of several species of fish and that AGA concentration is much higher in those species and tissues of fish that have significant levels of CPSase III and urea cycle activity; for example, the levels of AGA are higher in liver of adult gulf toadfish (*Opsanus beta*) and spiny dogfish shark, both of which have high CPSase III activity, than in largemouth bass or rainbow trout (*Oncorhynchus mykiss*), which have much lower or no CPSase III activity, respectively. In the gulf toadfish the levels of AGA in liver are considerably higher in the fed than in the fasting state, as is observed in liver in mammalian species; in addition, the level of AGA increases when the gulf toadfish are confined (stressed), which has been shown to induce a ureotelic response (see section below). The observations that AGA is present and that the levels of AGA correlate with parameters related to urea cycle activity suggest that in most fish where ureotelism has been documented AGA probably plays a role in regulation of the urea cycle similar to that observed for amphibian and mammalian species (Julsrud et al., 1998).

III. GLUTAMINE SYNTHETASE AND AMMONIA DETOXIFICATION

Glutamine synthetase catalyzes the first of the reactions shown below, but is usually assayed by the second transferase reaction, because the product γglutamyl hydroxamate is easily detected colorimetrically (Webb and Brown, 1976; Shankar and Anderson, 1985):

$$\text{Glutamate}^{1-} + \text{NH}_3 + \text{ATP}^{4-} + \text{H}^{1+} \xrightarrow{\text{Mg}^{2+}} \text{glutamine} + \text{ADP}^{3-} + \text{Pi}^{1-}, \quad (4)$$

$$\text{Glutamine} + \text{NH}_2\text{OH} + \text{ADP}^{3-} \xrightarrow{\text{Mg}^{2+}, \text{ASO}_4^{3-}} \gamma\text{-glutamyl hydroxamate} + \text{ADP}^{3-} + \text{NH}_3. \quad (5)$$

However, the colorimetric transferase assay is not very sensitive and it is difficult to draw conclusions about units of biosynthetic GSase activity present in tissue extracts because the ratio of transferase to biosynthetic activity, normally about 15:1 (Shankar and Anderson, 1985; Anderson and Walsh, 1995), can be quite variable from one species to another (Walsh, 1996). In addition, assessment of physiologic functions that may involve adaptations reflected in altered kinetic properties requires use of the biosynthetic reaction (Shankar and Anderson, 1985; Walsh, 1996). Studies on the species and tissue distribution of GSase in fish have generally shown high levels of GSase in brain, very high levels in liver of ureoosmotic elasmobranchs, and little or no activity in liver of most teleosts, except in those species that are ureogenic or ureotelic, that is, in association with the glutamine-dependent CPSase III (Webb and Brown, 1976, 1980; Table 1 in Campbell and Anderson, 1991; Cao et al., 1991; Anderson and Walsh, 1995). In addition, the subcellular distribution in liver varies—mitochondrial, cytosolic, or both.

A. Elasmobranchs

The unique feature of CPSase III, when compared to CPSase I, is its ability to use glutamine as the nitrogen-donating substrate. Conversion of ammonia to carbamoyl phosphate for urea synthesis by spiny dogfish liver mitochondria involves obligatory intermediate formation of glutamine, catalyzed by the sequential action of GSase and CPSase III (Anderson and Casey, 1984) (Fig. 3). That this is the sole function of GSase in elasmobranch liver is suggested by the following observations. Given the high K_m of 11 mM for glutamate, the units of GSase activity under *in vivo* conditions are probably not in large excess over the units of CPSase III activity (Shankar and Anderson, 1985). GSase is localized exclusively in the mitochondrial matrix along with CPSase III (Casey and Anderson, 1983), but most biosynthetic reactions requiring glutamine as substrate are localized in the cytosol. CPSase II activity, which requires glutamine as substrate, and the remainder of the pyrimidine nucleotide biosynthetic pathway enzymes, present in liver in virtually all other animal species, is absent in elasmobranch liver. This suggests that other pathways that require glutamine as substrate besides the pyrimidine pathway, such as the purine nucleotide biosynthetic pathway, are also likely absent in elasmobranch liver, requiring biosynthesis elsewhere and transport of the end products or intermediates (e.g., purine and pyrimidine nucleosides) to liver (Hong et al., 1995).

The obligatory formation of glutamine as the first step in the conversion of ammonia to urea in elasmobranchs likely results in a greater efficiency for sequestering low concentrations of ammonia for urea synthesis than that obtained by converting ammonia directly to carbamoyl phosphate catalyzed by CPSase I. This is due to the unusually low K_m (≈ 3 μM) (Shankar and Anderson, 1985) for ammonia exhibited by the mitochondrial GSase and the lower K_m (≈ 0.2 mM) for glutamine exhibited by CPSase III relative to the higher K_m (≈ 5 mM) for ammonia exhibited by CPSase I (Anderson, 1981). Isolated mitochondria from spiny dogfish liver have been shown to efficiently convert low concentrations of ammonia to citrulline, involving the sequential catalytic action of mitochondrial GSase, CPSase III, and OCTase (Casey and Anderson, 1985).

The only elasmobranch GSase that has been highly purified and characterized is that from spiny dogfish shark liver (Shankar and Anderson, 1985). Other than the unusually low K_m for ammonia, the structural and kinetic properties of elasmobranch GSases are generally similar to those of mammalian GSases (Campbell and Anderson, 1991). However, whereas GSase in mammalian species is localized in the cytosol in all tissues (Campbell and Anderson, 1991), the subcellular localization of GSase in spiny dogfish shark is tissue specific, exclusively mitochondrial in kidney and liver (presumably related to urea synthesis) and exclusively cytosolic in brain and other tissues (presumably related to biosynthetic or ammonia detoxification functions) (Campbell and Anderson, 1991). The available evidence indicates that spiny dogfish shark have only one gene for GSase (Campbell and Anderson, 1991; Laud and Campbell, 1994), suggesting that the compartmental isozymes arise either from separate mRNAs generated by differential transcription or by differential translation of a single mRNA. Laud and Campbell (1994) have shown that the GSase mRNA has two start codons and have proposed that this provides a mechanism for obtaining the two isozymes (mitochondrial and cytosolic) consistent with the observation that the subunit molecular weight of the mitochondrial GSase is higher than the molecular weight of the cytosolic GSase (estimated by SDS–PAGE as $\approx 47,000$ and $\approx 45,000$, respectively, the difference being ≈ 2000). Translation beginning at the second start codon would give a product with a predicted molecular weight of 41,869, presumably the cytosolic GSase. Translation beginning at the first start codon would give a product that has 29 additional amino acids (predicted molecular weight of 45,406), the sequence of which has the properties of a mitochondrial targeting sequence with two putative cleavage sites; if processed in this way, the "mature" product would have a molecular weight of 43,680, presumably the mitochondrial GSase, giving a difference of ≈ 2000.

B. Teleosts

The subcellular localization of GSase in the ureotelic freshwater air-breathing teleost *H. fossilis* is tissue specific, analogous to elasmobranchs (Chakravorty *et al.,* 1989). However, in most teleosts that are ureotelic or ureogenic, GSase in liver is usually localized primarily in the cytosol or in both the cytosol and mitochondria. The only GSases from teleosts that have been isolated and characterized are the mitochondrial and cytosolic GSases from liver of gulf toadfish (Walsh, 1996). The structural and kinetic properties of the purified enzymes are similar to those from spiny dogfish shark and to each other; the two do differ, however, in their ratio of biosynthetic to transferase activities, pH optimum, and inhibition by methionine sulfoximine, which may imply a mechanistic difference of physiologic significance between the two (Walsh, 1997). Like elasmobranchs, there appears to be only one gene, and sequence analysis of the gulf toadfish GSase mRNA has shown that the mechanism of expression of isozymes targeted to different compartments, cytosolic and mitochondrial, appears to be analogous to that in spiny dogfish (Walsh *et al.,* 1999). Just how this allows differential regulation of one or the other, which must occur as noted in a later section, remains to be elucidated, however.

IV. UREA CYCLE IN ELASMOBRANCHS

Many of the known properties and unique aspects of the urea cycle in elasmobranchs have been reviewed previously (Anderson, 1991, 1995a; Campbell and Anderson, 1991; Ballantyne, 1997) and several are alluded to above in the context of discussions of CPSase III and GSase (see Fig. 3). The role of urea transporters is discussed in Chapter 8.

Elasmobranchs are ureoosmotic, retaining high concentrations of urea in their tissues for the purpose of osmoregulation, which is accomplished by low permeability of the gills and by reabsorption mechanisms in the kidney (Wood *et al.,* 1995a; see also Chapters 1 and 8). As noted by Wood *et al.* (1995a), a system designed to retain urea may not be designed to excrete urea. Nevertheless, their studies clearly confirm the commonly held view that elasmobranchs are also ureotelic, excreting the majority of their nitrogen via the gills as urea ($>97\%$ as urea under resting conditions). Low permeability of ammonia as well as low urea permeability at the gills may be important for maintaining urea levels for osmoregulation during periods of fasting; ammonia is the major form whereby nitrogen is transported in the circulatory system, and loss at the gills would preclude its use for urea synthesis in the liver (Wood *et al.,* 1995a). The high affinity of liver mitochondria for ammonia for use in citrulline and, ultimately, urea formation, as noted above, may contribute to the strategy of minimizing loss of ammonia as a means of detoxification (Casey and Anderson, 1985). Regulation of urea synthesis for these two different purposes may primarily be related to osmoregulation, excess ammonia simply being detoxified by conversion to urea and urea excretion being regulated in response to the plasma urea levels required for osmoregulation.

Prominent unique features of the urea cycle in elasmobranchs include (1) the synergistic effect of the binding of AGA and glutamine to CPSase III (Anderson, 1981), (2) localization of arginase in the mitochondrial matrix in liver (Casey and Anderson, 1982, 1985), and (3) localization of GSase in the mitochondrial matrix in liver and the obligatory intermediate formation of glutamine from ammonia as the first step in the synthesis of citrulline (and urea) from ammonia (Anderson and Casey, 1984). The presence of high levels of AGA in liver mitochondria and the significant effect of AGA on ATP binding as well as glutamine binding by CPSase III would suggest that AGA plays a role in urea cycle regulation analogous to that in mammalian species (Anderson, 1981; Casey and Anderson, 1983). However, this remains to be demonstrated experimentally and may be characterized by features uniquely related to the role of urea synthesis in osmoregulation. Efforts to characterize the enzyme responsible for synthesis of AGA have not been successful (L. Freiburger and P. M. Anderson, unpublished observations). This enzyme (AGA synthase) in mammalian species is localized in the mitochondria and is regulated by arginine, a positive allosteric effector; the mitochondrial localization of the high levels of arginase in elasmobranch liver would seem to preclude this mode of regulation in elasmobranchs if AGA synthase is also localized in the mitochondria (Meijer et al., 1990; Meijer, 1995).

The metabolic significance, if any, of the mitochondrial localization of arginase is also unknown. This does not appear to play a role in regulating ornithine availability (Casey and Anderson, 1985). However, CPSase III and GSase are both progressively inhibited by increasing concentrations of urea at levels that are physiologically significant, suggesting that urea may act as a classical end-product feedback inhibitor; this effect may be rendered more sensitive by the fact that urea is produced by the action of arginase in the mitochondria rather than in the cytosol (Anderson, 1981; Shankar and Anderson, 1985). This hypothesis has received support from the observation that mitochondrial formation of citrulline from ammonia, glutamate, and bicarbonate, with succinate as the energy source, is inhibited by urea, but under the same conditions mitochondrial respiration is not inhibited (Anderson, 1986).

As noted above, CPSase II and the other enzymes of the pyrimidine pathway are not expressed in elasmobranch liver tissue. Anderson (1989) suggested that this may be due to the fact that glutamine is formed only in the mitochondria in liver and this glutamine may be utilized exclusively for citrulline synthesis and not be available as substrate for a cytosolic CPSase II, or for other cytosolic amidotransferase biosynthetic enzymes, such as glutamine phosphoribosylpyrophosphate amidotransferase and CTP synthetase (Zalkin, 1993). In the case of pyrimidine and purine biosynthesis, for example, this would necessitate a substantial transport of nucleosides from other tissues to liver and, probably, a very active salvage pathway in liver. It has been suggested that this situation in liver, together with the unusually low levels of glutamine in plasma, may have a causative relationship to the observed unusually low incidence of neoplasms in elasmobranchs

(Anderson, 1989). Ballantyne (1997) has recently advanced this idea in more detail in the context of other aspects of metabolism attributed to the high concentrations of urea in elasmobranchs.

Like exon I for the CPSase I gene for rat and frog (*R. catesbeiana*) (Lagacé *et al.*, 1987; Chen, 1995), the spiny dogfish CPSase III exon 1 contains the coding sequence for the mitochondrial signal peptide (38 amino acids) and four N-terminal amino acids of the mature enzyme (Hong *et al.*, 1996). Thus, the junction of exon 1 and intron 1 in the rat and frog CPSase I genes is concordant with that of the spiny dogfish shark CPSase III gene. This is consistent with the view that the CPSase I gene evolved or was adapted from the CPSase III gene (Campbell and Anderson, 1991; Hong *et al.*, 1994; Anderson, 1995b).

A unique feature of the TATA box (TACAAA) in the promoter for the spiny dogfish shark CPSase III gene is the existence of C (rather than the much more common T) at the third position, suggesting that, as in other species where this occurs, it only functions as a weak promoter (Hong *et al.*, 1996). CPSase III mRNA is very abundant in spiny dogfish shark liver (Hong *et al.*, 1995), as is the protein (Anderson, 1981; Anderson and Casey, 1984). For a gene with a weak TATA box promoter sequence to be transcribed in the liver at a high rate but not to be expressed in other tissues, it is reasonable to assume that a tissue-specific transcription factor(s) may be needed to bind to an enhancer(s) of the promoter. Consensus sequences for binding C/EBPα, an enhancer factor implicated in liver-specific gene expression, have been identified in the promoter of the spiny dogfish shark CPSase III gene (Hong *et al.*, 1996). Moreover, because similar consensus sequences have been identified in the promoter of the amphibian frog (*R. catesbeiana*) and rat CPSase I genes (Chen, 1995; Chen and Atkinson, 1997; Goping *et al.*, 1992) and because C/EBP-binding proteins, in concert with other transcription factors, have been implicated in regulating the expression of CPSase I (Goping *et al.*, 1992; Lagacé *et al.*, 1992; Chen *et al.*, 1994), it seems reasonable to assume that this factor also plays a similar role in regulating the expression of the CPSase III gene in spiny dogfish shark.

Consensus sequences for heat-shock elements have also been identified in the spiny dogfish shark CPSase III gene promoter as well as in the rat CPSase I promoter, suggesting that these genes might be activated by elevations in environmental temperatures and/or by other stressful conditions such as changes in salinity (Hong *et al.*, 1996). Perhaps there is an evolutionary advantage for an organism that can produce urea under stressful conditions. Precedence for this possibility has been reported; gulf toadfish, which has high levels of CPSase III and the other urea cycle enzymes, is known to switch to ureogenesis when subjected to the stress of physical confinement (Walsh, 1997).

It has been pointed out that studies on the mechanism of gene regulation can provide insights into the evolutionary relationships between the different CPSases (Takiguchi *et al.*, 1989; Farnham and Kollmar, 1990; Goping *et al.*, 1992). This would appear to be particularly true for the CPSase III gene in fish in view of the diversity of expression and function of CPSase III activity among different species. In addition to providing insights into the function(s) of the urea cycle in fish and the evolutionary origin of the urea cycle and of CPSase I in terrestrial vertebrates, such studies should prove to be important as models for understanding the regulation of tissue-specific transcription of eukaryotic genes (Hong *et al.*, 1996).

The pathway of urea synthesis and the basic kinetic properties of the enzymes involved in elasmobranchs are fairly well understood. However, considerable work remains if we are to understand the regulatory relationships between the urea cycle, the synthesis, transport, and utilization of glutamine, and the pathways of ammonia formation outside the liver in the context of urea synthesis for osmoregulation versus urea synthesis for ammonia detoxification.

V. UREA CYCLE IN TELEOSTS: ADAPTATION TO UNIQUE ENVIRONMENTAL CIRCUMSTANCES

Teleost fishes cannot be easily categorized in the context of expression of the urea cycle. As noted above, urea cycle enzymes are nearly undetectable and there is little evidence for urea cycle activity in most adult teleosts. However, some are ureogenic but the role of the urea cycle is not evident, and some are constitutively ureotelic or become ureotelic in response to various environmental factors. This section focuses primarily on the few teleost species where recent studies have provided insight into expression of the urea cycle in teleost fish.

A. Largemouth Bass (*Micropterus salmoides*), Rainbow Trout (*Oncorhynchus mykiss*), and Other Nonureotelic Teleosts: Expression of CPSase III and OCTase in Liver and Muscle

The presence of CPSase III in fish was first described in largemouth bass liver (Anderson, 1976) and the properties of highly purified and characterized largemouth bass CPSase III (Casey and Anderson, 1983), as well as the spiny dogfish shark CPSase III (Anderson, 1981), have served as reference points for studying CPSase III from other species. The CPSase IIIs from these two species are quite similar, except that the spiny dogfish shark CPSase III is much more sensitive to inhibition by urea and activation by trimethyamine oxide, probably reflecting its role in osmoregulation, as noted above.

The level of CPSase III activity in largemouth bass liver is much lower than in elsmobranchs, but low levels of all of the other enzymes required for urea synthesis, including GSase, are also present in liver of adult largemouth bass, establishing the possibility of a functioning urea cycle (Table I). However, several observations suggest that the urea cycle pathway and CPSase III in adult largemouth bass may not have a physiologically significant role. In contrast to elasmobranchs, GSase in largemouth bass is localized in the cytosol where glutamine is required for other amidotransferase reactions, including the cytosolic CPSase II (Cao *et al.*, 1991). Thus, glutamine may not be readily available for carbamoyl phosphate synthesis and, consequently, urea formation, catalyzed by the mitochondrial CPSase III. The role of glutamine-dependent CPSase III activity in liver may also be limited by the fact that the units of biosynthetic GSase activity in liver are about equal to or less than the units of CPSase III activity (Kong *et al.*, 1998). As would be expected, isolated mitochondria do not synthesize citrulline from glutamate plus ammonia as do isolated mitochondria from spiny dogfish shark liver, but neither do they appear to synthesize citrulline from glutamine (Cao *et al.*, 1991), as occurs in elasmobranchs.

Largemouth bass do excrete a somewhat higher percentage of their total nitrogen as urea (about 30%) compared to other nonureogenic teleosts, perhaps reflecting their ureogenic nature (Kong *et al.*, 1998). However, exposure to elevated levels of external ammonia, which results in a large increase in plasma ammonia levels as expected, does not result in an alteration of the rate of urea excretion and does not alter the level of expression of CPSase III or other urea cycle enzyme activities (Kong *et al.*, 1998). In contrast, exposure to elevated levels of external ammonia resulting in an increased rate of urea excretion has been observed in other fish known to have the full complement of urea cycle enzymes (Saha and Ratha, 1986, 1989, 1990, 1994; Wood *et al.*, 1989; Walsh *et al.*, 1990). Thus, largemouth bass, while ureogenic, are primarily ammonotelic and remain so even when exposed to high levels of external ammonia.

Although all urea cycle enzymes are present in largemouth bass liver, assuming that the urea excreted arises primarily from the urea cycle pathway, the total units of CPSase III activity in liver are not sufficient to account for the urea that is excreted. This led Kong *et al.* (1998) to analyze the levels of CPSase III activity in other tissues. Surprisingly, CPSase III as well as OCTase activities were found to be present in skeletal muscle (Tables I and II). The level of activity in muscle is very low, but given the fact that muscle accounts for more than 50% of total mass, the total units of CPSase III in muscle were found to be more than sufficient to account for the observed rate of urea excretion. Ribonuclease protection assays have confirmed the expected presence of CPSase III mRNA in muscle and liver, and have also established that CPSase III mRNA is expressed in kidney, spleen, and intestine (Kong *et al.*, 1998). Interestingly, the level of expression of CPSase III mRNA in liver appears to be less than in any of the other tissues.

In mammalian species the only tissue besides liver that has both CPSase I and OCTase activity is the intestinal mucosa (Jones *et al.*, 1961; Raijman, 1974); however, argininosuccinate synthase and argininosuccinate lyase are not present in the intestinal mucosa, and the citrulline formed is transported to other tissues for conversion to urea (Windmueller and Spaeth, 1981; Meijer et al., 1990). A similar kind of interorgan urea cycle pathway may occur in largemouth bass with muscle, as well as intestine. Given that CPSase III appears to be expressed in many tissues other than liver, establishing the presence or absence of all urea cycle enzymes in many tissues besides liver (e.g., kidney) is needed, as are related metabolic studies (e.g., demonstrating the conversion of [^{14}C]CO_2 into [^{14}C]urea), before a significant role for the urea cycle in adult largemouth bass can be ruled out.

The observation that there is sufficient CPSase III activity in muscle to account for the urea excreted does not establish that the urea cycle is the major source of urea, however. In fact, the possible significance of CPSase III in muscle is rendered problematical given the observations that (1) the level of GSase in muscle is very low and argininosuccinate synthetase and argininosuccinate lyase activities could not be detected (the latter activities could not be detected in intestinal extracts, either) (Kong et al., 1998) and (2) the levels of argininosuccinate synthetase and argininosuccinate lyase activities in liver are also very low and cannot account for the urea that is excreted even if an interorgan transfer of citrulline from muscle to liver occurred (Cao et al., 1991).

CPSase III, OCTase, and GSase are transiently expressed at high levels during embryogenesis in rainbow trout (see below). Perhaps this also occurs during embryogenesis in largemouth bass; the low level of expression of CPSase III and the other urea cycle enzymes in the adult may simply represent a basal level of transcription that is not of physiologic importance. In other teleosts closely related to largemouth bass, such as crappies and bluegills, where these enzymes cannot be detected in the adult liver (Cao et al., 1991), transcription may be more tightly regulated. The possibilities that other yet undiscovered environmental, dietary, or life-cycle circumstances may induce an increased expression of CPSase III and the other urea cycle enzymes in largemouth bass or that a low level of urea synthesis via the urea cycle may have a physiologic function should continue to be a consideration.

In contrast to largemouth bass, CPSase III and other urea cycle enzymes cannot be detected in liver of rainbow trout. However, as in largemouth bass, low levels of CPSase III and OCTase are present in muscle extracts (Tables I and II) (Korte et al., 1997). As shown in Fig. 5, while CPSase II mRNA is expressed in all tissues analyzed, CPSase III mRNA is expressed only in muscle, which is consistent with the observation that there is virtually no CPSase III activity in liver. The prominent band of CPSase III mRNA derived from muscle RNA seems at odds with the observed exceedingly low levels of CPSase III activity in muscle. As with largemouth bass, the level of GSase activity is also exceedingly low in muscle, as are argininosuccinate synthetase and argininosuccinate lyase activities (Korte et al., 1997). Whole-animal experiments carried out by Chiu et al. (1986) showed that juvenile rainbow trout have the ability to convert [^{14}C]ornithine to [^{14}C]arginine, indicating an active urea cycle. If a physiologically significant urea

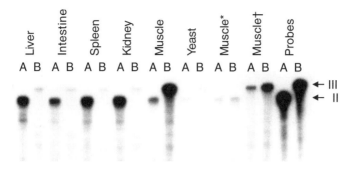

Fig. 5. Expression of CPSase II and CPSase III mRNA in different tissues of rainbow trout fingerlings measured by ribonuclease protection assays. Lanes A and B correspond to rainbow trout-specific CPSase II and CPSase III probes, respectively. (Adapted from Korte et al., 1997, with permission.)

cycle does operate in adult rainbow trout, then it would appear that at least the first two steps probably occur predominately in muscle.

Why would CPSase III and OCTase be present in muscle tissue when these enzymes have no known function outside of the urea cycle? A possible explanation may be that they serve to detoxify excess ammonia released in muscle from the deamination of AMP catalyzed by AMP deaminase as a component of the purine nucleotide cycle during exercise (Lowenstein, 1990; Mommsen and Hochachka, 1988). The significantly elevated urea excretion rates in rainbow trout relative to ammonia observed during high-speed compared to low-speed aerobic swimming (Lauff and Wood, 1996, 1997; Alsop and Wood, 1997) may be related to ammonia detoxification via an interorgan urea cycle pathway; note that in these studies total nitrogen excretion remains unchanged and urea nitrogen excretion remains less than ammonia nitrogen excretion. Citrulline formed in muscle from ammonia catalyzed by CPSase III and OCTase could be transported to liver for processing to urea, since the remaining urea cycle enzyme activities are present in liver (Korte et al., 1997).

Marble goby (*Oxyeleotris marmoratus*) (Jow et al., 1999) and midshipman (*Porichthys notatus*) (Anderson, 1980; Anderson and Walsh, 1995; T. Mommsen and P. M. Anderson, unpublished observations) are both ureogenic but they are primarily ammonotelic, not ureotelic, analogous to largemouth bass. In addition to liver, CPSase III and OCTase activity is present at low levels in muscle of both species (A. Ip and P. M. Anderson, unpublished observations, and T. Mommsen and P. M. Anderson, unpublished observations, respectively), similar to largemouth bass and rainbow trout.

Felskie et al. (1998) recently reevaluated previous reports of urea cycle-related CPSase and related enzyme activities in liver of several ammonotelic teleost species (common carp, *Cyprinus carpio;* goldfish, *Carassius auratus;* channel catfish, *Ictalurus punctatus*) and a holostean fish (bowfin, *Amia calva*). In this study kidney, intestinal, muscle, and liver extracts as well as the mitochondrial and cytosolic fractions from liver were analyzed. CPSase II activity was present in liver of all species, but there was no definitive evidence for CPSase III activity. A low level of CPSase III activity was tentatively identified in bowfin muscle and perhaps intestine and in common carp muscle; low levels of OCTase were present in all tissues of all four species except common carp. Another study aimed at measuring CPSase III activity in tissues where CPSase II activity is present showed that CPSase II could be readily detected in liver extracts of bullhead (Ameiurus family), bluegills and crappies (like largemouth bass, members of the sunfish family), but that CPSase III activity could not be detected (Cao *et al.*, 1991). Terjesen *et al.* (2000), using gel filtration chromatography to separate CPSase II from CPSase III, have definitively established that, like rainbow trout, CPSase III activity is absent in liver and kidney, but is present in muscle of Atlantic halibut (*Hippoglossus hippoglossus* L.). These studies illustrate the difficulty of ascertaining whether CPSase III is present when higher levels of CPSase II are present, and that even if present, the levels are usually very low and the physiologic significance, if any, as with rainbow trout or largemouth bass as noted above, remains to be determined.

The presence of CPSase III and OCTase activity in muscle has now been reported in several species of fish, but the generality of these observations and the physiologic significance, if any, remain to be determined. Only two observations suggest that muscle can play a physiologically significant role in urea synthesis in some fish. The observation that the AGA concentration in gulf toadfish muscle increases when the fish are fed provides indirect evidence that the presence of these first two enzymes of the urea cycle in muscle may be physiologically significant (Julsrud *et al.*, 1998). Secondly, as discussed in the next section, one species of fish has adapted to life in a highly alkaline environment by expressing high levels of urea cycle enzymes in muscle and excreting nitrogen wastes as urea. Studies are clearly needed to determine if an interorgan pathway between muscle or other tissues occurs as a general characteristic of fish, analogous to the interorgan pathway between intestine, kidney, and liver for mammalian species, as noted above.

B. Alkaline Lake-Adapted Tilapia (*Oreochromis alcalicus grahami*): Ureotelism in Muscle as an Adaptation to High pH

At water pH values as high as 10, branchial excretion of ammonia in fish is severely inhibited (Wright and Wood, 1985; Randall and Wright, 1989; Lin and Randall, 1990; Wilkie and Wood, 1991; Yesaki and Iwama, 1992). Yet, the alkaline lake-adapted tilapia thrives as the only resident fish in Lake Magadi, Kenya, a highly alkaline lake where the water pH is 10–10.5. This species of fish has adapted to life in this environment by expressing high levels of all urea cycle enzymes and excreting virtually all nitrogen wastes as urea, that is, by becoming ureotelic (Randall *et al.*, 1989; Wood *et al.*, 1989). Moreover, although all urea cycle enzymes are present in liver of this species, much higher concentrations of CPSase III and OCTase (Tables I and II) as well as all other urea cycle enzymes are expressed in muscle where the majority of urea is apparently synthesized (Lindley *et al.*, 1999). Analysis for urea cycle enzyme activities in muscle was prompted by the observations that the levels of urea cycle enzyme activities in liver are not sufficient to account for the observed rate of urea excretion by this species (Randall *et al.*, 1989; Wood *et al.*, 1989, 1994) and the reports noted above of low levels of CPSase III and OCTase activities in muscle of some teleosts. The level of these enzymes (Tables I and II) as well as the other urea cycle enzyme activities in muscle are exceedingly high, comparable to the levels in liver of ureoosmotic elasmobranchs and the ureogenic gulf toadfish (next section), and are more than sufficient to account for the observed rates of urea excretion (Lindley *et al.*, 1999). This adaptive expression of urea cycle enzymes in muscle is in sharp contrast to a closely related tilapia (*O. nilotica*) species that lives in freshwater a few kilometers away; this species has virtually no urea cycle enzyme activity, excretes only 20% of the total nitrogen as urea, and does not respond to ammonia loading by increased urea excretion (Randall *et al.*, 1989; Wood *et al.*, 1989; Wright, 1993).

Several aspects of this adaptation are important to note in the context of assessing the possible significance and the nature of expression of the urea cycle in muscle in fish noted above. Although there is considerable GSase activity in liver, there is little GSase activity in muscle of this species, an observation that seems to be consistent with observations noted above for other species. In addition, in liver the GSase is localized in the cytosol, not the mitochondria, and the total units of GSase activity in muscle and liver fall far short of the level needed to support urea synthesis if glutamine was an obligatory intermediate as it is in elasmobranch liver (Anderson and Casey, 1984). The apparent lack of sufficient glutamine as substrate may be reflected in several kinetic properties of the alkaline lake-adapted tilapia CPSase III that are uniquely different from spiny dogfish shark or largemouth bass CPSase III. Most notably, (1) the V_{max} for the alkaline lake-adapted tilapia CPSase III with ammonia as substrate is just as high as that with glutamine; (2) perhaps related to property 1, the binding of glutamine and AGA are not synergistic (AGA does affect the binding of ATP, however, as it does in other CPSase IIIs); (3) the K_m for glutamine is quite high (2 mM compared to 0.2 mM for other CPSase IIIs); and (4) the enzyme has substantial activity in the absence of AGA (i.e., AGA has little effect on activity with ammonia). Thus, adaptations may include kinetic changes that favor ammonia as the primary substrate. A unique modification of the glutamine binding site that might contribute to this adaptation is noted in the comparative alignments in Fig. 4.

The adaptation that led to expression of the urea cycle enzymes in muscle might appear to require a rather complex alteration in the gene regulatory mechanisms that control differential tissue expression. Alternatively, fish may have in place control elements that normally target muscle as a site for expression under certain environmental or life-cycle circumstances, so that adaptation would have involved a quantitative change rather than a qualitative change in expression. The presence of low levels of CPSase III and OCTase in muscle of several teleost species noted above supports this possibility. The presence of all urea cycle enzymes, not just CPSase III and OCTase, in muscle of the alkaline lake-adapted tilapia, suggests the possibility that the apparent absence of argininosuccinate synthetase and argininosuccinate lyase in muscle of other teleosts may simply reflect levels too low to detect, thus not necessitating the proposal of an interorgan transfer of urea cycle intermediates to convey significance to the presence of CPSase III and OCTase in muscle. The necessity for the adaptive localization of the urea cycle in muscle in the alkaline lake-adapted tilapia is suggested by the fact that the catalytic turnover rate of CPSase III is quite low and that it is simply not possible to package enough enzyme into the mitochondria in the relatively small liver of these fish to accomplish the observed rate of urea formation (Lindley *et al.*, 1999). In addition, the anatomical arrangements required for efficient swimming would not easily allow an increased liver mass.

As noted in a recent review by Wilkie and Wood (1996), excretion of urea as a result of expression of urea cycle enzymes is apparently not a universal mechanism in fish for adapting to an alkaline aqueous environment. A study of four teleost species of fish native to alkaline (pH 9.5) Pyramid Lake in the United States (Nevada) showed that the processes for nitrogenous waste excretion were the same as those of other teleost species, that is, the four species were primarily ammonotelic. Although the percent of total nitrogen excreted as urea was found to be somewhat higher than in other species, the authors concluded that this was likely due to uricolysis, since high levels of the uricolytic enzymes and very low levels of urea cycle enzymes were present in liver (McGeer *et al.*, 1994). Challenging one of the four species (cutthroat trout, *Oncorhynchus clarki henshawi*) to higher alkalinity (pH 10) (Wilkie *et al.*, 1993) resulted in increased GSase activity and urea excretion rate, but this was not accompanied by an increase in urea cycle enzyme activities. This species appears to adapt to an alkaline environment by dramatically lowering its rate of nitrogen metabolism. Similar results have been reported for a teleost fish from Lake Van in eastern Turkey, where the pH is 9.8 (Danulat and Kempe, 1992). Like the cutthroat trout, this fish has little or no urea cycle capability, but adaptation appears to be simply a high ammonia tolerance. It would be of interest to measure the level of CPSase III and OCTase in muscle of these species. Walsh and colleagues (P. J. Walsh, personal communication) have noted that a major factor in whether or not a species becomes obligately ureotelic when faced with alkaline waters may be the buffering capacity of the water. Well-buffered water, as in Lake Magadi, may prevent acidification of the local boundary layer of water at the gill surface to assist ammonia excretion, as occurs in freshwater rainbow trout (Wright *et al.,* 1989).

C. Ureotelism and Other Responses to Long-Term Air Exposure

Many air-breathing fish are capable of surviving out of water for days or weeks. Under such circumstances where the normal route of excretion across the gills into an aqueous environment is severely restricted, it might be reasonable to suggest that storage of ammonia may occur by conversion to urea or amino acids or other alternative nitrogen-containing component until reimmersion or elimination of ammonia via other routes may occur. This topic has been reviewed briefly by Anderson (1995a) and more recently and more extensively by Graham (1997) and is a topic in several chapters in this volume. In general it can be stated that a clear relationship between amphibious air-breathing capability and one specific mechanism of ammonia detoxification as an alternative to ammonotelism is not apparent. An increased rate of urea excretion is invoked in some species on emersion, but not in many others, and it is not clear in the former where the urea comes from. Alternative routes of detoxification/excretion include uricolysis, synthesis and storage of glutamine and/or other amino acids, and elimination of ammonia through the skin (see Chapter 4). In this section, recent studies that have contributed in a definitive way to a clearer understanding of the metabolic mechanisms of ammonia detoxification/excretion are discussed, in particular as related to urea cycle capability.

Saha and Ratha and coworkers have reported an extensive series of studies in recent years with several air-breathing teleost species from the Indian subcontinent, reporting that although ammonotelic, most of these species have a full complement of measurable urea cycle enzyme activities in liver and other tissues. In some, if not all, of these species transition from ammonotelism to ureotelism occurs when exposed to high concentrations of ambient ammonia concentrations, extended periods of time out of water (air exposure), or semiarid conditions inside mud during habitat drying as a normal yearly environmental circumstance (Saha and Ratha, 1986, 1987, 1989, 1990, 1994; Saha *et al.,* 1988, 1995, 1997, 1999, 2000; Saha and Das, 1999; Ratha *et al.,* 1995; Das *et al.,* 1991; Dkhar *et al.,* 1991; Chakravorty *et al.,* 1989; see also Ramaswamy and Reddy, 1983). Much of this earlier work has been reviewed by Anderson (1995a) and more recently and comprehensively by Saha and Ratha (1998).

The features of the air-breathing amphibious Singhi catfish *H. fossilis,* which is typical of the air-breathing teleost species from the Indian subcontinent, can be summarized in the context of nitrogen excretion as follows: (1) Relatively high levels of a full complement of the urea cycle enzymes are present in liver and kidney, and three of these enzyme activities (CPSase, OCTase, and arginase) are also present, at lower levels, in brain and muscle (Tables I and II) (Saha and Ratha, 1987, 1989); (2) subcellular localization of the urea cycle enzymes and GSase is analogous to that of elasmobranchs (i.e., GSase and arginase are both localized in the mitochondria in liver and kidney and GSase is localized in the cytosol in brain) (Chakravorty et al., 1989; Dkhar et al., 1991); (3) a difference from elasmobranchs is that the pyrimidine-related CPSase II is present in liver cytosol (Saha et al., 1997); (4) the fish is primarily ammonotelic in water (Saha et al., 1988; Saha and Ratha, 1989); (5) the fish switches to ureotelism when exposed to air for 24 h (urea accumulates in tissues at a faster rate than ammonia and is accompanied by a 1.5- to 2-fold induction of all urea cycle enzyme activities except arginase) (Ratha et al., 1995); and (6) the fish is extremely tolerant to high concentrations of ambient ammonia, surviving exposure to 75 mM ammonium chloride for several weeks and apparently becoming ureotelic (urea excretion rate increases 3- to 4-fold after 7 days, accompanied, as in air exposure, by a 1.5- to 3-fold induction of all urea cycle enzyme activities except arginase) (Saha and Ratha, 1986, 1990, 1994).

Interestingly, despite the high ureogenic capability of this species, the percent of total nitrogen excreted as urea under normal conditions is quite low, about 15% (Saha and Ratha, 1989). If liver and muscle account for 1.5 and 50% of total body weight, respectively, the total units of CPSase III activity per gram of fish is about 0.07 and 0.24 μmol/g fish/h at 37°C, respectively (Saha and Ratha, 1987, 1989). Thus, the level of CPSase activity in liver (plus the lesser amount in kidney) seems barely sufficient to account for the rate of urea excreted [≈0.05 μmol/g fish/h at 20°C (Saha and Ratha, 1989)], but the total CPSase activity in muscle is more than sufficient. Because the levels of argininosuccinate synthetase and lyase activities in liver are quite high (equivalent to 0.42 μmol/g fish/h, a little lower in kidney), but are not detectable in muscle (Saha and Ratha, 1987, 1989), a significant function for CPSase III and OCTase in muscle would require an interorgan transfer of citrulline from muscle to liver (or kidney), as discussed above in the context of the significance of the presence of CPSase III and OCTase in muscle.

Perfusion of whole liver with different concentrations of ammonium chloride over a period of 1 h has revealed several unusual features of the urea cycle in *H. fossilis* (Saha et al., 1995). Urea output, which was <0.05 μmol/g/min with no ammonium chloride in the perfusate, increased dramatically with increasing concentrations of ammonium chloride; the urea excretion rate (and also ammonia uptake) was found to be a saturable process, reaching a maximum of 0.42 μmol urea excreted/g/min at a rate of ammonium chloride addition of 1.18 μmol/g/min. The levels of all urea cycle enzyme activities, except arginase, increased during this 1 h of perfusion; induction of activity was proportional to ammonium chloride concentration in the perfusate, reaching a maximum of 2- to 3.5-fold at a rate of ammonium chloride addition of 0.6 μmol/g/min. This result is striking given the short time period, suggesting that this apparent regulatory process may reflect covalent modification or other form of enzyme activation rather than induction of an increased rate of gene expression. Another interesting observation is that the maximum induced CPSase activity, 0.15 μmol/min/g liver at 30°C, is not sufficient to account for the maximum observed rate of urea excretion (0.42 μmol urea excreted/g/min), suggesting that there may be another source of urea. The liver does have high levels of the uricolytic enzymes (Saha and Ratha, 1987), more than sufficient to account for this rate of urea excretion if flux of ammonia through the purine nucleotide pathway provides uric acid at this rate. The level of GSase activity in liver (1.0 μmol/min/g liver at 30°C) would be just sufficient to provide the one nitrogen of urea that originates from glutamine via either pathway (Chakravorty *et al.*, 1989).

A surprising feature of the mitochondrial CPSase in liver of *H. fossilis* is that, like the CPSase III from the alkaline lake-adapted tilapia, maximal activity with ammonia as substrate is actually higher (1.5-fold) than that with glutamine; unlike the alkaline lake-adapted tilapia CPSase III activity, however, AGA greatly stimulates activity, and, more interestingly, activity with both ammonia and glutamine present is nearly additive (Saha *et al.*, 1997). One possible explanation for this observation is that this fish has genes for both CPSase I and CPSase III and both are expressed. The presence of a gene for both CPSase I and CPSase III would not seem likely within the context of our current understanding of the structural relationships between these two enzymes, the known species distribution of the two enzymes, and the view that CPSase I evolved from CPSase III (Campbell and Anderson, 1991; Mommsen and Walsh, 1991; Hong *et al.*, 1994; Anderson, 1995b). Molecular phylogenetic analysis of the CPSase sequences now available is consistent with this current thinking about the evolution of CPSase I, that is, that CPSase I arose from CPSase III, not by gene duplication and divergence, but simply as an adaptation of CPSase III (D. Hewett-Emmet and P. M. Anderson, unpublished observations). It may be that this property reflects an adaptation to accommodate the unique environment of this species, either as a result of an isolated gene duplication event where one gene underwent structural changes resulting in a CPSase I-like activity or the CPSase III gene is modified, resulting in expression of a CPSase with separate sites for ammonia and glutamine (Saha *et al.*, 1997). Biochemical studies aimed at isolation and characterization will likely provide an answer to this question.

Very similar results to those discussed above for *H. fossilis* have been reported recently for the walking catfish, *Clarius batrachus,* another Indian air-breathing ureogenic fish (Saha and Das, 1999; Saha *et al.*, 2000).

It is important in the context of this section to point out that ureotelism is not the universal response of air-breathing teleosts to long-term air exposure. Marble goby is a facultative air-breather that can tolerate continuous air exposure for up to a week and has low levels of a full complement of the urea cycle enzymes. However, this fish does not increase urea excretion during long-term air exposure. Instead, GSase activity increases significantly, perhaps as a result of hormonally induced covalent modification of the enzyme, and ammonia is converted to glutamine, which accumulates in the tissues (Jow et al., 1999). Perfusion with ammonium chloride results in significant increases in free amino acids as well as urea in liver of *C. batrachus*, which is accompanied by increases in the levels of the amino acid metabolism-related enzymes glutamate dehydrogenase, aspartate aminotransferase, and GSase (Saha et al., 2000). The mudskippers *Periophthalmodon schlosseri* and *Boleophthalmus boddaerti* (Ip et al., 1993; Lim et al., 2000) and perhaps also *Periophthalmus cantonensis* (Iwata et al., 1981; Iwata, 1988) adapt to terrestrial conditions by accumulating amino acids in their tissues and/or greatly reducing protein and amino acid catabolic rates (see Chapter 4). Another strategy involves loss of ammonia across the skin (see Chapter 4).

D. Gulf Toadfish (*Opsanus beta*) and Related Species:
   A Switch from Ammonotelism to Ureotelism
   Induced by Stress

Because earlier work with gulf toadfish has been reviewed by Anderson (1995a) and in more detail by Walsh (1997), only those earlier results needed to place recent studies in context are discussed here.

Gulf toadfish is a marine teleost species that has a full complement of all urea cycle enzymes, including GSase, in liver at levels comparable to those in marine elasmobranchs (Mommsen and Walsh, 1989; Anderson and Walsh, 1995). Isolated hepatocytes have a high capacity for [^{14}C]urea synthesis when incubated with [^{14}C]bicarbonate and other appropriate substrates, and isolated mitochondria are capable of synthesizing citrulline formation from glutamine, analogous to elasmobranchs (Walsh et al., 1989; Henry and Walsh, 1997). Nevertheless, the gulf toadfish is only facultatively ureotelic, being primarily ammonotelic under conditions of minimal stress (Walsh et al., 1990; Walsh and Milligan, 1995). Ureotelism is not linked to osmoregulation or regulation of acid–base balance (Walsh et al., 1990; Barber and Walsh, 1993). Ureotelism and a high rate of urea excretion is induced by feeding (Walsh and Milligan, 1995) and by long-term air exposure or exposure to ammonia in the water (Walsh et al., 1990, 1994). The view that ureotelic capability in gulf toadfish may be related to air exposure in an intertidal environment during the tidal cycle has recently been discounted by Hopkins et al. (1999). This and a related study by Hopkins et al. (1997), however, suggest that an important factor in triggering gulf toadfish to excrete urea naturally may be the apparently high levels of ammonia encountered by these fish in the rhizome environment of the seagrasses that they inhabit in the subtidal zone.

Gulf toadfish also become ureotelic when they are either crowded or are confined individually in small chambers (Walsh et al., 1994; Walsh and Milligan, 1995). This effect appears to be stress related and may be physiologically significant as a means of predator avoidance and/or nitrogen retention when the fish are confined for periods of time to small spaces for shelter or during breeding (Walsh et al., 1994; Walsh, 1997).

Unlike elasmobranchs, liver GSase in gulf toadfish is localized in both the mitochondria ($\approx$30% of the total) and in the cytosol, and the level of GSase activity in the mitochondria is not sufficient to supply glutamine for maximal CPSase III activity or the observed rates of urea excretion; the additional capacity of the cytosolic GSase is just sufficient to account for these rates (Anderson and Walsh, 1995). These and related observations suggest that glutamine supply for CPSase III activity and ureogenesis likely requires supplementation by cytosolic GSase and this may be a major regulatory site (Anderson and Walsh, 1995).

When gulf toadfish are subjected to confinement/crowding, their pattern of nitrogen excretion is altered such that ammonia excretion is greatly curtailed and urea accounts for >90% of the total nitrogen excreted; the actual rate of urea excretion does not actually increase (Walsh and Milligan, 1995). The level of CPSase III activity does not change, but the level of cytosolic (but not mitochondrial) GSase activity increases more than five-fold (Walsh et al., 1994; Walsh and Milligan, 1995; Julsrud et al., 1998). These observations indicate that the transition to apparent ureotelism actually represents a shutting down of ammonotelism rather than an activation of ureogenesis. GSase mRNA also increases about five-fold under similar conditions of confinement/crowding, as does the GSase protein concentration as determined by Western blots, clearly identifying GSase gene expression as a major regulatory control point (Kong et al., 2000). The increase in GSase activity during confinement correlates with a surge in plasma cortisol, making cortisol an obvious potential candidate for a role in the transcriptional regulation of GSase expression (Hopkins et al., 1995). A putative glucocorticoid response element has been identified in the promoter region of the gulf toadfish GSase gene (P. J. Walsh, personal communication) and preliminary studies with isolated gulf toadfish hepatocytes indicate that treatment with cortisol induces GSase mRNA expression (P. M. Anderson and P. J. Walsh, unpublished observations).

Interestingly, the crowding/confinement of gulf toadfish that induces expression of GSase mRNA accompanied by an increase in total GSase activity also induces expression of CPSase III mRNA approximately eight-fold (Kong et al., 2000), even though total CPSase III activity does not increase. An explanation for this observation remains to be elucidated, but Kong et al. (2000) have suggested that this may reflect an increased expression of CPSase III needed to offset a possible general increased rate of protein degradation needed to support cortisol-induced gluconeogenesis from amino acids (Mommsen et al., 1999).

Isolated gulf toadfish liver mitochondria catalyze citrulline synthesis from glutamine with succinate as an energy donor at a rate that is higher than that needed to account for the highest rates of urea excretion from hepatocytes or whole fish (Henry and Walsh, 1997). Glutamine transport into the mitochondria is also subject to regulation, since the rate of transport is increased ten-fold and the K_m is increased four-fold (from 7 to 22 mM) in mitochondria from fish subject to confinement (Henry and Walsh, 1997). However, the rates of transport are quite high and do not appear to be rate limiting for urea production, even with the increase in K_m for glutamine for transport observed with mitochondria from confined fish. Inhibition of mitochondrial carbonic anhydrase results in a decrease in the rate of citrulline synthesis by isolated mitochondria [and of urea synthesis by isolated hepatocytes (Walsh *et al.*, 1989)], indicating that carbonic anhydrase activity is required for catalyzing bicarbonate formation from CO_2 required for carbamoyl phosphate formation catalyzed by the CPSase III. This carbonic anhydrase activity is not induced by crowding, however, and, though required, is in large excess and is not rate limiting under normal *in vivo* conditions. Henry and Walsh (1997) conclude, therefore, that the rates of urea synthesis in gulf toadfish are regulated, in part, by a balance between glutamine production by cytoplasmic GSase and glutamine consumption as substrate for mitochondrial CPSase III, as proposed by Anderson and Walsh (1995). As noted earlier, hepatic CPSase III and, therefore, urea formation also appear to be regulated by changing levels of AGA, a positive allosteric effector for CPSase III; the levels of AGA in liver increase more than two-fold in confined (fasted) versus control (fasted) gulf toadfish and increase more than ten-fold in fed versus fasted gulf toadfish (Julsrud *et al.*, 1998).

An unusual feature that accompanies ureotelism associated with confinement/crowding is the pulsatile nature of urea excretion; urea is stored in various body tissues and is all excreted within a 3-h time period approximately once a day (Wood *et al.*, 1995b; see also Chapter 8). This is not due to a pulsatile production mechanism (urea is synthesized continuously and the urea concentration in plasma and other body compartments increases steadily between pulsatile excretion) and is not triggered by a specific plasma urea concentration threshold (Wood *et al.*, 1997). The pulse of excreted urea occurs through the gills (Pärt *et al.*, 1999). Among several hypotheses investigated to account for pulsatile excretion (Gilmour *et al.*, 1998; Perry *et al.*, 1998), recent studies implicate periodic activation of a facilitated urea transporter in the gills, similar to the vasopressin-regulated urea transporter in mammalian kidney (Wood *et al.*, 1998).

These confinement/crowding studies all indicate that urea excretion remains relatively constant and it is the rate of ammonia excretion that changes greatly. As pointed out by Wood *et al.* (1995b), induction of GSase activity in liver and other tissues (as the result of an increase in plasma cortisol levels that occurs in response to stress such as crowding) serves to prevent ammonia excretion by converting it to glutamine. The amide group of glutamine can, in turn, be converted and stored as urea, which can then be released rapidly as pulses, perhaps for reasons related to minimizing predation or nitrogen conservation, as noted above.

Studies on the regulation of the urea cycle in this interesting fish in the context of fed versus fasting states or exposure to air or high levels of ambient ammonia, all of which, in contrast to confinement/crowding, markedly increase the rate of urea excretion, have not been carried out in detail and may be as instructive as those related to confinement/crowding. Like rainbow trout and largemouth bass, CPSase III mRNA and activity can be demonstrated in muscle and other extrahepatic tissues of gulf toadfish (Julsrud et al., 1998; Kong et al., 2000). An extrahepatic requirement for CPSase III activity to sustain general ureotelic capability does not seem likely, since the units of CPSase III in liver [$\approx$0.6 μmol/min/g liver (Walsh and Milligan, 1995), representing a potential capacity of 6 μmol/min/kg fish, assuming that liver represents 1% of the body weight (P. J. Walsh, personal communication)] are sufficient to account for the observed rates of urea excretion; the highest reported rates of urea excretion (fed state) are $\approx$0.3 μmol/min/kg fish (Walsh et al., 1990; Wood et al., 1995b, 1997; Walsh and Milligan, 1995).

Wang and Walsh (2000) recently reported that three members of the Batrachoididae family (the toadfish *O. tau* and *O. beta* and the midshipman *Porichthys notatus*) have unusually high tolerances to ammonia exposure and note that this tolerance seems to be exhibited by those species of fish that are ureotelic. However, this tolerance appears to correlate more with the level of GSase activity, especially levels of GSase in the brain, and levels of glutamine that accumulate in tissues than to urea cycle capability. The authors suggest that mechanisms other than or in addition to urea synthesis for ammonia detoxification must be considered to explain the exceptionally high ammonia tolerance by these species.

E. Embryogenesis and the Urea Cycle

Studies with three different species have now demonstrated induction of key urea cycle enzymes during early embryogenesis [rainbow trout embryos (Wright et al., 1995; Korte et al., 1997), Atlantic cod, *Gadus morhua* L. (Chadwick and Wright, 1999), and Atlantic halibut (Terjesen et al., 2000)]. These results seem to clearly suggest a physiologic role for the urea cycle during early stages of development that may be a characteristic feature of all fish. The topic of expression of the urea cycle during embryogenesis is covered in detail in Chapter 5.

VI. SUMMARY

The focus of this chapter is on recent developments in our understanding of the expression of the urea cycle and the roles of glutamine synthetase (GSase) and glutamine-dependent carbamoyl-phosphate synthetase (CPSase III) in ammonia detoxification in fish, with emphasis on teleosts. Marine elasmobranch fishes have an active urea cycle, synthesizing and retaining urea for the purpose of osmoregulation. Well-characterized biochemical features uniquely different from the urea cycle in ureotelic terrestrial vertebrates that have served as a reference for studies on teleost fishes are reviewed. The majority of teleost fishes are ammonotelic, that is, ammonia is simply excreted directly across the gills into the surrounding aqueous environment. But, a functional urea cycle and ureotelism have been documented in a few adult species as adaptations to unusual environmental circumstances (stress, air exposure, high pH, exposure to high concentrations of ammonia). These adaptations are reflected in altered (1) CPSase III amino acid sequences and kinetic properties, (2) regulation of the levels of gene expression, and (3) specificity of expression with respect to organ localization (e.g., expressed in muscle instead of liver) and life-cycle stage. The emerging view is that all fish likely have genes for the urea cycle enzymes; teleost fish normally do not express the urea cycle enzymes (except perhaps during embryogenesis); fish are opportunistic in variations of adapting expression of the urea cycle to local environments; and fish are highly individualistic in the mechanisms they employ for adapting to varying environmental challenges, that is, expression of the urea cycle is only one of several possible strategies.

ACKNOWLEDGMENTS

The author expresses his appreciation to the National Science Foundation for research grant support of work cited in this review and that has made possible his participation in bringing this book to print. He also thanks Drs. Pat Walsh and Pat Wright for critically reviewing the manuscript for this chapter.

REFERENCES

Alsop, D. H., and Wood, C. M. (1997). The interactive effects of feeding and exercise on oxygen consumption, swimming performance and protein usage in juvenile rainbow trout (*Oncorhynchus mykiss*). *J. Exp. Biol.* **200,** 2337–2346.

Anderson, P. M. (1976). A glutamine- and N-acetyl-L-glutamate-dependent carbamyl phosphate synthetase activity in the teleost *Micropterus salmoides*. *Comp. Biochem. Physiol.* **54B,** 261–263.

Anderson, P. M. (1980). Glutamine- and N-acetylglutamate-dependent carbamyl phosphate synthetase in elasmobranchs. *Science* **208,** 291–293.

Anderson, P. M. (1981). Purification and properties of the glutamine- and N-acetyl-L-glutamate-dependent carbamoyl phosphate synthetase from liver of *Squalus acanthias*. *J. Biol. Chem.* **256,** 12228–12238.

Anderson, P. M. (1986). Effects of urea, trimethylamine oxide, and osmolality on respiration and citrulline synthesis by isolated hepatic mitochondria from *Squalus acanthias*. *Comp. Biochem. Physiol.* **85B**, 783–788.

Anderson, P. M. (1989). Glutamine-dependent carbamoyl-phosphate synthetase and other enzyme activities related to the pyrimidine pathway in spleen of *Squalus acanthias* (spiny dogfish). *Biochem. J.* **261**, 523–529.

Anderson, P. M. (1991). Glutamine-dependent urea synthesis in elasmobranch fishes. *Biochem. Cell Biol.* **69**, 317–319.

Anderson, P. M. (1995a). Urea cycle in fish: Molecular and mitochondrial studies. *In* "Fish Physiology" Volume 14, "Ionoregulation: Cellular and Molecular Approaches to Fish Ionic Regulation" (C. M. Wood and T. J. Shuttleworth, eds.), Chap. 3, pp. 57–83. Academic Press, New York.

Anderson, P. M. (1995b). Molecular aspects of carbamoyl phosphate synthesis. *In* "Nitrogen Metabolism and Excretion" (P. J. Walsh, and P. A. Wright, eds.), Chap. 9, pp. 33–50. CRC Press, Boca Raton, FL.

Anderson, P. M., and Casey, C. A. (1984). Glutamine-dependent synthesis of citrulline by isolated hepatic mitochondria from *Squalus acanthias*. *J. Biol. Chem.* **259**, 456–462.

Anderson, P. M., and Walsh, P. J. (1995). Subcellular localization and biochemical properties of the enzymes of carbamoyl phosphate and urea synthesis in the Batrachoidid fishes *Opsanus beta*, *Opsanus tau*, and *Porichthys notatus*. *J. Exp. Biol.* **198**, 755–766.

Atkinson, D. E. (1992). Functional roles of urea synthesis in vertebrates. *Physiol. Zool.* **65**, 243–267.

Ballantyne, J. S. (1997). Jaws: the inside story. The metabolism of elasmobranch fishes. *Comp. Biochem. Physiol.* **118B**, 703–742.

Barber, M. L., and Walsh, P. J. (1993). Interactions of acid–base status and nitrogen excretion and metabolism in the ureogenic teleost *Opsanus tau*. *J. Exp. Biol.* **185**, 87–105.

Brown, G. W. Jr., James, J., Henderson, R. J., Thomas, W. N., Robinson, R. O., Thompson, A. L., Brown, E., and Brown, S. G. (1966). Uricolytic enzymes in liver of the dipnoan *Protopterus aethiopicus*. *Science* **153**, 1653–1654.

Campbell, J. W. (1991). Excretory nitrogen metabolism. *In* "Environmental and Metabolic Animal Physiology" (C. L. Prosser, ed.), Chap. 7, pp. 277–324. Wiley-Liss, New York.

Campbell, J. W., and Anderson, P. M. (1991). Evolution of mitochondrial enzyme systems in fish: the mitochondrial synthesis of glutamine and citrulline. *In* "Biochemistry and Molecular Biology of Fishes" (P. W. Hochachka, and T. P. Mommsen, eds.), Vol. 1, pp. 43–76. Elsevier, Amsterdam.

Cao, X., Kemp, J. R., and Anderson, P. M. (1991). Subcellular localization of two glutamine-dependent carbamoyl-phosphate synthetases and related enzymes in liver of *Micropterus salmoides* (largemouth bass) and properties of isolated liver mitochondria: comparative relationships with elasmobranchs. *J. Exp. Zool.* **258**, 24–33.

Casey, C. A., and Anderson, P. M. (1982). Subcellular location of glutamine synthetase and urea cycle enzymes in liver of spiny dogfish (*Squalus acanthias*). *J. Biol. Chem.* **257**, 8449–8453.

Casey, C. A., and Anderson, P. M. (1983). Glutamine- and N-acetyl-L-glutamate-dependent carbamoyl phosphate synthetase from *Micropterus salmoides*. Purification, properties, and inhibition by glutamine analogs. *J. Biol. Chem.* **258**, 8723–8732.

Casey, C. A., and Anderson, P. M. (1985). Submitochondrial localization of arginase and other enzymes associated with urea synthesis and nitrogen metabolism, in liver of *Squalus acanthias*. *Comp. Biochem. Physiol.* **82B**, 307–315.

Chadwick, T. D., and Wright, P. A. (1999). Nitrogen excretion and expression of urea cycle enzymes in the Atlantic cod (*Gadus morhua* L.): a comparison of early life stages with adults. *J. Exp. Biol.* **202**, 2653.

Chakravorty, J., Saha, N., and Ratha, B. K. (1989). A unique pattern of tissue distribution and subcellular localization of glutamine synthetase in a freshwater air-breathing teleost, *Heteropneustes fossilis* (Bloch). *Biochem. Int.* **19**, 519–527.

Chen, Y. (1995). Transcriptional regulation of the urea cycle enzyme genes. Ph.D. thesis, University of Western Ontario, London, Canada.

Chen, Y., and Atkinson, B. G. (1997). Role for the *Rana catesbeiana* homologue of C/EBP alpha in the reprogramming of gene expression in the liver of metamorphosing tadpoles. *Dev. Genet.* **20,** 152–162.

Chen, Y., Hu, H., and Atkinson, B. G. (1994). Characterization and expression of C/EBP-like genes in the liver of *Rana catesbeiana* tadpoles during spontaneous and thyroid hormone-induced metamorphosis. *Dev. Genet.* **15,** 366–377.

Chiu, Y. N., Austic, R. E., and Rumsey, G. L. (1986). Urea cycle activity and arginine formation in rainbow trout (*Salmo gairdneri*). *J. Nutr.* **116,** 1640–1650.

Cohen, P. P. (1976). Evolutionary and comparative aspects of urea biosynthesis. *In* "The Urea Cycle" (S. Grisolia, R. Báguena, and F. Mayor, eds.), pp. 21–38. John Wiley, New York.

Danulat, E., and Kempe, S. (1992). Nitrogenous waste and accumulation of urea and ammonia in *Chalcalburnus tarichi* (Cyprinidae), endemic to the extremely alkaline Lake Van (Eastern Turkey). *Fish Physiol. Biochem.* **9,** 377–386.

Das, J. R., Saha, N., and Ratha, B. K. (1991). Tissue distribution and subcellular localization of glutamate dehydrogenase in a freshwater air-breathing teleost, *Heteropneustes fossilis*. *Biochem. Systematics Ecol.* **19,** 207–212.

Dkhar, J., Saha, N., and Ratha, B. K. (1991). Ureogenesis in a freshwater teleost: an unusual subcellular localization of ornithine–urea cycle enzymes in the freshwater air-breathing teleost *Heteropneustes fossilis*. *Biochem. Int.* **25,** 1061–1069.

Evans, D. E. (1986). CAD, a chimeric protein that initiates *de novo* pyrimidine biosynthesis in higher eukaryotes. *In* "Multidomain Proteins–Structure and Function" (D. G. Hardie, ed.), Chap. 9, pp. 283–331. Elsevier, Amsterdam.

Farnham, P. J., and Kollmar, R. (1990). Characterization of the 5' end of the growth-regulated Syrian hamster CAD gene. *Cell Growth Diff.* **1,** 179–189.

Felskie, A. K., Anderson, P. M., and Wright, P. A. (1998). Expression and activity of carbamoyl phosphate synthetase III and ornithine–urea cycle enzymes in various tissues of four fish species. *Comp. Biochem. Physiol. B* **119,** 355–364.

Gilmour, K. M., Perry, S. F., Wood, C. M., Henry, R. P., Laurent, P., Pärt, P., and Walsh, P. J. (1998). Nitrogen excretion and the cardiorespiratory physiology of the gulf toadfish, *Opsanus beta*. *Physiol. Zool.* **71,** 492–505.

Goping, I. S., Lagacé, M., and Shore, G. C. (1992). Factors interacting with the rat carbamoyl phosphate synthetase promoter in expressing and nonexpressing tissues. *Gene* **118,** 283–287.

Graham, J. B. (1997). "Air-Breathing Fishes: Evolution, Diversity, and Adaptation." Academic Press, San Diego.

Griffith, R. W. (1991). Guppies, toadfish, lungfish, coelacanths and frogs: a scenario for the evolution of urea retention in fishes. *Environ. Biol. Fish.* **32,** 199–218.

Hayashi, S., Fujiwara, S., and Noguchi, T. (1989). Degradation of uric acid in fish liver peroxisomes. Intraperoxisomal localization of hepatic allantoicase and purification of its peroxisomal membrane-bound form. *J. Biol. Chem.* **264,** 3211–3215.

Helbing, C. C., and Atkinson, B. G. (1994). 3,5,3'-Triiodothyronine-induced carbamyl-phosphate synthetase gene expression is stabilized in the liver of *Rana catesbeiana* tadpoles during heat shock. *J. Biol. Chem.* **269,** 11743–11750.

Henry, R. P., and Walsh, P. J. (1997). Mitochondrial citrulline synthesis in the ureagenic toadfish, *Opsanus beta*, is dependent on carbonic anhydrase activity and glutamine transport. *J. Exp. Zool.* **279,** 521–529.

Hong, J., Salo, W. L., Lusty, C. L., and Anderson, P. M. (1994). Carbamoyl phosphate synthetase III, an evolutionary intermediate in the transition between glutamine-dependent and ammonia-dependent carbamoyl phosphate synthetases. *J. Mol. Biol.* **243,** 131–140.

Hong, J., Salo, W. L., and Anderson, P. M. (1995). Nucleotide sequence and tissue-specific expres-

sion of the multifunctional protein carbamoyl-phosphate synthetase-aspartate transcarbamoylase-dihydroorotase (CAD) mRNA in *Squalus acanthias. J. Biol. Chem.* **270**, 14130–14139.
Hong, J., Salo, W. L., Chen, Y., Atkinson, B. G., and Anderson, P. M. (1996). The promoter region of the carbamoyl-phosphate synthetase III gene of *Squalus acanthias. J. Mol. Evol.* **43**, 602–609.
Hopkins, T. E., Wood, C. M., and Walsh, P. J. (1995). Interactions of cortisol and nitrogen metabolism in the ureogenic gulf toadfish, *Opsanus beta. J. Exp. Biol.* **198**, 2229–2235.
Hopkins, T. E., Serafy, J. E., and Walsh, P. J. (1997). Field studies on the ureogenic gulf toadfish, in a subtropical bay. II. Nitrogen excretion physiology. *J. Fish Biol.* **50**, 1271–1284.
Hopkins, T. E., Wood, C. M., and Walsh, P. J. (1999). Nitrogen metabolism and excretion in an intertidal population of the gulf toadfish (*Opsanus beta*). *Mar. Freshw. Behav. Physiol.* **33**, 21–34.
Huggins, A. K., Skutsch, G., and Baldwin, E. (1969). Ornithine–urea cycle enzymes in teleostean fish. *Comp. Biochem. Physiol.* **28**, 587–602.
Ip, Y. K., Lee, C. Y., Chew, S. F., Low, W. P., and Peng, K. W. (1993). Differences in the responses of two mudskippers to terrestrial exposure. *Zool. Sci.* **10**, 511–519.
Iwata, K. (1988). Nitrogen metabolism in the mudskipper, *Periophthalmus cantonensis:* changes in free amino acids and related compounds in carious tissues under conditions of ammonia loading with reference to its high ammonia tolerance. *Comp. Biochem. Physiol.* **91A**, 499–508.
Iwata, K., Kakuta, M., Ikeda, G., Kimoto, S., and Wada, N. (1981). Nitrogen metabolism in the mudskipper, *Periophthalmus cantonensis:* a role of free amino acids in detoxification of ammonia produced during its terrestrial life. *Comp. Biochem. Physiol.* **68A**, 589–596.
Jones, M. E. (1980). Pyrimidine nucleotide biosynthesis in animals: genes, enzymes, and regulation of UMP biosynthesis. *Ann. Rev. Biochem.* **49**, 253–280.
Jones, M. E., Anderson, A. D., Anderson, C., and Hodes, S. (1961). Citrulline synthesis in rat tissues. *Arch. Biochem. Biophys.* **95**, 499–507.
Jow, L. Y., Chew, S. F., Lim, C. B., Anderson, P. M., and Ip, Y. K. (1999). The marble goby *Oxyeleotris marmoratus* activates hepatic glutamine synthetase and detoxifies ammonia to glutamine during air exposure. *J. Exp. Biol.* **202**, 237–245.
Julsrud, E. A., Walsh, P. J., and Anderson, P. M. (1998). *N*-acetyl-L-glutamate and the urea cycle in gulf toadfish (*Opsanus beta*) and other fish. *Arch. Biochem. Biophys.* **350**, 55–60.
Kong, H., Edberg, D. D., Salo, W. L., Wright, P. A., and Anderson, P. M. (1998). Nitrogen excretion and expression of carbamoyl-phosphate synthetase III activity and mRNA in extrahepatic tissues of largemouth bass (*Micropterus salmoides*). *Arch. Biochem. Biophys.* **350**, 157–168.
Kong, H., Kahatapitiya, N., Kingsley, K., Salo, W. L., Anderson, P. M., Wang, Y. S., and Walsh, P. J. (2000). Induction of carbamoyl phosphate synthetase III and glutamine synthetase mRNA during confinement stress in gulf toadfish (*Opsanus beta*). *J. Exp. Biol.* **203**, 311–320.
Korsgaard, B., Mommsen, T. P., and Wright, P. A. (1995). Nitrogen excretion in teleostean fish: adaptive relationships to environment, ontogenesis, and viviparity. *In* "Nitrogen Metabolism and Excretion" (P. J. Walsh and P. Wright, eds.), pp. 259–287. CRC Press, Boca Raton, FL.
Korte, J. J., Salo, W. L., Cabrera, V. M., Wright, P. A., Felskie, A. K., and Anderson, P. M. (1997). Expression of carbamoyl-phosphate synthetase III mRNA during the early stages of development and in muscle of adult rainbow trout *(Oncorhynchus mykiss)*. *J. Biol. Chem.* **272**, 6270–6277.
Lagacé, M., Howell, B. W., Burak, R., Lusty, C. J., and Shore, G. C. (1987). Rat carbamoyl phosphate synthetase I gene: promoter sequence and tissue-specific transcriptional regulation *in vitro. J. Biol. Chem.* **262**, 10415–10418.
Lagacé, M., Goping, I. S., Muller, C. R., Lazzaro, M., and Shore, G. C. (1992). The carbamoyl phosphate synthetase promoter contains multiple binding sites for C/EBP-related proteins. *Gene* **118**, 231–238.
Laud, P. R., and Campbell, J. W. (1994). Genetic basis for tissue isozymes of glutamine synthetase in elasmobranchs. *J. Mol. Evol.* **39**, 93–100.

Lauff, R. F., and Wood, C. M. (1996). Respiratory gas exchange, nitrogenous waste excretion, and fuel usage during aerobic swimming in juvenile rainbow trout. *J. Comp. Physiol. B* **166,** 501–509.

Lauff, R. F., and Wood, C. M. (1997). Effects of training on respiratory gas exchange, nitrogenous waste excretion, and fuel usage during aerobic swimming in juvenile rainbow trout (*Oncorhynchus mykiss*). *Can. J. Fish. Aquat. Sci.* **54,** 566–571.

Lim, C. B., Chew, S. F., Anderson, P. M., and Ip, Y. K. (2001). Mudskippers (*Periophthalmodon schlosseri* and *Boleophthalmus boddaerti*) reduce the rates of protein and amino acid catabolism to slow down internal ammonia build up during aerial exposure in constant darkness. *J. Exp. Biol.* **204,** 1605–1614.

Lin, H. L., and Randall, D. J. (1990). The effect of varying water pH on the acidification of expired water in rainbow trout. *J. Exp. Biol.* **149,** 149–160.

Lindley, T. E., Scheiderer, C. L., Walsh, P. J., Wood, C. M., Bergman, H. L., Bergman, A. L., Laurent, P., Wilson, P., and Anderson, P. M. (1999). Muscle as the primary site of urea cycle enzyme activity in an alkaline lake-adapted tilapia, *Oreochromis alcalicus grahami*. *J. Biol. Chem.* **274,** 29858–29861.

Lowenstein, J. M. (1990). The purine nucleotide cycle revized. *Int. J. Sports Med.* **11,** S37-S46.

Marshall, M. (1976). Carbamyl phosphate synthetase I from frog liver. In "The Urea Cycle" (S. Grisolia, R. Báguena, and F. Mayor, eds.), pp. 133–142. John Wiley, New York.

McGeer, J. C., Wright, P. A., Wood, C. M., Wilkie, M. P., Mazur, C. F., and Iwama, G. K. (1994). Nitrogen excretion in four species of fish from an alkaline lake. *Trans. Am. Fish. Soc.* **123,** 824–829.

Meijer, A. J. (1995). Urea synthesis in mammals. In "Nitrogen Metabolism and Excretion" (P. J. Walsh and P. A. Wright, eds.), Chap. 12, pp. 193–204. CRC Press, Boca Raton, FL.

Meijer, A. J., Lamers, W. H., and, Chamuleau, R. A. F. M. (1990). Nitrogen metabolism and ornithine cycle function. *Physiol. Rev.* **70,** 701–748.

Mommsen, T. P., and Hochachka, P. W. (1988). The purine nucleotide cycle as two temporally separated metabolic units: a study on trout muscle. *Metabolism* **37,** 552–556.

Mommsen, T. P., and Walsh, P. J. (1989). Evolution of urea synthesis in vertebrates: the piscine connection. *Science* **243,** 72–75.

Mommsen, T. P., and Walsh, P. J. (1991). Urea synthesis in fishes: evolutionary and biochemical perspectives. In "Biochemistry and Molecular Biology of Fishes" (P. W. Hochachka and T. P. Mommsen, eds.), Vol. 1, pp. 137–163. Elsevier, Amsterdam.

Mommsen, T. P., and Walsh, P. J. (1992). Biochemical and environmental perspectives on nitrogen metabolism in fishes. *Experientia* **48,** 583–593.

Mommsen, T. P., Vijayan, M. M., and Moon, T. W. (1999). Cortisol in teleosts: dynamics, mechanisms of action, and metabolic regulation. *Rev. Fish Biol. Fish.* **9,** 211–208.

Nyunoya, H., Broglie, K. E., Widgren, E. E., and Lusty, C. J. (1985). Characterization and derivation of the gene coding for mitochondrial carbamyl phosphate synthetase I on rat. *J. Biol. Chem.* **260,** 9346–9356.

Pärt, P., Wood, C. M., Gilmour, K. M., Perry, S. F., Laurent, P., Zadunaisky, J., and Walsh, P. J. (1999). Urea and water permeability in the ureotelic gulf toadfish (*Opsanus beta*). *J. Exp. Zool.* **283,** 1–12.

Perlman, D. F., and Goldstein, L. (1988). Nitrogen metabolism. In "Physiology of Elasmobranch Fishes" (T. J. Shuttleworth, ed.), pp. 253–276. Springer-Verlag, Berlin.

Perry, S. F., Gilmour, K. M., Wood, C. M., Pärt, P., Laurent, P., and Walsh, P. J. (1998). The effects of arginine vasotocin and catecholamines on nitrogen excretion and the cardio-respiratory physiology of the gulf toadfish, *Opsanus beta*. *J. Comp. Physiol. B* **168,** 461–472.

Raijman, L. (1974). Citrulline synthesis in rat tissues and liver content of carbamoyl phosphate and ornithine. *Biochem J.* **138,** 225–232.

Ramaswamy, M., and Reddy, T. G. (1983). Ammonia and urea excretion in three species of air-breathing fish subjected to aerial exposure. *Proc. Indian Acad. Sci. (Anim. Sci.)* **92**, 293–297.
Randall, D. J., and Wright, P. A. (1989). The interaction between carbon dioxide and ammonia excretion and water pH in fish. *Can. J. Zool.* **67**, 2936–2942.
Randall, D. J., Wood, C. M., Perry, S. F., Bergman, H., Maloiy, G. M. O., Mommsen, T. P., and Wright, P. A. (1989). Urea excretion as a strategy for survival in a fish living in a very alkaline environment. *Nature* **337**, 165–166.
Ratha, B. K., Saha, N., Rana, R. K., and Choudhury, B. (1995). Evolutionary significance of metabolic detoxification of ammonia to urea in an ammoniotelic freshwater teleost, *Heteropneustes fossilis* during temporary water deprivation. *Evol. Biol.* **8/9**, 107–117.
Ratner, S. (1973). Enzymes of arginine and urea synthesis. *Adv. Enzymol.* **39**, 1–90.
Saha, N., and Das, L. (1999). Stimulation of ureogenesis in the perfused liver of an Indian air-breathing catfish, *Clarias batrachus,* infused with different concentrations of ammonium chloride. *Fish Physiol. Biochem.* **21**, 303–311.
Saha, N., and Ratha, B. K. (1986). Effect of ammonia stress on ureogenesis in a freshwater air-breathing teleost, *Heteropneustes fossilis.* In "Contemporary Themes in Biochemistry" (O. L. Kon, M. C. Chung, P. L. H. Hwang, S. Leong, K. H. Loke, P. Thiyagaraja, and P. T. Wong, eds.), Vol. 6, pp. 432–343. Cambridge University Press, Cambridge.
Saha, N., and Ratha, B. K. (1987). Active ureogenesis in a freshwater air-breathing teleost, *Heteropneustes fossilis. J. Exp. Zool.* **241**, 137–141.
Saha, N., and Ratha, B. K. (1989). Comparative study of ureogenesis in freshwater, air-breathing teleosts. *J. Exp. Zool.* **252**, 1–8.
Saha, N., and Ratha, B. K. (1990). Alterations in excretion pattern of ammonia and urea in a freshwater air-breathing teleost, *Heteropneustes fossilis* (Bloch) during hyper-ammonia stress. *Indian J. Exp. Biol.* **28**, 597–599.
Saha, N., and Ratha, B. K. (1994). Induction of ornithine–urea cycle in a freshwater teleost, *Heteropneustes fossilis,* exposed to high concentrations of ammonium chloride. *Comp. Biochem. Physiol.* **108B**, 315–325.
Saha, N., and Ratha, B. K. (1998). Ureogenesis in Indian air-breathing teleosts: adaptation to environmental constraints. *Comp. Biochem. Physiol. A* **120**, 195–208.
Saha, N., Chakravorty, J., and Ratha, B. K. (1988). Diurnal variation in renal and extra-renal excretion of ammonia-N and urea-N in a freshwater air-breathing teleost, *Heteropneustes fossilis* (Bloch). *Proc. Indian Acad. Sci. (Anim. Sci.)* **97**, 529–537.
Saha, N., Dkhar, J., and Ratha, B. K. (1995). Induction of ureogenesis in perfused liver of a freshwater teleost, *Heteropneustes fossilis,* infused with different concentrations of ammonium chloride. *Comp. Biochem. Physiol.* **112B**, 733–741.
Saha, N., Dkhar, J., Anderson, P. M., and Ratha, B. K. (1997). Carbamyl phosphate synthetases in an air-breathing teleost, *Heteropneustes fossilis. Comp. Biochem. Physiol.* **116B**, 57–63.
Saha, N., Das, L., and Dutta, S. (1999). Types of carbamyl phosphate synthetases and subcellular localization of urea cycle and related enzymes in air-breathing walking catfish, *Clarias batrachus* infused with ammonium chloride: a strategy to adapt under hyperammonia stress. *J. Exp. Zool.* **283**, 121–130.
Saha, N., Dutta, S., and Häussinger, D. (2000). Changes in free amino acid synthesis in the perfused liver of an air-breathing walking catfish, *Clarias batrachus* infused with ammonium chloride: a strategy to adapt under hyperammonia stress. *J. Exp. Zool.* **286**, 13–23.
Shankar, R. A., and Anderson, P. M. (1985). Purification and properties of glutamine synthetase from liver of *Squalus acanthias. Arch. Biochem. Biophys.* **239**, 248–259.
Simmer, J. P., Kelly, R. E., Rinker, A. G., Scully, J. L., and Evans, D. R. (1990). Mammalian carbamyl phosphate synthetase (CPS). cDNA sequences and evolution of the CPS domain of the Syrian hamster multifunctional protein CAD. *J. Biol. Chem.* **265**, 10395–10402.

Takiguchi, M., Matsubasa, T., Amaya, Y., and Mori, M. (1989). Evolutionary aspects of urea cycle enzyme genes. *Bioessays* **10**, 163–166.

Terjesen, B. F., Ronnestad, I., Norberg, B., and Anderson, P. M. (2000). Detection of basic properties of carbamoyl phosphate synthetase III during teleost ontogeny: a case study in the Atlantic halibut (*Hippoglossus hippoglossus* L.). *Comp. Biochem. Physiol.* **126B**, 521–535.

Thoden, J. B., Wesenberg, G., Raushel, F. M., and Holden, H. M. (1997). Carbamoyl phosphate synthetase: closure of the B-domain as a result of nucleotide binding. *Biochemistry* **38**, 2347–2357.

Trammel, P. R., and Campbell, J. W. (1970). Carbamyl phosphate synthesis in a land snail, *Strophocheilus oblongus*. *J. Biol. Chem.* **245**, 6634–6641.

Trammel, P. R., and Campbell, J. W. (1971). Carbamyl phosphate synthesis in invertebrates. *Comp. Biochem. Physiol.* **40B**, 395–406.

Walsh, P. J. (1996). Purification and properties of hepatic glutamine synthetases from the ureotelic gulf toadfish, *Opsanus beta*. *Comp. Biochem. Physiol.* **115B**, 523–532.

Walsh, P. J. (1997). Evolution and regulation of ureogenesis and ureotely in (batrachoidid) fishes. *Ann. Rev. Physiol.* **59**, 299–323.

Walsh, P. J., and Milligan, C. J. (1995). Effects of feeding on nitrogen metabolism and excretion in the gulf toadfish (*Opsanus beta*). *J. Exp. Biol.* **198**, 1559–1566.

Walsh, P. J., Parent, J. J., and Henry, R. P. (1989). Carbonic anhydrase supplies bicarbonate for urea synthesis in toadfish (*Opsanus beta*) hepatocytes. *Physiol. Zool.* **62**, 1257–1272.

Walsh, P. J., Danulat, E., and Mommsen, T. P. (1990). Variation in urea excretion in the gulf toadfish (*Opsanus beta*). *Mar. Biol.* **106**, 323–328.

Walsh, P. J., Tucker, B. C., and Hopkins, T. E. (1994). Effects of confinement and crowding on ureogenesis in the gulf toadfish *Opsanus beta*. *J. Exp. Biol.* **191**, 195–206.

Walsh, P. J., Handel-Fernandez, M. E., and Vincek, V. (1999). Cloning and sequencing of glutamine synthetase cDNA from liver of the ureotelic gulf toadfish (*Opsanus beta*). *Comp. Biochem. Physiol. B* **124**, 251–259.

Wang, X., and Walsh, P. J. (2000). High ammonia tolerance in fishes of the family batrachoididae (toadfish and midshipmen). *Aquat. Toxicol.* **50**, 205–221.

Webb, J. T., and Brown, G. W. (1976). Some properties and occurrence of glutamine synthetase in fish. *Comp. Biochem. Physiol. B* **54**, 171–175.

Webb, J. T., and Brown, G. W., Jr. (1980). Glutamine synthetase: an assimilatory role in liver as related to urea retention in marine chondrichthyes. *Science* **208**, 293–295.

Wilkie, M. P., and Wood, C. M. (1991). Nitrogenous waste excretion, acid–base regulation, and ionoregulation in rainbow trout (*Oncorhynchus mykiss*) exposed to extremely alkaline water. *Physiol. Zool.* **64**, 1069–1086.

Wilkie, M. P., and Wood, C. M. (1996). The adaptations of fish to extremely alkaline environments. *Comp. Biochem. Physiol.* **113B**, 665–673.

Wilkie, M. P., Wright, P. A., Iwama, G. K., and Wood, C. M. (1993). The physiological responses of the Lahontan cutthroat trout (*Oncorhynchus clarki henshawi*), a resident of highly alkaline Pyramid Lake (pH 9.4), to challenge at pH 10. *J. Exp. Biol.* **175**, 173–194.

Windmueller, H. G., and Spaeth, A. E. (1981). Source and fate of circulating citrulline. *Endocrinol. Metab.* **4**, E473–E480.

Wood, C. M. (1993). Ammonia and urea metabolism and excretion. *In* "The Physiology of Fishes" (D. H. Evans, ed.), pp. 379–425. CRC Press, Boca Raton, FL.

Wood, C. M., Perry, S. F., Wright, P. A., Bergman, H. L., and Randall, D. J. (1989). Ammonia and urea dynamics in the Lake Magadi tilapia: a teleost fish adapted to an extremely alkaline environment. *Respir. Physiol.* **77**, 1–20.

Wood, C. M., Bergman, H. L., Laurent, P., Maina, J. N., Narahara, A., and Walsh, P. J. (1994). Urea production, acid–base regulation and their interactions in the Lake Magadi tilapia, a unique teleost adapted to a highly alkaline environment. *J. Exp. Biol.* **189**, 13–36.

Wood, C. M., Pärt, P., and Wright, P. A. (1995a). Ammonia and urea metabolism in relation to gill function and acid–base balance in a marine elasmobranch, the spiny dogfish (*Squalus acanthias*). *J. Exp. Biol.* **198,** 1545–1558.

Wood, C. M., Hopkins, T. E., Hogstrand, C., and Walsh, P. J. (1995b). Pulsatile urea excretion in the ureagenic toadfish *Opsanus beta:* an analysis of rates and routes. *J. Exp. Biol.* **198,** 1729–1741.

Wood, C. M., Hopkins, T. E., and Walsh, P. J. (1997). Pulsatile urea excretion in the toadfish (*Opsanus beta*) is due to a pulsatile excretion mechanism, not a pulsatile production mechanism. *J. Exp. Biol.* **200,** 1039–1046.

Wood, C. M., Gilmour, K. M., Perry, S. F., Pärt, P., Laurent, P., and Walsh, P. J. (1998). Pulsatile urea excretion in gulf toadfish (*Opsanus beta*): evidence for activation of a specific facilitated diffusion transport system. *J. Exp. Biol.* **201,** 805–817.

Wright, P. A. (1993). Nitrogen excretion and enzyme pathways for ureagenesis in freshwater tilapia (*Oreochromis niloticus*). *Physiol. Zool.* **66,** 881–901.

Wright, P. A. (1995). Nitrogen excretion: three end products, many physiological roles. *J. Exp. Biol.* **198,** 273–281.

Wright, P. A., and Wood, C. M. (1985). An analysis of branchial ammonia excretion in the freshwater rainbow trout: effects of environmental pH change and sodium uptake blockage. *J. Exp. Biol.* **114,** 329–353.

Wright, P. A., Randall, D. J., and Perry, S. F. (1989). Fish gill boundary layer: a site of linkage between carbon dioxide and ammonia excretion. *J. Comp. Physiol. B* **158,** 627–635.

Wright, P. A., Iwana, G. K., and Wood, C. M. (1993). Ammonia and urea excretion in Lahontan cutthroat trout (*Oncorhynchus clarki henshawi*) adapted to highly alkaline Pyramid Lake (pH 9.4). *J. Exp. Biol.* **175,** 153–172.

Wright, P. A., Felskie, A. K., and Anderson, P. M. (1995). Induction of ornithine–urea cycle enzymes and nitrogen metabolism and excretion in rainbow trout (*Oncorhynchus mykiss*) during early life stages. *J. Exp. Biol.* **198,** 127–135.

Yesaki, T. Y., and Iwama, G. K. (1992). Some effects of water hardness on survival, acid–base regulation, ion regulation and ammonia excretion in rainbow trout in highly alkaline water. *Physiol. Zool.* **65,** 763–787.

Zalkin, H. (1993). The amidotransferases. *Adv. Enzymol. Relat. Areas Mol. Biol.* **66,** 203–309.

Zalkin, H., and Dixon, J. E. (1992). *De novo* purine nucleotide biosynthesis. *Prog. Nucleic Acid Res. Mol. Biol.* **42,** 259–287.

8

UREA TRANSPORT

P. J. WALSH AND C. P. SMITH

I. Introduction
II. Urea Transport in Mammalian Systems
   A. Background
   B. Urea Transporters
   C. Physiologic Role of Urea Transporters in the Mammalian Kidney
   D. Physiologic Role of Urea Transporters in Mammalian Hepatocytes and Red Blood Cells
   E. Active Urea Transport
III. Physiology of Urea Transport in Fish
   A. Elasmobranchs
   B. Teleosts
IV. Prospects for Future Research and Evolutionary Perspectives
References

I. INTRODUCTION

Although a great deal of attention has been focused for decades on the importance of urea retention as an osmolyte in cartilagenous fishes (and the coelacanth) from the perspective of the evolutionary fitness of the intracellular milieu (see Chapter 9), oddly, the literature on the actual transport mechanisms for urea in fishes is not voluminous. In the older literature, the major highlights of the few mechanistic studies that were performed include, of course, (1) the classic studies of Homer Smith in the 1930s (Smith, 1931a,b, 1936) demonstrating that the elasmobranch kidney has incredible power to reabsorb urea; (2) the studies of Schmidt-Nielsen and coworkers (e.g., Schmidt-Nielsen *et al.,* 1972) demonstrating that this resorptive power was linked to sodium resorption; and (3) the studies of Boylan (1967) demonstrating that the gills of elasmobranchs likewise had low effective urea permeability. Further progress on urea transport in elasmobranchs was probably limited by the incredibly complex architecture of the elasmobranch kidney (and to some extent the gill). Historically, urea transport in teleosts was considered a nonissue. Many ichthyologists, like most biologists in general, probably worked under the assumption that lipid bilayers are rather permeable to urea, and that urea fluxes to the environment and between intracellular compartments in fish (even in ureoosmotic fish) were mostly the consequence of unregulated "leak." Examination of the early literature on urea permeabilities shows, however, a rather limited ability, compared to ammonia, of urea to cross lipid bilayers unaided (Collander, 1937; Galluci et al., 1971). More recently, detailed studies of the physiology of mammalian kidney segments clearly demonstrated a wide range of normal permeabilities to urea, considerably greater than would be expected if urea were to permeate cells simply by lipid-phase diffusion. In addition, permeability was found to be modulated by hormones and change in osmolality (reviewed by Knepper and Chou, 1995). Most recently, a suite of specific protein transporters for urea has been discovered in mammalian tissues and ultimately in fish (see below).

The emerging view that we hope unfolds in this chapter is that permeability and transport of urea are tightly controlled in fish, even in those species not utilizing urea as a major osmolyte. Although information on urea transporters in taxa more primitive to fish is still somewhat lacking, enough information exists to posit that some forms of urea transporters predate the appearance of fish. So, the broader evolutionary question is not "Did urea transporters first appear in fish?" but, more likely, "What evolutionary pressures led to the elaboration of urea transporters in fish?" Whereas these pressures and needs are somewhat clear for fish using urea as a major osmolyte, the picture is cloudier for other fish. In these species urea may serve other roles and hence the key to potentially understanding its transport physiology is not to treat urea simply as a "waste" molecule like ammonia. For the most part, ammonia (especially as NH_3) is highly permeable across membranes in fish, and it would be difficult for fish to retain it by standard transport mechanisms, if some aspect of their ecology, development, reproduction, and so on, required nitrogenous solute retention (toxicity considerations aside). Ammonia retention would require specialized membrane lipids (see Kikeri et al., 1989) or metabolic trapping (e.g., by glutamine synthetase or by low pH) in epithelial tissues, specializations perhaps not compatible with other transport processes. On the other hand, because urea is *not* especially permeable when unaided by carriers, its compartmentation *can* be readily controlled in circumstances when the retention of so-called "waste" nitrogen is selectively advantageous. Although roles for urea other than as an osmolyte remain speculative, we expect that further study of urea transport in noncartilaginous fish will begin to elucidate novel additional roles for urea.

Because the molecular basis of urea transport is far better understood in mammals, we begin with a brief review of mammalian urea transport as a backdrop to what is known for fish.

II. UREA TRANSPORT IN MAMMALIAN SYSTEMS

A. Background

Prior to the molecular characterization of the urea transporter proteins, two cell types were the favored test beds for experiments addressing urea transport. These were the red blood cell (RBC) and the terminal inner medullary collecting duct ($IMCD_t$) of the kidney. Using these cell types and what can now be termed classic physiological, biochemical, and biophysical techniques, the following key characteristics were identified (reviewed in Knepper and Rector, 1991): (1) Transport of urea across the IMCD epithelia or red blood cell membrane was facilitative and as such was incapable of transporting urea against a concentration gradient. (2) The pathway taken by urea was separate to that taken by water. This issue had been debated for some time mainly due to the fact that urea and water are similar in size and share certain physicochemical properties. (3) Urea transport was found to be inhibitable by millimolar concentrations of the aglycon phloretin, a characteristic that was later made use of to clone the first urea transporter cDNA. Phloretin inhibited IMCD urea transport regardless of on which side of the epithelia it was placed. (4) The K_m and V_{max} of urea transport were very high, indicating low affinity and high turnover number transport. (5) The antidiuretic hormone vasopressin caused a dramatic increase in the urea permeability of the IMCD, but not in the RBC.

B. Urea Transporters

On the shoulders of this sound knowledge, the first urea transporter cDNA was isolated from rabbit inner medullar using an expression cloning approach that utilized heterologous expression of mRNA in *Xenopus* oocytes (You *et al.*, 1993). This cDNA encoded a 397-amino-acid glycoprotein termed UT-A2 (originally known as UT2). Expression in *Xenopus* oocytes confirmed that the protein translocated urea in a facilitative manner and was inhibited by phloretin. Soon after this breakthrough, a homologous cDNA was isolated from bone marrow that encoded a urea transporter that shared the characteristics of UT-A2, but only 60% amino acid identity (Olivès *et al.*, 1994). It has now transpired that these two proteins are the products of two genes, known as UT-A and UT-B. In mouse and human these genes are arranged in tandem on chromosome 18 (Olivès *et al.*, 1995, 1996; Fenton *et al.*, 1999). The proximity of the genes to each other and high degree of homology shared by the gene products have led to the suggestion that they arose from a common ancestral gene (Fenton *et al.*, 1999).

Both genes give rise to multiple transcripts. At least four transcripts are derived from UT-A by differential splicing and two from UT-B by differential use of polyadenylation signals. The UT-A transcripts encode proteins of different size, whereas the UT-B transcripts encode the same protein. The largest UT-A transcript so far characterized is 4.0 kb and codes for UT-A1, a 927-amino-acid protein (Shayakul *et al.*, 1996). A feature of this and the other UT gene products is that they are comprised of a high number of hydrophobic residues relative to other membrane proteins. Considerable energy has been expended by various researchers to model the UT structure based on computer analysis (Sands *et al.*, 1997), however, it is difficult to discern whether the protein forms membrane-spanning α-helices and, if so, how many or whether they are composed of amino acids arranged in a β-sheet confirmation. This area of research requires empirical testing and should be a fruitful avenue to pursue. Simply for convenience we have chosen to represent the basic structural module as a four-membrane-spanning α-helical structure (Fig. 1). We see in Fig. 1 that UT-A1 is composed of UT-A3 and UT-A2 linked by a span of hydrophilic residues. Within this linking sequence and the reasonably long N- and C-terminal residues are the concensi for protein

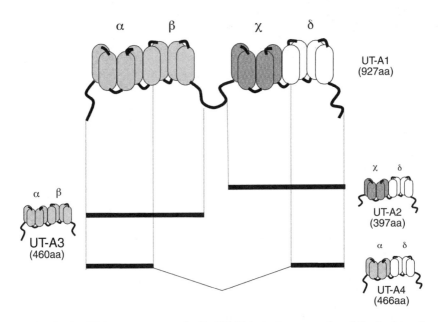

Fig. 1. The UT-A urea transporter family. UT-A1 is the largest member of the family so far characterized. It consists of four homologous domains designated, α, β, χ, and δ. The other UT-A gene products UT-A2, UT-A3, and UT-A4 are also shown. Thick black lines delineate UT-A1 domains, which serve to make up other family members: UT-A2 consists of domains χ and δ; UT-A3 consists of domains α and β; UT-A4 consists of α and δ spliced together. Numbers in parentheses denote the predicted number of amino acids (aa) that make up a given protein.

kinase A (PKA) and protein kinase C (PKC) sites. There are also potential glycosylation sites in regions predicted to project extracellularly.

The protein can be divided into similar sized and homologous "cassettes" (Fig. 1). For example UT-A1 is composed of cassettes α, β, χ, and δ, whereas UT-A3 is composed of cassettes α and β and UT-A2 consists of cassettes χ and δ. Although no equivalent to UT-A1 derived from the UT-B gene has been identified, the UT-B protein, like UT-A2, UT-A3, and UT-A4, is also composed of two structurally similar cassettes that instill in the molecule considerable internal homology.

The molecular biology of urea transporters continues to advance rapidly, allowing further advances in the physiologic aspects, all of which are summarized in several review articles (Hediger *et al.,* 1996; Sands *et al.,* 1997; Bankir and Trinh-Trang-Tan, 1998; Sands, 1999). Most of the physiologic and molecular data are for rat, but human genes have also been cloned. Across mammalian species and tissue types, transporters within either the UT-A or UT-B groups show 75–95% amino acid identity, whereas identities between UT-A and UT-B fall in the range of 50–70%. To date, most of the kinetic studies of how the transporters function have been performed with erythrocytes, although *Xenopus* oocyte expression of cDNA clones has also added to these data. These data do not conform to a carrier-like translocation, which alternates the urea binding site to either membrane face, which would characteristically have a realtively slow turnover number (V_{max}). However, the very high turnover number suggests a channel-based type of function and more recent data for kidney lend credence to this notion (Kishore *et al.,* 1997).

As might be surmised from the above discussion of collecting duct water versus urea permeabilities, the UT transporters are highly selective for urea and are generally held not to transport water when expressed at levels representing those found *in vivo* (Sidoux-Walter *et al.,* 1999). Similarly, the aquaporin water channels (AQPs) are typically highly specific for water (Borgnia *et al.,* 1999), although several do transport urea.

Both UT-A and UT-B family members are expressed in the kidney. UT-A is differentially expressed along the nephron (Shayakul *et al.,* 1997), whereas UT-B is expressed in blood vessel of the descending vasa recta (Xu *et al.,* 1997). Products of both genes have been detected in several other tissues including liver, brain, and colon (You *et al.,* 1993; Olivès *et al.,* 1996). The physiologic role these proteins play in some of these tissues is very much open to speculation and requires careful consideration. The fact that these proteins *are* expressed in tissues not associated with osmoregulation is further evidence in support of urea serving multiple functions; however, these currently remain speculative and we will restrict our discussion to the well-established role urea plays in the mammalian kidney.

C. Physiologic Role of Urea Transporters in the Mammalian Kidney

The major role of the mammalian kidney is the removal of metabolic waste from the plasma while maintaining water balance. Although urea is not considered a metabolically useful molecule, between 50 and 70% of urea filtered at the glomerulus is reabsorbed. Why half or more of the urea filtered is reabsorbed is yet unknown and calls into question the classification of urea as a "waste" by-product of catabolism. It has however, been known for many years that urea contributes to the high medullary osmolality that is responsible for urinary concentration. As part of this process, the mammalian kidney is capable of concentrating urea from plasma/glomerular filtrate values in the range of 4–10 mM to approximately 600–800 mM in urine. According to Knepper and Chou (1995), the transport of urea out of the mammalian nephron is largely, if not exclusively, by passive permeation (faciliated diffusion), although recent evidence suggests that under certain circumstances an ion couple component may contribute (see below). Values for urea permeability in the mammalian nephron range from 0.4 to 69 $\times$ 10^{-5} cm·s^{-1} (Knepper and Rector, 1991), values that are typically higher than the basal permeabilities of, for example, the gills of teleosts studied (see below). Importantly, mammalian kidney segments are *differentially* permeable to urea. This fact, when coupled with the complex architecture of the mammalian kidney, and the temporal variation in urea permeability that is possible *within* some segments, allows for large fluxes and recycling of urea within the kidney, and the osmotic extraction of water from the urine. Several key points of mammalian kidney physiology as it relates to urea movement are summarized from Knepper and Chou (1995).

1. KIDNEY ANATOMY

The anatomy of the kidney, with distinct zones and a countercurrent arrangement of tubules and blood vessels allows for the development of concentration gradients of solutes within the kidney. Fig. 2 shows a schematic drawing of mammalian nephron segments, particularly one long-looped and one short-looped nephron. The glomerulus is where blood plasma undergoes ultrafiltration, and the resultant ultrafiltrate is subject to modification as it passes along the various nephronal segments. For the purposes of our discussion, note particularly the proximity of the thin limbs of Henle's loops with the inner medullary portion of the collecting duct (Fig. 2).

The second anatomic feature of importance is the renal vasculature, which carries blood to and from the renal medulla, the *vasa recta* (not pictured). Descending vasa recta deliver blood to the capillaries of each level of the medulla, which then feed into ascending vasa recta. Descending and ascending vasa recta pass in proximity to each other in the inner stripe of the outer medulla, and the counterflow arrangement of the vessels facilitates countercurrent exchange of sol-

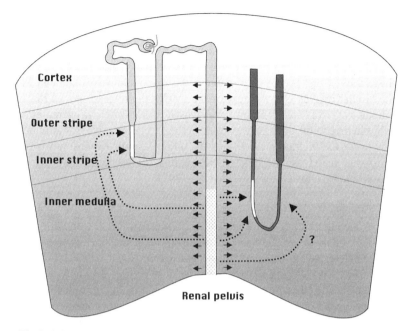

Fig. 2. Schematic diagram of mammalian kidney showing a short (left) and a long (right) looped nephron. Small arrows represent water movement out of the collecting duct. Urea (dotted arrows) moves out of the terminal inner medullary collecting duct via UT-A1 (white speckled shading) and is recycled into descending loops of Henle via UT-A2 (white open area). Urea may also enter long loops in the thin ascending limbs (arrow with question mark) although UT-A proteins have not been identified in this nephron segment.

utes and water. This countercurrent exchange allows a reduced effective blood flow but maintains a high absolute perfusion rate, leading to the preservation of solute concentration gradients that are central to urinary concentration. UT-B has been shown to be expressed on the descending vasa recta and is proposed to allow urea to diffuse into the blood vessels and in so doing be carried down into the medulla. In other words, blood flow is arranged so as to maintain high urea concentrations in the medullary interstitium, and it is this compartment of relative hyperosmolality that is responsible for water movement out of the nephron and consequently concentration of the urine (see below).

2. PERMEABILITIES

For the most part, the permeabilities of nephron segments fall in the range of 0.4 to 3.5×10^{-5} cm·s^{-1}. However, the thin descending and ascending portions of Henle's loop in the inner medulla have a permeability approximately 10-fold higher (13.5 to 19×10^{-5} cm·s^{-1}). The second exceptional segment is the final one-third of the collecting duct, the so-called $IMCD_t$, which has a permeability of 69×10^{-5} cm·s^{-1} (Knepper and Chou, 1995). This permeability increases to an extremely high value in response to the nonapeptide hormone vasopressin (antidiuretic hormone). The high permeability of the thin descending limbs and $IMCD_t$ is due to the presence of UT-A proteins in the apical and basolateral membranes of the tubular epithelia. In the case of the $IMCD_t$ the transporters on the apical membrane (UT-A1) are activated by increased cAMP levels triggered by vasopressin binding to the V_2 vasopressin receptor. In contrast, UT-A2 is in part responsible for the high permeability of the thin descending limbs; however, this protein is not acutely activated by cAMP, but its level has been shown to increase in response to sustained thirst (Smith *et al.*, 1995).

3. MOVEMENT OF UREA

How does the kidney concentrate urea to massive levels without obligating very much urinary water? In many of the earlier segments, the movement of urea can best be described as passively following osmotic gradients, and it is unclear if this is a urea transporter-mediated process because none of the known urea transporters localizes to these segments. It is not known if this movement is via the lipid bilayer, via nonspecific movement through other transporters, or via other "leak" components. However, in distal segments differential permeability instilled by urea transporter proteins comes into play in concentrating urea. In the thick ascending limb of Henle's loop, vigorous active NaCl extrusion leads to a high NaCl concentration in the interstitium of the outer medulla.

Importantly, this segment has a very low urea and water permeability. In the cortical and outer medullary collecting duct, urea permeability remains relatively low, but water permeability is high due to the presence of aquaporin proteins; water is drawn from the initial collecting duct segments leading to the concentration of urea in the collecting duct fluid. In the $IMCD_t$, the urea permeability is high and urea rapidly moves from the IMCD fluid to the inner medullary interstitium, leading to a high concentration of urea in the medullary interstitium. This creates an osmotic gradient allowing more water to be extracted and thus increases the solute concentration of the urine. The net result is an inner medulla with urea concentrations approaching 1 *M*, and a tubular urine high in solute and low in volume (water). Therefore, "waste" solutes can be voided from the body with minimal water loss.

Because of the high urea permeability of the thin limbs and the concentration gradient that exists between the medullary interstitium and the fluid in the thin limbs, urea moves into the thin limbs and is thus "recycled" back along the distal tubule and collecting duct (Fig. 2). This recycling is a central feature of the countercurrent multiplication system and is dependent on the differential urea permeabilities conferred by urea transporters.

Modulation of the system is achieved by coordinated activation of UT-A1 transporters and AQPs by vasopressin. For example, if vasopressin leves are low, water and urea permeabilities remain low in the collecting duct, resulting in the excretion of relatively large volumes of dilute urine.

Numerous other factors have been found to acutely regulate tubular urea permeability. These include hyperosmolality and glucagon (Sands *et al.*, 1997), and no doubt more will be discovered in the near future. Unfortunately, knowledge of the molecular events that bring about changes in urea permeability is lacking. In addition to being acutely regulated, urea transporters are regulated in the long term at the gene and possibly the protein level in response to dietary protein content, water availability, hypercalcemia, and glucocorticoids among other factors (Smith *et al.*, 1995; Sands *et al.*, 1997; Naruse *et al.*, 1999). The end point of these regulatory changes is to affect the number of urea transporter molecules and in so doing increase/decrease urea permeability. Regulation at this level is most likely to involve modulation of gene transcription, mRNA stability, translation, or protein degradation or possibly several of these processes. With so many factors influencing urea transporter expression and activity, it becomes clear that regulation of urea permeabilities is extremely complex and is not a static process, but one that responds to the prevailing status of the animal.

D. Physiologic Role of Urea Transporters in Mammalian Hepatocytes and Red Blood Cells

Urea is synthesized in the mammalian liver and the presence of urea transporters has been predicted in this tissue to allow exit of urea into the circulation. However, initial studies to isolate a liver urea transporter homologous to UT-A or UT-B failed and it was left to the expression cloners to come up with an explanation of how urea exits hepatocytes. The answer arrived in the form of a cDNA encoding a novel member of the aquaporin family, aquaporin 9, which, in addition to transporting water, also transports urea and is inhibited by phloretin. This discovery caused researchers to question the need for a specific urea transporter in liver. However, despite the failure of earlier efforts, recent studies have reported the presence of UT-A transporters in liver (Klein *et al.*, 1999).

The rationale for UT-B expression in the kidney has been discussed above, so we now turn our attention to UT-B expression in RBCs. It has been suggested that the physiologic role of urea transporters in RBCs is to prevent detrimental volume change as they pass through the vasa recta and experience vast changes in the concentrations of urea (Macey and Yousef, 1988).

Interestingly, there is a human condition in which, due to a mutation in the UT-B gene, UT-B protein is not expressed. People carrying this mutation, known as Jknull, are therefore devoid of UT-B protein. If UT-B was important in maintaining RBC integrity, then one would expect evidence of damaged RBCs and hemolysis, yet Jknull individuals are asymptomatic and only when challenged by thirsting do they show a mild reduction in their ability to concentrate urine (Sands et al., 1992). A more likely function of UT-B is that its presence prevents the rapid washout of the medullary urea gradient by circulating RBCs. The fact that UT-B has asymmetric transport properties, in that urea moves out of RBCs (efflux) more effectively than it moves in (influx), supports this hypothesis (Macey and Yousef, 1988). From this it can be predicted that animals such as fish that lack the ability to concentrate their urine may have no requirement for RBC urea transporters.

E. Active Urea Transport

Although the bulk of researchers' attention has been focused on facilitated diffusion of urea in the mammalian kidney, at least three measurable components of urea flux appear to be active. Rats fed a low-protein diet for several weeks might experience a physiologic need to conserve nitrogen as urea, which can then be recycled by gut flora (see e.g., Wolfe, 1981). Under these condiditons, a sodium-dependent secondary active Na^+-urea cotransport is expressed in the IMCD. Urea transport is inhibited when Na^+,K^+-ATPase is inhibited and by removing Na^+ from the lumen (but not the bath) and is thus hypothesized to be in the apical membrane. It is not inhibited by phloretin, not stimulated by vasopressin, and is encoded for by an mRNA that is longer than UT-A1 (Isozaki et al., 1993, 1994a,b; Sands et al., 1996).

A second active component is expressed in the initial IMCD from furosemide-treated rats (Kato and Sands, 1998a). It is inhibited by removing sodium from the bath, but not the lumen, is stimulated by vasopressin, and inhibited by both phloretin and oubain, suggesting a basolateral sodium-urea countertransporter. Lastly, Kato and Sands (1998b) have also described an apical sodium-urea countertransporter in $IMCD_3$ which is inhibited by removing sodium from the lumen but not the bath, is inhibited by phloretin and ouabain, and is stimulated by vasopressin.

There are also reports of several types of active urea transport in amphibians, but these appear to be of a different variety, typically being proton dependent rather than sodium dependent or, in fact, being a primary active transport (Ussing and Johnansen, 1969; Garcia-Romeu et al., 1981; Lacoste et al., 1991; Rapoport et al., 1988; Dykto et al., 1993). Active urea transport has also been reported in yeast (Cooper and Sumrada, 1975; Pateman et al., 1982) and bacteria (Jahns et al., 1988). Interestingly, the yeast transporter gene (*DUR3*) has been cloned and its amino acid sequence (ElBerry et al., 1993) is completely unlike other UTs (P. J. Walsh, unpublished observations). Perhaps its sequence will be useful for cloning active urea transporters from vertebrates. It is possible that use of elasmobranch tissues where mRNA for such transporters might be more abundant (see below) than in mammalian kidney might prove useful in this regard.

III. PHYSIOLOGY OF UREA TRANSPORT IN FISH

A. Elasmobranchs

Although elasmobranchs do retain urea as an osmolyte, they appear to be ureotelic, at least under limited laboratory circumstances. A study by Wood *et al.* (1995b) on *fasted* spiny dogfish shark, *Squalus acanthias* measured rates of J_{Urea} of approximately 500–700 μmol urea-N $kg^{-1} \cdot h^{-1}$, while ammonia excretion rates were only about 2–5% of the total nitrogen excreted under these conditions. The proportion of urea and ammonia excretion has only been measured simultaneously in very few species of elasmobranchs (reviewed by Perlman and Goldstein, 1988; Shuttleworth, 1988; Wood, 1993), but appears to follow the same general pattern as that of Wood *et al.* (1995b), with urea predominating over ammonia.

However, data on ammonia infusion in this same study suggest that the effect of feeding on patterns of nitrogen excretion by elasmobranchs should be systematically examined. When spiny dogfish were infused with ammonia, most of the ammonia load was rapidly excreted as ammonia (J_{Amm} rose from approximately 30 to 1000 μmol N $kg^{-1} \cdot h^{-1}$), while urea excretion peaked hours later at a lower incremental value (i.e., an additional over baseline J_{Urea} of approximately 500–700 μmol-urea N $kg^{-1} \cdot h^{-1}$) (Wood *et al.*, 1995b). It would be of interest to determine if a similar ammonia excretion pattern resulted during the natural nitrogen loading of feeding, that is, to test the hypothesis that urea synthesis (an energetically costly pathway) would be attuned to only produce enough urea to osmotically balance unavoidable losses. These experiments might also be designed to further test the hypothesis of metabolic ammonia trapping by gill glutamine synthetase proposed by Wood *et al.* (1995b). Finally, as Wood *et al.* (1995b) pointed out, the data of Goldstein and Palatt (1974) indicate that trimethylamine oxide (TMAO) losses in elasmobranchs may be an important component of nitrogen excretion, contributing 10–20% of urea losses. Given the uncertainty surrounding the dietary versus synthetic origins of TMAO in elasmobranchs (Goldstein and Palatt, 1974), TMAO excretion measurements in feeding studies would also be informative.

Because elasmobranchs appear to be ureotelic, we now turn our attention to the routes and mechanisms of ureotely. Several studies have documented that the dominant route of urea loss is via the gills, whereas the kidney is highly efficient at limiting urea losses through a powerful reabsorption mechanism. For example, in spiny dogfish, the kidney and gills account for 6.5 and 91.2% of urea excretion, respectively (Wood *et al.*, 1995b). We examine these mechanisms separately in the next sections.

1. KIDNEY

It was known since Städeler and Frérichs (1858) that elasmobranch blood and tissues contained an unusually high concentration of urea, and since Denis (1913) and Scott (1913) that the urine contained a much lower level of urea than plasma. In classic studies, Homer Smith (Smith, 1931a,b; 1936) extended these observations to numerous elasmobranch species (both freshwater and saltwater), and also measured other major osmolytes. He then formulated the model of osmoregulation in these species from which current researchers still begin. Much of the other earlier literature is reviewed by Hickman and Trump (1969). The salient findings of this earlier literature and selected recent studies can be summarized as follows:

1. A substantial amount of urea is filtered across the glomeruli and most is reabsorbed in the tubules (e.g., 70–99.5% in the smooth dogfish, *Mustelus canis*; Kempton, 1953).
2. There is a constant, albeit small, residuum of urea in the urine that does not vary with the filtered load or absolute resorption of urea, leading Kempton (1953) to speculate that this residuum is obligately linked to water excretion (necessary since elasmobranchs have a slightly higher osmolality than seawater and face passive water gain across the gills).
3. The analogs methylurea and acetamide are both reabsorbed by the kidney nearly as well as urea, but thiourea is not (in a wide variety of elasmobranch species; Schmidt-Neilsen and Rabinowitz, 1964).
4. Urea reabsorption is correlated with sodium reabsorption in a ratio of 1.6 urea:Na^+, over a very wide range of urine flow rates and urea reabsorption values (30–6600 μmol·kg^{-1}·h^{-1}) induced by exposure to diluted seawater and expansion of extracellular fluid volume by various infusions (in the spiny dogfish; Schmidt-Nielsen *et al.*, 1972).
5. Phloretin, administered intravenously, increased the fractional excretion of urea by fourfold, and led to a proportional decrease (~30%) in both urea and Na^+ reabsorption (in the spiny dogfish; Hays *et al.*, 1977). In the same studies chromate injection (believed to interfere with cAMP production) decreased urea reabsorption, but had less effect on Na^+ reabsorption, thus slightly reducing the urea:Na^+ ratio from 1.6 to between 1.22 and 1.34. These authors suggest that both chromate and phloretin permeated basolateral and apical sides, so no inferences regarding polarity of action can be inferred.
6. The architecture of the elasmobranch kidney is complex, but various studies (of *Raja erinacea*; Lacy and Reale, 1991; Lacy *et al.*, 1985) reveal several important features (Fig. 3): (a) an arrangement of nephron loops in a countercurrent fashion; (b) the inclusion of some of the loops (fully or partly depending on the segment) in either a peritubular sheath or a blood sinus, that is, distinct anatomic zones; and (c) an extensive capillary net-

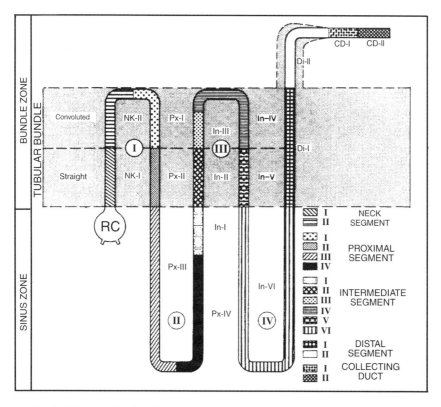

Fig. 3. Schematic drawing of the course of the elasmobranch nephron into four loops, distal tubule, and the collecting duct. Loops I and III and the early distal tubule as well as the late distal tubule leaving the bundle (countercurrent system) are wrapped by the peritubule sheath. The subdivisions of each tubule segment are indicated by the symbols and abbreviations. RC, renal corpuscle. (Modified from "Fine structure of the elasmobranch renal tubule." E. R. Lacy and E. Reale. *Am. J. Anat.* © 1991. With permission of Wiley-Liss, Inc., a subsidiary of John Wiley & Sons, Inc.)

work in countercurrent arrangment to many of the nephron segments. Overall, these authors suggest that a complex renal countercurrent multiplier system may be important in fluid regulation in elasmobranchs.

7. In the first identification of a urea transporter in fish, Smith and Wright (1999) cloned by homology to mammalian UT-A2 a gene from dogfish shark encoding a protein (ShUT) belonging to the UT-A family. By Northern analysis, it is clear that the gene is transcribed in kidney and brain of dogfish shark and this gene, or a related one, is suggested in several other tissues including gill under lower stringency hybridization conditions.

The above data can fit either a model of passive or active reabsorption of urea, and definitive proof for either model is lacking. However, a passive model appears to have been favored to date. This model was originally proposed in its simplest form by Boylan (1972), and later elaborated by Friedman and Hebert (1990), both based on the complicated architecture of the kidney, and the assumption/prediction that there is differential urea permeability in the various nephron segments. By active transport of sodium (chloride) from the tubule lumen to the interstitium, and the maintenance of high interstitial [NaCl] by the anatomic/circulatory aspects, favorable conditions for passive (diffusional) movement of urea could be established in much the same way as the mammalian kidney uses differential permeabilities to generate a urea gradient for the extraction of water from the urine.

The passive model for urea reabsorption in elasmobranch kidney incorporates the dependence on sodium, as well as the existence of a facilitated urea transporter (Fig. 4a). Notably, it would require a urea transporter on both basolateral and apical membranes of tubular cells as is indicated in the mammalian kidney (Knepper and Chou, 1995).

As Smith and Wright (1999) caution, and in light of the recent data for Na^+-coupled urea transport in the mammalian kidney by Sands and coworkers, however, secondary active urea reabsorption cannot be ruled out. For example, given the very tight coupling between urea and sodium reabsorption under many circumstances (Schmidt-Nielsen *et al.*, 1972), it is tempting to speculate that urea is transported by a Na^+-coupled urea cotransporter (analogous to the first of the active transporters described above for mammalian kidney). In this arrangement (Fig. 4b), an inwardly directed sodium gradient would be established by the basolateral Na^+,K^+-ATPase, and urea movement from the tubule into the cell at the apical surface would be directly coupled to the movement of sodium. Urea would then exit the cell at the basolateral membrane via a facilitated transporter. Assuming that this Na^+-urea cotransporter is not phloretin sensitive like its mammalian counterpart, this active model cannot be ruled out on the basis of the phloretin inhibition studies of Hays *et al.* (1977), as this inhibition could simply be on the last step of the transport pathway (e.g., the facilitated diffusion via a basolateral UT-A or -B protein), rather than on the first, Na^+-coupled, step. An alternative active scenario might account for the phloretin sensitivity by including a Na^+-urea cotransporter on the apical membrane, and a Na^+-urea countertransporter on the basolateral membrane (Fig. 4c). Interestingly, both of these scenarios would be electrogenic, necessitating perhaps another facet of the transport process to be understood, as well as another component that could be manipulated experimentally in voltage-clamp-type studies. Both models (Figs. 4b and 4c) could also include metabolic urea generated in the kidney tubule cells to augment (or perhaps even generate) urea gradients as suggested for the gill by Wood *et al.* (1995b); see below.

It is somewhat dissappointing to note that roughly a century since the initial observations of the large urea gradient from plasma to urine in elasmobranchs, mechanisms of urea reabsorption are still so poorly understood, relative to the

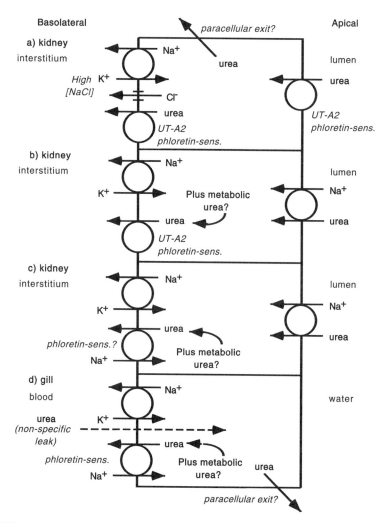

Fig. 4. *Hypothetical* arrangements of urea transporters in the elasmobranch kidney (a–c) and gill (d). Depicted are kidney reabsoprtion in (a) Urea transport driven by osmotic movement linked to NaCl transport and a high [NaCl] in the insterstitium in which passage across the basolateral membrane is either by UT-2A or nonspecific paracellular pathways; (b) secondarily active urea transport linked to sodium entry at the apical membrane and exit at the basolateral membrane via UT-2A; and (c) secondarily active urea transport linked to both sodium entry at the apical (cotransport) and basolateral (countertransport) membranes. Depicted in (d) is a model for gill extrusion/backtransport of urea. In models (b), (c), and (d), metabolically produced urea by the gill could account for part or all of the urea gradient.

function of the mammalian kidney. Clearly, further work applying micropuncture techniques, isolated tubule studies, perfused kidney preparations, electrophysiologic approaches, and molecular probes will shed light on these mechanisms.

2. Gills

Although the early studies of Homer Smith recognized that the gills of elasmobranchs must be rather impermeable to urea for the ureoosmotic strategy to be feasible, gill urea permeability was not really addressed experimentally until the studies of *Squalus acanthias* by Boylan and colleagues in the 1960s, summarized in Boylan (1967) where he prefaces his work with "The gill, an epithelial structure one to two cells in thickness is perfused on its inner surface by blood having a urea concentration of 2 g percent (~ 330 mM) and on its outer surface by sea water virtually free of urea."

In these early studies, Boylan and colleagues examined the movement of ^{14}C-labeled urea and thiourea from blood to water, in what we might describe now as an *in vivo* perfused head preparation. They established that gill urea permeability values were some 100-fold lower than for water (7.5×10^{-8} cm·s^{-1}), and some 35-fold lower than for urea through toad bladder (Maffly and Leaf, 1959). Furthermore, although as noted above the dogfish kidney rejects thiourea (Schmidt-Nielsen and Rabinowitz, 1964), thiourea permeabilities at the gill were about equal to urea (Boylan, 1967), an early clue that gill transport mechanisms may differ from those of the kidney.

Study of the urea transport dynamics of the elasmobranch gill then drops from the literature for some 26 years, an amazing hiatus given the incredible nature of the gradient being held back by a one- to two-cell layer epithelium, seemingly a transport physiologist's dream system!

Work on this problem resumed recently. Wood *et al.* (1995b) used an *in vivo* approach with *S. acanthias* which, as noted above, through the use of urinary catheters established firmly that the bulk of ureotely occurs through the gill. Next, through a series of experiments that followed urea *and ammonia* excretion during infusions of acids, bases, ammonia, urea, and urea analogs, they were able to reach several conclusions. First, the low net or effective permeability values of Boylan (1967) were reconfirmed. Second, dogfish gills were seen to have an ammonia permeability some 22-fold lower than for teleosts (see below). Although Boylan initially speculated that urea permeability was low due to structural features, this conclusion is difficult to reconcile with the apparent normal permeabilities of the gases CO_2 and O_2 and the lower permeability of a third gas, ammonia. Wood *et al.* (1995b) speculated that the low ammonia permeability may be the result of a metabolic trapping system by gill glutamine synthetase. Third, infusion of thiourea or acetamide (at levels equivalent to only about 15% of blood urea) resulted in increased urea excretion. This observation is consistent with Wood *et al.*'s hypothesis of a "backtransporter" at the gill that normally returns urea "leak" to the plasma; inhibition of the transporter with thiourea or acetamide would allow the "leak" to continue unabated leading to the observation of increased net excretion.

Surprisingly, ammonia infusion markedly increased urea efflux despite no change in plasma urea excretion. This was taken as generally indicating a decoupling of urea excretion rates from plasma urea concentration, although activation of alternative, less specific pathways for urea movement (e.g., paracellular movement) was not ruled out. In contrast, urea infusion caused an initial brief *decrease* in urea excretion rates followed by a return to normal (Wood *et al.,* 1995b). Both of these results suggest that processes at the gill epithelium control the rate of excretion, rather than simply being responsive to plasma urea concentrations per se.

Pärt *et al.* (1998) extended this line of thought and experimention by hypothesizing that the net excretion across the gill epithelium would be a balance between passive influx from the blood, the function of a basolateral urea transporter facing the blood, and possibly metabolic urea production in the gill tissue, which would effectively cause a gradient from gill to blood. Using an isolated perfused head preparation (IPHP) from *S. acanthius,* they confirmed earlier values for urea permeability and showed that urea efflux was not strongly dependent on the perfusate urea concentration in the range of 175–300 mM. In contrast to the *in vivo* studies of Wood *et al.* (1995b), thiourea perfusion had no effect on urea efflux in the IPHP. However, phloretin addition to the perfusate nearly doubled urea excretion rates. A washout experiment in which urea-free perfusate was used allowed simultaneous measurement of urea flux to both the water and the perfusate, and the calculation that the basolateral membrane is about 14 times more permeable to urea than the apical membrane.

Last, the addition of adrenaline to the perfusate caused a three-fold increase in urea efflux, with little effect on water flux. These data in sum suggest a specific pathway for urea movement through gill and the possibility of the backtransport mechanism proposed by Wood *et al.* (1995b). The arrangements suggested for kidney in Figa. 4a–c would require slight modification to account for these observations (Fig. 4d) and a phloretin-sensitive basolateral Na^+-urea countertransporter would seem to be the most parsimonious explanation for the current observations. Urea exit across the apical membrane of the gill would not need a facilitated urea transporter of the UT-type, because "leak" either cellularly or paracellularly from the massive gradient might be enough to account for ureotely. However, Smith and Wright (1999) showed evidence for a UT homolog in gill tissue (under low stringency) although the location of this mRNA expression (i.e., gill epithelia or blood vessel) remains unknown, and it is possible that the signal is from a related Na^+-urea countertransporter.

Fines *et al.* (2000) have performed studies with isolated basolateral membranes of dogfish shark gills and have demonstrated the presence of a phloretin-sensitive, Na^+-urea countertransporter, with orientation of urea transport toward the serosal side as required by this model. However, in addition to this timely finding, they also measured a rather high cholesterol content in the shark gill membrane lipids, and suggest that it is this high cholesterol content that actually accounts for the bulk of the low permeability of the gill to urea. These observations follow closely from other research on elasmobranch membranes, which show unusual properties of the lipids (e.g., Barton et al., 1999; Glemet and Ballantyne, 1996). Clearly more research is required with this fascinating system. In addition to continued work on the IPHP and isolated gill vesicles, valuable information could be obtained from immunocytochemistry when appropriate probes are available, and from ornithine–urea cycle (O-UC) enzymology on the gill tissue to test the hypothesis of urea production by gill cells.

3. Hepatocytes and Red Blood Cells

Studies of urea transport in hepatocytes and RBCs of elasmobranchs are relatively limited. Examination of urea transport in elasmobranch hepatocytes is limited to one study that supports a passive mode of urea transport (Walsh et al., 1994). Considerably more work has been carried out to address urea transport across RBC membranes. Initial studies by Murdaugh et al. (1964), measuring time to lysis of *S. acanthius* RBCs in several artificial solutions, suggested the presence of urea transporter proteins. However, this hypothesis was rejected later by Rabinowitz and Gunther (1973) using more stringent experimental tests on *Squalus suckleyi*, including the use of ^{14}C-tracer methodology. Kaplan et al. (1974) similarly reported no evidence for facilitated transport in erythrocytes from *S. acanthias* and *R. erinecea*. Walsh et al. (1994) extended this conclusion by showing no effect of phloretin or acetamide on the influx of urea into RBCs of the lesser spotted dogfish *Scyliorhinus caniculata*. In a study of skate erythrocytes, Carlson and Goldstein (1997) seem to have established that urea in fact permeates via the lipid phase of the membrane. The evolutionary implications of the apparent lack of specialized urea transport in elasmobranch RBCs are addressed in Section IV. It is tempting to conclude that, with the obvious exception of possible gradients within specialized transport tissues (gills and kidney) aimed at major transepithelial movement, urea concentrations within the elasmobranch body are largely uniform and that urea movement responds slowly to the overall poise set by synthesis and transport/leak at the animal–environment interface. Clearly, further molecular studies are required to determine more precisely the presence/absence of transporters in various tissues.

4. Urea Dynamics in Euryhaline Adaptation

We have known since the early work of Smith (1931a) that urea plays an important role in elasmobranch adaptation to lower salinity. In this study, four freshwater elasmobranchs had 70% lower plasma urea contents; this pattern was reconfirmed for freshwater bull shark (*Carcharhinus leucas*) by Urist (1962) and Thorson (1967). Furthermore, Smith (1931a) presents evidence that this was due to an increased urinary output and a diminished fractional tubular reabsorption.

Price (1967) and Price and Creaser (1967) demonstrated that for clear-nose skate, *Raja eglanteria,* a species that can be brackish, serum urea varied directly with salinity.

These early observations were followed by several more mechanistic studies in the late 1960s and 1970s, *generally* establishing that gill permeabilities to urea did not appear to change much during low salinity adaptation (although ultimately urea fluxes at the gills might have decreased due to the eventually lowered gradient of plasma to water urea), and that urinary urea voiding always increased (Goldstein *et al.,* 1968; Goldstein and Forster, 1971; Forster *et al.,* 1972; deVlaming and Sage, 1973; Payan *et al.,* 1973; Wong and Chan, 1977). Only one study addressed the role of the rectal gland (Wong and Chan, 1977) and found it to be minimal. These generalizations "make sense" in that it would be expected that in more dilute environments increases in the already passive water movement across the gills would lead to a water load that would be handled by the kidney in parallel to urea handling. Note, however, that although the kidney is the organ that responds by *increasing* net urea excretion, when measured, the gill still accounts for half of the urea losses in dilute salinities (e.g., Wong and Chan, 1977). Not much examination of this phenomenon has taken place in the interim, and future studies are very much in order.

The generalizations listed above are far from being firmly established conclusions for several reasons: (1) At least six different elasmobranch species were examined in the above studies with very different measurements performed on each; (2) the above species represent a range of "stenohaline marine" and "euryhaline" lifestyles; (3) the above studies varied greatly in whether acute or acclimatory salinity effects were being examined; (4) there are some differences in species responses (but perhaps due to differing protocols) with regard to whether the increase in urinary urea output was due to changes in glomerular filtration rate, urea reabsorption, or both; and (5) potential changes in metabolic urea output were not always addressed, and responses of metabolism were varied when studied (decreasing in some cases and not changing in others).

Future in-depth studies would do well to (1) concentrate on fewer species (perhaps one representative stenohaline and one euryhaline species) where natural encounter rates with changes in salinity are well known from, for example, movement data from tracking devices; (2) address both acute and acclimatory effects (so that realistic predictions can be made about strategies employed by animals when encountering salinity changes in nature); and (3) carefully and simultaneously measure branchial, renal, and metabolic components.

B. Teleosts

Urea excretion mechanisms in teleost fish have not been extensively examined for the obvious reason that most teleosts are predominantly ammoniotelic. For many species, for example, it is not known if branchial or renal pathways predominate in urea excretion. In one systematic study using a divided chamber approach, Sayer and Davenport (1987) demonstrated that anterior pathways accounted for at least 50% of urea excretion in six out of seven teleost species. In a more recent study, Wright et al. (1995) demonstrated that about 75% of urea excretion occurred at the anterior end of the tidepool sculpin (*Oligocottus maculosus*). A similar finding of 80% was reported by Wood et al. (1994) for the obligately ureotelic Lake Magadi tilapia (*Alcolapia grahami*). It would appear as if both gills and kidneys deserve attention in future studies. By far, the most extensive data set comes from recent work with the gulf toadfish, *Opsanus beta*. These data confirm that branchial pathways predominate in this species, so we will begin our analysis of teleosts with this tissue and species.

1. GILLS

Gulf toadfish are among a small group of adult teleost fishes that excretes urea as a predominant waste product (see Chapter 1). The urea excretion event is highly pulsatile in *O. beta*, with virtually all of the daily urea load being excreted in a single pulse through the gills lasting from 0.5 to 3 h (reviewed by Walsh, 1997; Wood et al., 1995a, 1997, 1998; Gilmour et al., 1998; Pärt et al., 1999). Gill pavement cells, but not the chloride cells, undergo massive morphologic changes during a pulse, including elaboration of the apical membrane and an increase in the number of cytosolic vesicles and their fusion with the apical membrane, suggesting that these events may be linked to the increase in urea exit (Laurent et al., 2001). In parallel, a 1.8-kb gill cDNA (tUT) with high homology to the mammalian kidney UT-A2 and the dogfish shark kidney (ShUT) has been cloned and found to encode a UT-like protein (Smith et al., 1998; Walsh et al., 2000; GenBank Accession AF165893). Of the tissues examined on Northern blots using a tUT cDNA probe (gill, liver, RBC, kidney, skin, intestine), only gill yields a positive signal (Smith et al., 1998; Walsh et al., 2000).

Considerable physiologic evidence that excretion of urea is by a facilitated diffusion pathway in toadfish gill has been obtained. Plasma urea concentrations fall dramatically at the time when urea appears in the water (Wood et al., 1997). Because urea appears to be in equilibrium between tissues and plasma, plasma concentration changes have been used to predict whole-body urea loss during a pulse, and these estimates nearly exactly match the amount of urea appearing in the bath water (Wood et al., 1997). *In vivo*, the pathway is inhibited by the urea analogs thiourea and acetamide, and is not saturable at concentrations up to 60 mM urea (Wood et al., 1998). However, recent studies by McDonald et al. (2000) demonstrate that only acetamide appears to be transported by the pathway, at about 35–50% the effectiveness of urea; thiourea was transported less efficiently (15%) and thiourea effects on transport appeared to be via decreases in the number of pulses rather than the pulse size. The following is the most dramatic representation of the facilitated diffusion nature of the pathway. When *bath* urea concentrations are elevated above typical plasma levels of urea (i.e., 60 vs. 15 mM) such that the gradient was reversed to outward → inward, and unidirectional influx and efflux of urea were measured (using injected ^{14}C-urea), an *increase* in total plasma [urea] was observed during a pulse, with kinetics that directly mirror the timing of the unidirectional efflux. When efflux and influx rates are normalized to the respective urea concentration (i.e., plasma and bath), there is a highly correlated 1:1 correspondence, indicative of a facilitated diffusion transporter whose net flux is dependent on the gradient size and direction (Wood *et al.*, 1998).

Toadfish gill permeabilities to urea (as determined both *in vivo* and *in vitro* with the IPHP) *during nonpulsing periods* fall into the range of 2 to 4 × 10^{-7} cm·s^{-1} (Pärt *et al.*, 1999). During pulse periods, these permeability values increase by about 35-fold (Pärt *et al.*, 1999; Wood *et al.*, 1998). One hypothesis being tested for the mechanism of this massive increase in permeability involves targeted vesicular shuttling of transporter containing vesicles as has been described for the increase in water permeability triggered by vasopressin in the mammalian collecting duct (Inoue *et al.*, 1999). In the toadfish, we hypothesize that transporters are recruited to the pavement cell membrane by fusion with the plasma membrane of intracellular vesicles rich in tUT transporter. The activation of the transporter appears to be preceded by an overall minimum in plasma cortisol levels, indicating hormonal involvment (Wood *et al.*, 1998). Furthermore, injections of arginine vasotocin (the piscine equivalent of vasopressin) appear to be able to induce pulses *in vivo* (Perry *et al.*, 1998), but not in the IPHP, indicating a possibly complex pattern of hormonal regulation.

Although the pulsatile release of urea across the gills is, as far as we know, a unique feature of *Opsanus* species, whether the gills of other teleosts are permeable to urea and whether there is evidence to support involvement of urea transporters are interesting questions that are only beginning to be addressed. Gill permeability to urea has been measured in a few other teleost species, and has been summarized by Isaia (1984) and Pärt *et al.* (1999). Values for teleost gill are generally in the region of 2 × 10^{-6} cm·s^{-1} and changes in salinity do not affect this value. For comparison, a typical value for permeability of urea through artificial bilayers is about 4 × 10^{-6} cm·s^{-1} (Galluci *et al.*, 1971), about the same value, and about an order of magnitude lower than for water permeability through teleost gills (1 × 10^{-5} cm·s^{-1}). Thus, on initial inspection, teleost gills appear to conform to a "passive" model for urea permeability that does not involve specific urea transporter proteins as we have discovered in the toadfish (see above), or with no special contingencies for the urea retention as seen in elasmobranchs where effective permeabilities are 3 to 8 × 10^{-8} cm·s^{-1} (Boylan, 1967; Wood *et al.*, 1995b; Pärt *et al.*, 1998).

However, Pärt *et al.* (1999) speculated that perhaps the true diffusive urea permeability of all fish gills is closer to the toadfish permeability values of 10^{-7} cm·s^{-1}. In this view, elasmobranch species would further decrease this permeability through a backtransport mechanism. Notably a backtransport mechanism has been ruled out in the toadfish (Wood *et al.*, 1998), and the presence of the facilitated urea transporter (ShUT or homolog) shows very weak expression in the dogfish gill (Smith and Wright, 1999). Furthermore if this view is correct, typical teleost urea permeabilities would be due to the presence of at least a modest number of facilitated diffusion urea transporters. In support of this view, we have recently obtained evidence by Northern analysis that mRNA for a tUT-like transporter is expressed in the gills of a wide variety of teleost species (P. J. Walsh and C. E. Campbell, unpublished data). However, in the study of Wright *et al.* (1995), little evidence for UT-like urea transport was found in the tidepool sculpin. Clearly, further studies are needed on this interesting problem.

2. KIDNEY

The handling of urea by the teleost fish kidney was not extensively studied until recent years. In the same study of gill transport characteristics in the toadfish by McDonald *et al.* (2000), it was discovered that the urine had urea concentrations about twice as high as plasma, and that much of this increase was achieved by urea secretion, rather than filtration and subsequent selective resorption of water. In fact, urea secretion clearance was some two- to threefold more effective for urea than for water or chloride. Furthermore, renal handling of the analogs thiourea and acetamide was opposite to that of the gill. Specifically, thiourea inhibited urea accumulation in the urine and was itself secreted as effectively as urea (relative to water and chloride), whereas acetamide was not concentrated in the urine and had no effect on urea concentration (McDonald *et al.*, 2000). Urine Na^+ levels were about one-tenth of plasma concentrations, leading these authors to speculate that perhaps urea is cotransported with Na^+ into the urine, with subsequent active resorption of Na^+ further down the tubule. They also point out that equally likely, however, is a mechanism similar to the deep $IMCD_3$ of the mammalian kidney, where Kato and Sands (1998a,b) have demonstrated a direct countertransport of urea with active Na^+ resorption.

In an earlier study of rainbow trout, McDonald and Wood (1998) demonstrated that urea movement in the opposite direction is in play in the freshwater rainbow trout. Urea is reabsorbed from urine to plasma, such that plasma concentration remains about double that in the urine. Moreover, on loading of trout with urea, urea reabsorption increased in direct proportion with the filtered urea load, with no evidence for saturation of the reabsorptive mechanism at up to fourfold above the normal range. Could the same transporter protein be responsible for observations made in saltwater and freshwater? Perhaps the same transporter working in opposite directions may be responsible for urea secretion/reabsorption in saltwater and freshwater teleosts. Clearly, it would be of interest to examine the kidneys of euryhaline species with a view toward answering this question.

3. HEPATOCYTES AND RED BLOOD CELLS

Studies of urea transport in RBCs and hepatocytes of teleosts are limited. Kaplan et al. (1974) showed the absence of facilitated transport of urea in erythrocytes of winter flounder (*Pseudopleuronectes americanus*) and goosefish (*Lophius piscatorius*). Walsh et al. (1994) discovered a small (25%) phloretin-sensitive component of urea uptake by redfish (*Scianops ocellatus*) RBCs, but a lack of inhibition in RBCs of *O. beta* by phloretin and acetamide. The latter finding agrees with the lack of a detectable UT homologous signal in *O. beta* RBC mRNA (Walsh et al., 2000). *O. beta* and *O. tau* (oyster toadfish) hepatocytes exhibited substantial inhibition of urea influx and efflux by phloretin (Walsh et al., 1994; Walsh and Wood, 1996). It is possible that this transport is due to the specialized modes of urea excretion in this family of fishes although northern analysis of *O. beta* liver did not detect UT homologs. Further evidence is needed.

IV. PROSPECTS FOR FUTURE RESEARCH AND EVOLUTIONARY PERSPECTIVES

As pointed out above, a number of areas are fertile for research. The mechanisms of urea reabsorption/retention at the elasmobranch kidney and gills are far from understood, and it seems logical to not only study these mechanisms in the "static" state of full-strength seawater, but also under conditions of lowered salinity. If urea transport in elasmobranch kidney (and gill) does turn out to be directly sodium coupled, given the "intensity" of transport in this tissue it is likely that there will be a high relative abundance of transporter message for such Na^+/urea transporters. Investigators in search of a Na^+/urea transporter gene in the mammalian kidney might best start molecular investigations with elasmobranchs.

Study of urea transport in teleosts may also offer systems that are in a sense "stripped down" from the elasmobranch situation, such that facilitated diffusion, rather than active transport, may predominate. Likewise, different tissues within teleosts also appear to offer heterogeneity of function and additional model systems. The toadfish system appears to offer an advantage in studies of the temporal dynamics of urea transport regulation. Whether this phenomenon is unique to *Opsanus* species remains to be determined. In addition, we have deliberately focused on urea transport in adult fish because urea transport during development is the subject of Chapter 5. Earlier stages in development may in fact continue to offer novel and simpler systems for the study of urea transport.

Finally, with the rapid progress in identifying urea transporters at the molecular level in both mammals and fish, phylogenetic perspectives of transporter evolution will be most informative. We can speculate that the mammalian UTs arose as a result of gene duplication of an ancestral gene. The urea transporters isolated from fish to date (ShUT, tUT, and sequence for the Magadi tilapia gill UT; Walsh *et al.,* 2001) appear to be of the UT-A type, rather than something ancestral to both UT-A and UT-B per se. Furthermore, it appears as if there are no special urea transport mechanisms in the RBCs of fish (see discussion above, as well as in Kaplan *et al.,* 1974), perhaps indicating the absence of UT-B in fish. Notably, tissue and plasma urea concentrations in elasmobranchs are already rather massive, suggesting that steep gradients within tissues (e.g., kidney) are not likely, perhaps explaining the lack of a need for "rapid" urea equilibration across elasmobranch RBCs. Did UT-B not evolve until land colonization, and the development of steep kidney urea gradients necessitated by enhanced water resorption? In light of the discovery of the "terrestrial" CPSase I isozymes in some teleosts (e.g., lungfish; see Chapter 1), it is appropriate to look for the appearance of UT-B genes in selected semiterrestrial teleosts as well as amphibians to answer this question. Although it is likely that both facilitated diffusion and active urea transport are ancient (e.g., see Cooper and Sumrada, 1975; Jahns *et al.,* 1988; Heller *et al.,* 1980), are these urea transporters related to those expressed in vertebrates, and are these transporters present in early animals? These and other interesting questions await further study.

REFERENCES

Bankir, L., and Trinh-Trang-Tanv, M.-M. (1998). Urea excretion revisited: physiology and transporters. *Adv. Nephrol. Necker. Hosp.* **28,** 83–135.

Barton, K. N., Buhr, M. M., and Ballantyne, J. S. (1999). Effects of urea and trimethylamine *N*-oxide on fluidity of liposomes and membranes of an elasmobranch. *Am. J. Physiol.* **276,** R397–R406.

Borgnia, M., Nielsen, S., Engel, A., and Agre, P. (1999). Cellular and molecular biology of the aquaporin water channels. *Annu. Rev. Biochem.* **68,** 425–458.

Boylan J. (1967). Gill permeability in *Squalus acanthias. In* "Sharks, Skates and Rays" (P. W. Gilbert, R. F. Mathewson, and D. P. Rall, eds.), pp. 197–206. Johns Hopkins University Press, Baltimore, MD.

Boylan, J. W. (1972). A model for passive urea reabsorption in the elasmobranch kidney. *Comp. Biochem. Physiol.* **42A,** 27–30.

Carlson, S. R., and Goldstein, L. (1997). Urea transport across the cell membrane of skate erythrocytes. *J. Exp. Zool.* **277,** 275–282.

Collander, R. (1937). The permeability of plant protoplasts to nonelectrolytes. *Trans. Faraday Soc.* **33,** 985–990.

Cooper, T. G., and Sumrada, R. (1975). Urea transport in *Saccharomyces cerevisiae. J. Bateriol.* **121,** 571–576.

Denis, W. (1913). Metabolism studies on cold-blooded animals. II. The blood and urine of fish. *J. Biol. Chem.* **16,** 389–393.

DeVlaming, V. L., and Sage, M. (1973). Osmoregulation in the euryhaline elasmobranch, *Dasyatis sabina. Comp. Biochem. Physiol.* **45A,** 31–44.

Dykto, G., Smith, P. L., and Kinter, L. B. (1993). Urea transport in toad skin (*Bufo marinus*). *J. Pharmacol. Exp. Ther.* **267,** 364–370.

ElBerry, H. M., Majumdar, M. L., Cunningham, T. S., Sumrada, R. A., and Cooper, T. G. (1993).

Regulation of the urea active transporter gene (DUR3) in *Saccharomyces cerevisiae. J. Bacteriol.* **175,** 4688–4698.
Fenton, R., Hewitt, J. E., Howorth, A., Cottingham, C. A., and Smith, C. P. (1999). The murine urea transporter genes Slc14a1 and Slc14a2 occur in tandem on chromosome 18. *Cytogen. Cellular Genet.* **87,** 95–96.
Fines, G. A., Ballantyne, J. S., and Wright, P. A. (2001). Active urea transport and an unusual basolateral membrane composition in the gills of a marine elasmobranch. *Am. J. Physiol.* **280,** R16–R24.
Forster, R. P., Goldstein, L., and Rosen, S. K. (1972). Intrarenal control of urea reabsorption by renal tubules of the marine elasmobranch, *Squalus acanthius. Comp. Biochem. Physiol.* **42A,** 3–12.
Friedman, P. A., and Herbert, S. C. (1990). Diluting sement in kidney of dogfish shark. I. Localization and characterization of chloride absorption. *Am. J. Physiol.* 258, R398–R408.
Galluci, E., Micelli, S., and Lippe, C. (1971). Non-electrolyte permeability across thin lipid membranes. *Arch. Int. Physiol. Biochim.* **79,** 881–887.
Garcia-Romeu, F., Masoni, A., and Isaia, J. (1981). Active urea transport through isolated skins of frog and toad. *Am. J. Physiol.* **241,** R114–R123.
Gilmour, K. M., Perry, S. F., Wood, C. M., Henry, R. P., Laurent, P., Pärt, P., and Walsh, P. J. (1998). Nitrogen excretion and the cardiorespiratory physiology of the gulf toadfish, *Opsanus beta. Physiol. Zool.* **71,** 492–505.
Glemet, H. C., and Ballantyne, J. S. (1996). Comparison of liver mitochondrial membranes from an agnathan (*Myxine glutinosa*), an elasmobranch (*Raja erinacea*) and a teleost fish (*Pleuronectes americanus*). *Mar. Biol.* **124,** 509–518.
Goldstein, L., and Forster, R. P. (1971). Osmoregulation and urea metabolism in the little skate *Raja erinacea. Am. J. Physiol.* **220,** 742–746.
Goldstein, L., and Palatt, P. J. (1974). Trimethylamine oxide excretion rates in elasmobranchs. *Am. J. Physiol.* **227,** 1268–1272.
Goldstein, L., Oppelt, W. W., and Maren, T. H. (1968). Osmotic regulation and urea metabolism in the lemon shark *Negaprion brevirostris. Am. J. Physiol.* **215,** 1493–1497.
Hays, R. M., Levine, S. D., Myers, J. D., Heinemann, H. O., Kaplan, M. A., Franki, N., and Berliner, H. (1977). Urea transport in the dogfish kidney. *J. Exp. Zool.* **199,** 309–316.
Hediger, M. A., Smith, C. P., You, G., Lee, W.-S., Kanai, Y., and Shayakul, C. (1996). Structure, regulation and physiological roles of urea transporters. *Kidney Int.* **49,** 1615–1623.
Heller, K. B., Lin, E. C. C., and Wilson, T. H. (1980). Substrate specificity and transport properties of the glycerol facilitator of *Escherichia coli. J. Bacteriol.* **144,** 274–278.
Hickman, C. P., and Trump, B. F. (1969). The Kidney. *In* "Fish Physiology" (W. S. Hoar and D. J. Randall, eds.), Vol. I, pp. 91–239. Academic Press, New York.
Inoue, T., Terris, J., Ecelbarger, C. A., Chou, C. L., Nielsen, S., and Knepper, M. A. (1999). Vasopressin regulates apical targeting of aquaporin-2 but not of UT1 urea transporter in renal collecting duct. *Am. J. Physiol.* **276,** F559–566.
Isaia, J. (1984). Water and nonelectrolyte permeation. *In* "Fish Physiology" (W. S. Hoar and D. J. Randall, eds.), Vol. 10B, pp. 1–38. Academic Press, Orlando, FL.
Isozaki, T., Verlander, J. W., and Sands, J. M. (1993). Low protein diet alters urea transport and cell structure in rat initial inner medullary collecting duct. *J. Clin. Invest.* **92,** 2448–2457.
Isozaki, T., Gillen, A. G., Swanson, C. E., and Sands, J. M. (1994a). Protein restriction sequentially induces new urea transport processes in rat initial IMCDs. *Am. J. Physiol.* 266, F756–F761.
Isozaki, T., Lea, J. P., Tumlin, J. A., and Sands, J. M. (1994b). Sodium-dependent net urea transport in rat initial inner medullary collecting ducts. *J. Clin. Invest.* **94,** 1513–1517.
Jahns, T., Zobel, A., Kleiner, D., and Kaltwasser, H. (1988). Evidence for carrier-mediated, energy-dependent uptake of urea in some bacteria. *Arch. Microbiol.* **149,** 377–383.

Kaplan, M. A., Hays, L., and Hays, R. M. (1974). Evolution of a facilitated diffusion pathway for amides in the erythrocyte. *Am. J. Physiol.* **226,** 1327–1332.

Kato, A., and Sands, J. M. (1998a). Active sodium-urea counter-transport is inducible in the basolateral membrane of rat renal initial inner medullary collecting ducts. *J. Clin. Invest.* **102,** 1008–1015.

Kato, A., and Sands, J. M. (1998b). Evidence for sodium-dependent active urea secretion in the deepest subsegment of the rat inner medullary collecting duct. *J. Clin Invest.* **101,** 423–428.

Kempton, R. T. (1953). Studies of the elasmobranch kidney. II. Reabsorption of urea by the smooth dogfish, *Mustelis canis. Biol. Bull.* **104,** 45–56.

Kikeri, D., Sun, A., Zeidel, M. L., and Hebert, S. C. (1989). Cell membranes impermeable to NH_3. *Nature* **339,** 478–480.

Kishore, B. K., Terris, J., Fernandez-Llama, P., and Knepper, M. A. (1997). Ultramicro-determination of vasopressin-regulated renal urea transporter protein in microdissected renal tubules. *Am. J. Physiol.* **272,** F531–F537.

Klein, J. D., Timmer, R. T., Rouillard, P., Bailey, J. L., and Sands, J. M. (1999). UT-A urea transporter protein expressed in liver: upregulation by uremia. *J. Am. Soc. Nephrol.* **10,** 2076–2083.

Knepper, M., and Chou, C.-L. (1995). Urea and ammonium transport in the mammalian kidney. *In* "Nitrogen Metabolism and Excretion" (P. J. Walsh and P. A. Wright, eds.), pp. 205–227. CRC Press, Boca Raton, FL.

Knepper, M. A., and Rector, F. C., Jr. (1991). Urinary concentration and dilution, *In* "The Kidney," pp. 445–455. W. B. Saunders, Philadelphia, PA.

Lacoste, I., Dunel-Erb, S., Harvey, B. J., Laurent, P., and Ehrenfeld, J. M. (1991). Active urea transport independent of H^+ and Na^+ transport in frog skin epithelium. *Am. J. Physiol.* **261,** R898–R906.

Lacy, E. R., and Reale, E. (1991). Fine structure of the elasmobranch renal tubule: neck and proximal segments of the little skate. *Am. J. Anat.* **190,** 118–132.

Lacy, E. R., Reale, E., Schlusselberg, D. S., Smith, W. K., and Woodward, D. J. (1985). A renal countercurrent system in marine elasmobranch fish: a computer aided reconstruction. *Science* **227,** 1351–1354.

Laurent, P., Wood, C. M., Wang, Y., Perry, S. F., Gilmour, K. M., Pärt, P., Chevalier, C., West, M., and Walsh, P. J. (2001). Intracellular vesicular trafficking in the gill epithelium of urea-excreting fish. *Cell Tissue Res.* **303,** 197–210.

Macey, R. I., and Yousef, L. W. (1988). Osmotic stability of red cells in renal circulation requires rapid urea transport. *Am. J. Physiol.* **254,** C669–674.

Maffly, R. H., and Leaf, A. (1959). Potential of water in mammalian tissues. *J. Gen. Physiol.* **42,** 1257–1275.

McDonald, M. D., and Wood, C. M. (1998). Reabsorption of urea by the kidney of the freshwater rainbow trout. *Fish Physiol. Biochem.* **18,** 375–386.

McDonald, M. D., Wood, C. M., Wang, Y. X., and Walsh, P. J. (2000). Differential branchial and renal handling of urea, acetamide, and thiourea in the gulf toadfish, *Opsanus beta:* evidence for two transporters. *J. Exp. Biol.* **203,** 1027–1037.

Murdaugh, H. V., Robin, E. D., and Hearn, C. D. (1964). Urea: apparent carrier-mediated transport by facilitated diffusion in dogfish erythrocytes. *Science* **144,** 52–53.

Naruse, M., Klein, J. D., Ashkar, Z. M., Jacobs, J. D., and Sands, J. M. (1999). Glucocorticoids downregulate the vasopressin-regulated urea transporter in rat terminal inner medullary collecting ducts. *J. Am. Soc. Nephrol.* **8,** 517–523.

Olivès, B., Neau, P., Bailly, P., Hediger, M. A., Rousselet, G., Cartron, J. P., and Ripoche, P. (1994). Cloning and functional expression of a urea transporter from human bone marrow cells. *J. Biol. Chem.* **269,** 31649–31652.

Olivès, B., Mattei, M. G., Huet, M., Neau, P., Martial, S., Cartron, J. P., and Bailly, P. (1995). Kidd

blood group and urea transport function of human erythrocytes are carried by the same protein. *J. Biol. Chem.* **270**, 15607–15610.
Olivès, B., Martial, S., Mattei, M. G., Matassi, G., Rousselet, G., Ripoche, P., Cartron, J. P., and Bailly, P. (1996). Molecular characterization of a new urea transporter in the human kidney. *FEBS Lett.* **386**, 156–160.
Pärt, P., Wright, P. A., and Wood, C. M. (1998). Urea and water permeability in dogfish (*Squalus acanthias*) gills. *Comp. Biochem. Physiol.* **119A**, 117–123.
Pärt, P., Wood, C. M., Gilmour, K. M., Perry, S. F., Laurent, P., Zadunaisky, J., and Walsh, P. J. (1999). Urea and water permeability in the ureotelic gulf toadfish (*Opsanus beta*). *J. Exp. Zool.* **283**, 1–12.
Pateman, J. A., Dunn, E., and Mackay, E. M. (1982). Urea and thiourea transport in *Aspergillus nidulans*. *Biochem. Genet.* **20**, 777–790.
Payan, P., Goldstein, L., and Forster, R. P. (1973). Gills and kidneys in ureosmotic regulation in euryhaline skates. *Am. J. Physiol.* **224**, 367–372.
Perlman D. F., and Goldstein, L. (1988). Nitrogen metabolism. *In* "Physiology of Elasmobranch Fishes" (T. J. Shuttleworth, ed.), pp. 253–276. Springer-Verlag, Berlin.
Perry, S. F., Gilmour, K. M., Wood, C. M., Pärt, P., Laurent, P., and Walsh, P. J. (1998). The effects of arginine vasotocin and catecholamines on nitrogen excretion and the cardiorespiratory physiology of the gulf toadfish, *Opsanus beta*. *J. Comp. Physiol. B.* **168**, 461–472.
Price, K. S. (1967). Fluctuations in two osmoregulatory components, urea and sodium chloride, of the clearnose skate, *Raja eglanteria* Bosc 1802. II. Upon natural variation of the salinity of the external medium. *Comp. Biochem. Physiol.* **23**, 77–82.
Price, K. S., and Creaser, E. P. (1967). Fluctuations in two osmoregulatory components, urea and sodium chloride, of the clearnose skate, *Raja eglanteria* Bosc 1802. I. Upon laboratory modification of external salinities. *Comp. Biochem. Physiol.* **23**, 65–76.
Rabinowitz, L., and Gunther, R. A. (1973). Urea transport in elasmobranch erythrocytes. *Am. J. Physiol.* **224**, 1109–1115.
Rapoport, J., Abuful, A., Chaimovits, C., Noeh, Z., and Hays, R. M. (1988). Active urea transport by the skin of *Bufo viridis:* amiloride- and phloretin-sensitive transport sites. *Am. J. Physiol.* **255**, F429–F433.
Sands, J. M. (1999). Regulation of renal urea transporters. *J. Am. Soc. Nephrol.* **10**, 635–646.
Sands, J. M., Gargus, J. J., Frohlich, O., Gunn, R. B., and Kokko, J. P. (1992). Urinary concentrating ability in patients with Jk(a-b-) blood type who lack carrier-mediated urea transport. *J. Am. Soc. Nephrol.* **2**, 1689–1696.
Sands, J. M., Martial, S., and Isozaki, T. (1996). Active urea transport in the rat initial inner medullary collecting duct: functional characterization and initial expression cloning. *Kidney Int.* **49**, 1611–1614.
Sands, J. M., Timmer, R. T., and Gunn, R. B. (1997). Urea transporters in kidney and erythrocytes. *Am. J. Physiol.* **273**, F321–F339.
Sayer, M. D. J., and Davenport, J. (1987). The relative importance of the gills to ammonia and urea excretion in five seawater and one freshwater teleost species. *J. Fish. Biol.* **31**, 561–570.
Schmidt-Nielsen, B., and Rabinowitz, L. (1964). Methylurea and acetamide: active reabsorption by elasmobranch renal tubules. *Science* **146**, 1587–1588.
Schmidt-Nielsen, B., Truniger, B., and Rabinowitz, L. (1972). Sodium-linked urea transport by the renal tubule of the spiny dogfish *Squalus acanthias*. *Comp. Biochem. Physiol.* **42A**, 13–25.
Scott, G. G. (1913). A physiological study of the changes in *Mustelus canis* produced by modifications in the molecular concentration of the external medium. *Ann. N.Y. Acad. Sci.* **23**, 1–75.
Shayakul, C., Steel, A., and Hediger, M. A. (1996). Molecular cloning and characterization of the vasopressin-regulated urea transporter of rat kidney collecting ducts. *J. Clin. Invest.* **98**, 2580–2587.

Shayakul, C., Knepper, M. A., Smith, C. P., DiGiovanni, S. R., and Hediger, M. A. (1997). Segmental localization of urea transporter mRNAs in rat kidney. *Am. J. Physiol.* **272,** F654–660.

Shuttleworth, T. J. (1988). Salt and water balance—extrarenal mechanisms. *In* "Physiology of Elasmobranch Fishes" (T. J. Shuttleworth, ed.), pp. 171–200. Springer-Verlag, Berlin.

Sidoux-Walter, F., Lucien, N., Olives, B., Gobin, R., Rousselet, G., Kamsteeg, E. J., Ripoche, P., Deen, P. M., Cartron, J. P., and Bailly, P. (1999). At physiological expression levels the Kidd blood group/urea transporter protein is not a water channel. *J. Biol. Chem.* **274,** 30228–30235.

Smith, C. P., and Wright, P. A. (1999). Molecular characterization of an elasmobranch urea transporter. *Am. J. Physiol.* **276,** R622–R626.

Smith, C. P., Lee, W. S., Martial, S., Knepper, M. A., You, G., Sands, J. M., and Hediger, M. A. (1995). Cloning and regulation of expression of the rat kidney urea transporter (rUT2). *J. Clin. Invest.* **96,** 1556–1563.

Smith, C. P., Heitz, M. J., Wood, C. M., and Walsh, P. J. (1998). Molecular identification of a gulf toadfish (*Opsanus beta*) urea transporter. *J. Physiol.* **511,** 33P.

Smith, H. W. (1931a). The absorption and excretion of water and salts by the elasmobranch fishes. I. Fresh water elasmobranchs. *Am. J. Physiol.* **98,** 279–295.

Smith, H. W. (1931b). The absorption and excretion of water and salts by the elasmobranch fishes. II. Marine elasmobranchs. *Am. J. Physiol.* **98,** 296–310.

Smith, H. W. (1936). The retention and physiological role of urea in the Elasmobranchii. *Biol. Bull.* **11,** 49–82.

Städeler, G., and Frérichs, F. T. (1858). Über das Vorkommen von Harnstoff, Taurin und Scyllit in den Organanen der Plagiostomen. *J. Prakt. Chem.* **73,** 48–55.

Thorson, T. B. (1967). Osmoregulation in fresh-water elasmobranchs. *In* "Sharks, Skates and Rays" (P. W. Gilber, R. F. Mathewson, and D. P. Rall, eds.), pp. 265–270. Johns Hopkins University Press, Baltimore, MD.

Urist, M. R. (1962). Calcium and other ions in the blood and skeleton of the Nicaraguan freshwater shark. *Science* **137,** 985–986.

Ussing, H., and Johnansen, B. (1969). Anomalous transport of sucrose and urea in toad skin. *Nephron* **6,** 317–328.

Walsh, P. J. (1997). Evolution and regulation of ureogenesis and ureotely in (barachoidin) fishes. *Ann. Rev. Physiol.* **59,** 299–323.

Walsh, P. J., and Wood, C. M. (1996). Interactions of urea transport and synthesis in hepatocytes of the gulf toadfish, *Opsanus beta*. *Comp. Biochem. Physiol.* **113B,** 411–416.

Walsh, P. J., Wood, C. M., Perry, S. F., and Thomas, S. (1994). Urea transport by hepatocytes and red blood cells of selected elasmobranch and teleost fishes. *J. Exp. Biol.* **193,** 321–335.

Walsh, P. J., Heitz, M., Campbell, C. E., Cooper, G. J., Medina, M., Wang, Y. S., Goss, G. G., Vincek, V., Wood, C. M., and Smith, C. P. (2000). Molecular identification of a urea transporter in gills of the ureotelic gulf toadfish (*Opsanus beta*). *J. Exp. Biol.* **203,** 2357–2364.

Walsh, P. J., Grosell, M., Goss, G. G., Laurent, P., Bergman, H. L., Bergman, A. N., Wilson, P., Alper, S., Smith, C. P., Kamunse, C., and Wood, C. H. (2001). Physiological and molecular characterization of urea transport by the gills of the Lake Magadi tilapia (*Alcolakia grahami*). *J. Exp. Biol.* **204,** 509–520.

Wolfe, R. R. (1981). Measurement of urea kinetics *in vivo* by means of a constant tracer infusion of di-^{15}N-urea. *Am. J. Physiol.* **240,** E428–E434.

Wong, T. M., and Chan, D. K. O. (1977). Physiological adjustments to dilution of the external medium in the Lip-shark *Hemiscyllium plagiosum* (Bennett). *J. Exp. Zool.* **200,** 85–96.

Wood, C. M. (1993). Ammonia and urea metabolism and excretion. *In* "The Physiology of Fishes" (D. H. Evans, ed.), pp. 379–425. CRC Press, Boca Raton, FL.

Wood, C. M., Bergman, H. L., Laurent, P. Maina, J. N., Narhara, A., and Walsh, P. J. (1994). Urea production, acid–base regulation, and their interactions in the Lake Magadi Tilapia, a unique teleost adapted to a highly alkaline environment. *J. Exp. Biol.* **189,** 13–36.

Wood, C. M., Hopkins, T. E., Hogstrand, C., and Walsh, P. J. (1995a). Pulsatile urea excretion in the ureagenic toadfish *Opsanus beta:* an analysis of rates and routes. *J. Exp. Biol* **198,** 1729–1741.

Wood, C. M., Pärt, P., and Wright, P. A. (1995b). Ammonia and urea metabolism in relation to gill function and acid–base balance in a marine elasmobranch, the spiny dogfish (*Squalus acanthias*). *J. Exp. Biol.* **198,** 1545–1558.

Wood, C. M., Hopkins, T. E., and Walsh, P. J. (1997). Pulsatile urea excretion in the toadfish (*Opsanus beta*) is due to a pulsatile excretion mechanism, not a pulsatile production mechanism. *J. Exp. Biol.* **200,** 1039–1046.

Wood, C. M., Gilmour, K. M., Perry, S. F., Pärt, P., Laurent, P., and Walsh, P. J. (1998). Pulsatile urea excretion in gulf toadfish (*Opsanus beta*): evidence for activation of a specific facilitated diffusion transport system. *J. Exp. Biol.* **201,** 805–817.

Wright, P. A., Pärt, P., and Wood, C. M. (1995). Ammonia and urea excretion in the tidepool sculpin (*Oligocottus maculosus*): sites of excretion, effects of reduced salinity and mechanisms of urea transport. *Fish Physiol. Biochem.* **14,** 111–123.

You, G., Smith, C. P., Kanai, Y., Lee, W.-S., Stelzner, M., and Hediger, M. A. (1993). Cloning and characterization of the vasopressin-regulated urea transporter. *Nature* **365,** 844–847.

Xu, Y., Olivès, B., Bailly, P., Fischer, E., Ripoche, P., Ronco, P., Cartron, J. P., and Rondeau, E. (1997). Endothelial cells of the kidney vasa recta express the urea transporter HUT11. *Kidney Int.* **51,** 138–146.

9

NITROGEN COMPOUNDS AS OSMOLYTES

PAUL H. YANCEY

I. Introduction
II. Types and Contents of Osmolytes
   A. Osmoconformers: Agnatha—Class Myxini
   B. Osmoconforming Hypoionic Regulators
   C. Hypoosmotic Hypoionic Regulators
III. Metabolism and Regulation of Osmolytes
   A. Osmoconformers: Myxini
   B. Osmoconforming Hypoionic Regulators: Chondrichthyes
   C. Hypoosmotic Hypoionic Regulators
IV. Properties and Functions of Nitrogen Osmolytes
   A. Inorganic Ions as Perturbants and Amino Acids as Compatibile Osmolytes
   B. Urea as Perturbing Osmolyte
   C. Methylamines as Stabilizing and Counteracting Osmolytes
   D. Urea and TMAO as Buoyancy Aids
   E. Urea, TMAO, and Taurine as Antioxidants
   F. Other Functions
V. Mechanisms of Compatibility and Counteraction: Osmolytes and Water Structure
VI. Evolutionary Considerations
VII. Practical Applications, Exceptions, and Unanswered Questions
   References

I. INTRODUCTION

The basic dissolved constituents of cells—potassium as the main cation, plus metabolites, proteins, and so on—typically yield an osmotic pressure of roughly 300 milliosmolar (mOsm) in most organisms. Because seawater is about 1000 mOsm, marine organisms must have adaptations to maintain cell volume in the face of this potentially dehydrating force. There appear to be three adaptive strategies (Kirschner, 1991), summarized in Fig. 1. First, most marine organisms (invertebrates, protists, bacteria) are "pure" *osmoconformers,* with internal osmotic pressures that are equal to or slightly higher than the environment. Thus, there is no tendency to lose water to the environment. Extracellular fluids typically re-

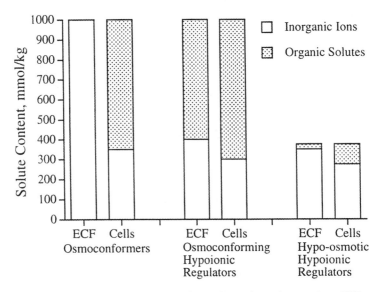

Fig. 1. Simplified schematic of major osmotic constituents in marine organisms. ECF, extracellular fluid.

semble seawater, at least in having similar concentrations of NaCl, and they typically change passively with external changes. Thus, there should be little cost to maintain internal ion balances, at least for extracellular fluids. Most marine osmoconformers appear to be stenohaline (unable to cope with large salinity changes), but some are euryhaline, able to adapt. Importantly, cellular osmotic pressure is not elevated greatly with inorganic ions but rather predominantly with certain small-molecular-weight *organic osmolytes* (Fig. 1, left). It is these that euryhaline species regulate to a greater extent than ions for cell volume maintenance. As extensively reviewed elsewhere (Yancey *et al.,* 1982; Yancey, 1994), across the spectrum of life these osmolytes fall into four broad categories: carbohydrates such as polyols, free amino acids such as glycine and taurine, methylammonium and methylsulfonium solutes, and urea. In marine animals, it is the nitrogen-based osmolytes, such as those shown in Fig. 2A, that dominate.

A second strategy, which also involves cellular osmolytes, is termed *osmoconforming hypoionic regulation*. These organisms, exemplified by chondrichthyean fishes, also have body fluid osmolalities equal to or slightly higher than the environment. Thus, again, there is no tendency to lose water and there may even be a modest gain. However, they also actively regulate their extracellular fluids to have considerably lower salt concentrations than the environment, with the osmotic difference balanced by extracellular (as well as intracellular) primarily nitrogenous organic osmolytes (Fig. 1, middle). In euryhaline species, these osmolytes are usually regulated to a greater extent than are inorganic ions for osmotic adjustments.

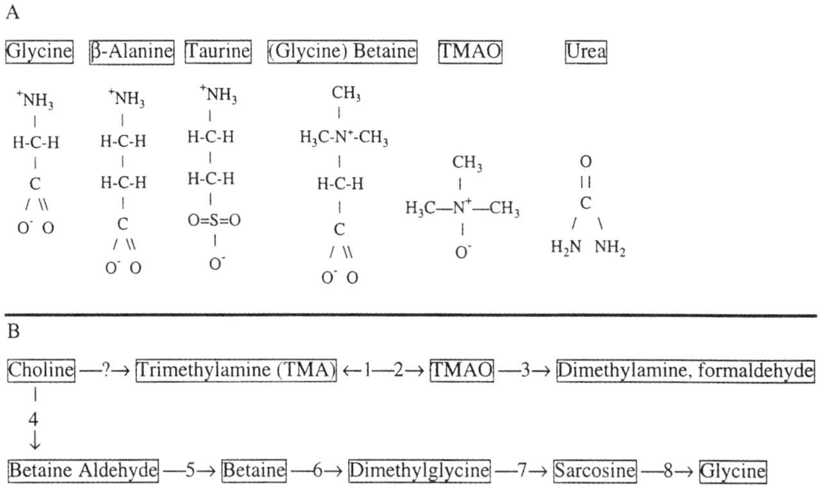

9. NITROGENOUS OSMOLYTES

A

Glycine   β-Alanine   Taurine   (Glycine) Betaine   TMAO   Urea

$$\begin{array}{cccccc}
^+NH_3 & ^+NH_3 & ^+NH_3 & CH_3 & & \\
| & | & | & | & & \\
H\text{-}C\text{-}H & H\text{-}C\text{-}H & H\text{-}C\text{-}H & H_3C\text{-}N^+\text{-}CH_3 & & \\
| & | & | & | & CH_3 & O \\
C & H\text{-}C\text{-}H & H\text{-}C\text{-}H & H\text{-}C\text{-}H & | & || \\
/\ \backslash\backslash & | & | & | & H_3C\text{-}N^+\text{-}CH_3 & C \\
O^-\ O & C & O{=}S{=}O & C & | & /\ \backslash \\
 & /\ \backslash\backslash & | & /\ \backslash\backslash & O^- & H_2N\ NH_2 \\
 & O^-\ O & O^- & O^-\ O & & \\
\end{array}$$

B

Choline —?→ Trimethylamine (TMA) ←1—2→ TMAO —3→ Dimethylamine, formaldehyde
|
4
↓

Betaine Aldehyde —5→ Betaine —6→ Dimethylglycine —7→ Sarcosine —8→ Glycine

Fig. 2. (A) Structures of typical nitrogen osmolytes found in marine animals, including fishes (TMAO, N-trimethylamine oxide). (B) Metabolic pathways for methylamines. (Modified from Van Waarde, 1988.) 1, TMAO reductase; 2, trimethylamine (TMA) oxidase, using NADPH and O_2; 3, TMAOase; 4, choline dehydrogenase, using FAD; 5, betaine aldehyde dehydrogenase, using NAD^+; 6, betaine homocysteine transmethylase, using homocysteine; 7, dimethylglycine oxidase, using FAD; 8, sarcosine oxidase, using FAD.

The third osmotic strategy is *hypoosmotic hypoionic regulation,* exhibited by most marine actinopterygian fishes. Body fluids in the fishes are usually maintained at 300–400 mOsm, even in euryhaline species (Lange and Fugelli, 1965). As a consequence, such fish tend to lose water constantly to any osmotic environment above this range. Cells in these animals require only low concentrations of organic osmolytes (Fig. 1, right). However, significant exceptions to this have recently been found, as discussed below.

A fourth strategy, *hyperosmotic regulation,* applies primarily to inhabitants of low salinity waters. In some cases these are the same (euryhaline) species that are hypoosmotic hypoionic regulators in seawater (e.g., anadromous fish). However, because this does not involve organic osmolytes extensively, we discussed it only briefly.

II. TYPES AND CONTENTS OF OSMOLYTES

A. Osmoconformers: Agnatha—Class Myxini

The two groups of the superclass Agnatha, hagfish and lampreys, fall into different adaptive categories. The latter are hypoosmotic regulators like most

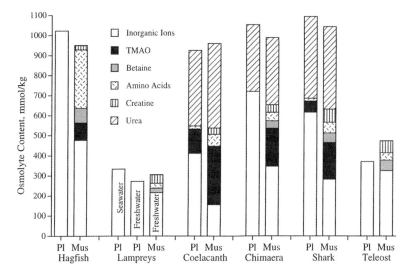

Fig. 3. Osmolyte contents of plasma (Pl) in mM and in muscles (Mus) estimated as mmol/kg cell water for marine fishes and freshwater lamprey. [Data from Robertson, 1976 (hagfish *Myxine glutinosa,* and *Chimaera monstrosa*); Evans, 1993 (plasma, lamprey *Petromyzon fluviatilis*); Robertson, 1984 (muscle, freshwater lamprey *Lampetra fluviatilis,* and teleost *Anguilla anguilla*); Lutz and Robertson, 1971 (coelacanth *Latimeria chalumnae*); and Robertson, 1989 (shark *Scyliorhinus canicula*).]

vertebrates (see Section II.C), but hagfish (Myxini) are unique among the fishes: They are strictly marine osmoconformers, with high levels of cellular organic osmolytes and with blood resembling seawater (though with some modest differences in ion levels; Evans, 1993). In one thoroughly analyzed species, the organic osmolytes were primarily free amino acids, with some methylamines, mostly *N*-trimethylamine oxide (TMAO) (Figs. 2 and 3), very similar to many marine invertebrates such as decapod shrimp (Carr *et al.,* 1996).

B. Osmoconforming Hypoionic Regulators

1. GNATHOSTOMATA—CLASS CHONDRICHTHYES

The marine cartilaginous fishes—elasmobranchs (sharks, skates, rays) and holocephalans (chimaeras)—have organic osmolytes in both extra- and intracellular fluids, dominated by urea (Fig. 2) (Smith, 1936). Thus, these fishes are sometimes termed *ureosmotic.* Urea concentrations are about equal in the two compartments (Fig. 3) because it equilibrates readily across most membranes via facilitated urea transporters (see Chapter 8) or, at least in skate erythrocytes, probably by simple diffusion through the lipid membrane (Carlson and Goldstein, 1997). These fish invariably also contain somewhat lesser amounts of methylamines and free amino acids (Fig. 3). These (charged) osmolytes are generally much higher inside cells due, in part, to low membrane permeability (Goldstein and Kleinzeller, 1987). TMAO is the most commonly found methylamine in these fishes, but in some cases betaine (also called glycine betaine or N-trimethylglycine; Fig. 2) dominates; for example, TMAO is the main methylamine in one species of holocephalan (Robertson, 1976), but betaine is in another (Bedford *et al.*, 1998). Basic metabolic pathways of these two methylamines are shown in Fig. 2B.

Sarcosine (N-methylglycine) (Fig. 2B) is also common in some skate tissues, and amino acids such as taurine and β-alanine (Fig. 2A) are also found at moderate levels, varying among species and tissues (King and Goldstein, 1983). All of these nitrogenous solutes are regulated during osmotic changes (see Section III).

As first noted and reviewed by Yancey and Somero (1980), estimated intracellular concentrations in analyzed chondrichthyean tissues reveal an approximately 2:1 to 3:2 ratio of urea to total methylamines, often roughly 400:200 mM (Fig. 3). In most species, TMAO is highest in muscle (approaching a 3:2 intracellular ratio with urea), whereas other tissues tend to have somewhat less (closer to a 2:1 ratio), as shown in Fig. 4A. In the latter tissues, other osmolytes such as taurine and certain (nonnitrogenous) polyols are elevated at the expense of TMAO; examples are shown in Fig. 4B. See Section IV.C.1 for an hypothesis regarding the 2:1 ratio.

TMAO is of concern to the fisheries industry because it can break down into

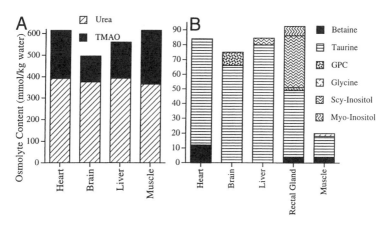

Fig. 4. (A) Contents of urea and TMAO in various organs of the shark *Mustelus manazo* (Suyama and Tokuhiro, 1954), calculated as mmol per total tissue water. Because urea is equal both intra- and extracellularly, while TMAO is much higher intracellularly, the urea:TMAO ratios inside cells will be lower than shown. (B) Contents of minor (nonurea, non-TMAO) osmolytes in organs of *Squalus acanthias*, as mmol/kg water (HPLC analysis; M. Starks and P. H. Yancey, unpublished).

harmful products, including formaldehyde, but especially trimethylamine (TMA), a volatile base (Fig. 2B). TMAO breakdown can be due to endogenous enzymes or to bacteria; the oxygen released can discolor cans (Hebard *et al.*, 1982; Van Waarde, 1988; Sotelo *et al.*, 1995). TMA not only has a strong odor of rotten fish, making food unpalatable, but is highly toxic. As an example, there is some evidence that TMA in Greenland shark meat has poisoned both humans and sled dogs, causing convulsions, vomiting, diarrhea, and even death (Anthoni *et al.*, 1991). Urea breaks down not only into harmful ammonia, but also isocyanate, which can carbamylate proteins (Stark, 1972). And as discussed extensively in Section IV.B, urea itself is a perturbing solute at 400 mM. However, the relatively high membrane permeability of urea makes it much easier to remove than TMAO during food preparation.

2. GNATHOSTOMATA—CLASS SARCOPTERYGII

Members of both living groups of sarcopterygian fishes—the coelacanth and lungfishes—exhibit urea retention. The coelacanth uses urea and methylamines as organic osmolytes much like chondrichthyans (Fig. 3). Interestingly, the total osmotic pressure reported is slightly less than seawater (Griffith and Pang, 1979), suggesting a tendency to lose water.

Lungfish (Dipnoi) that estivate (African and South American, but not Australian species) also build up large amounts of urea (reported up to 400 mM), presumably as a less toxic alternative to ammonia for storage during estivation (reviewed by Forster and Goldstein, 1969; Evans, 1993). Urea accumulation may also help retain water. There are no studies as to whether other organic osmolytes also accumulate during estivation, an interesting question for further investigation.

C. Hypoosmotic Hypoionic Regulators

1. AGNATHA—CLASS CEPHALASPIDOMORPHI

Lampreys, the other group of agnathans, regulate much like the majority of Actinopterygii (Fig. 3). They develop in freshwater, with some (euryhaline) species migrating to the oceans as adults. Recent studies have shown that although larval stages are ammonotelic they can produce urea, probably as a waste product only (Wilkie *et al.*, 1999).

2. GNATHOSTOMATA—CLASS ACTINOPTERYGII

a. Basic Hypoosmotic Patterns. According to dogma, all subclasses of actinopterygians—Chondrostei (sturgeons, etc.), Neopterygii (gars and bowfins), and Teleostei—have low levels of organic osmolytes because of their low internal osmotic pressures regulated at nearly constant levels, even in euryhaline species.

Teleosts often contain intracellular TMAO and/or betaines (including some unusual cyclic betaines such as homarine; Ito et al., 1994) as apparent or demonstrated organic osmolytes. However, until recently the total concentration of these organic osmolytes, usually dominated by TMAO, was reported to be less than 100 mmol/kg wet wt (Fig. 3), with gadiform fishes (cods and relatives) having the highest TMAO levels at about 70 mmol/kg wet wt (Hebard et al., 1982).

Taurine (Fig. 2A) is also common in teleost tissues, often as the most concentrated ninhydrin-positive solute (Van Waarde, 1988). As discussed in Section III.C.2, it is used as an osmolyte but may have other functions (see Section IV.E). Urea concentration can be high in some adult and embryonic teleosts under certain conditions, but probably for the purpose of detoxification of ammonia rather than for significant osmotic function (Wood, 1993). Recent research on urea in teleosts is reviewed elsewhere in this volume (Chapter 7; see also Chapters 1, 5, and 8).

Also contributing significantly to cell osmotic pressure in elasmobranchs and teleosts are the nitrogenous imidazole derivatives histidine, carnosine, and anserine. These can occur up to 50 mM in muscle, with (methylated) anserine being highest in elasmobranchs and salmon, and the others being highest in other teleosts. However, these compounds are not regarded as osmotic effectors but rather as proton buffers and transporters of metal ions (reviewed by Van Waarde, 1988). Similarly, creatine typically contributes significantly to cell osmotic pressure, especially in muscle (Fig. 3). It is known to participate in energy storage as creatine phosphate, but in mammals it can also act as an osmolyte in some tissues (Miller et al., 2000). Whether or not this occurs in fish is not clear, however. Thus, the imidazole derivatives and creatine are not considered further.

b. Exceptions. Significant exceptions have recently been found to the low-osmolality dogma for Actinopterygii. First, some arctic teleosts have several hundred millimolar plasma glycerol (Raymond, 1992), making them nearly isoosmotic with seawater. Second, consistently high levels of TMAO (83–288 mmol/kg muscle wet wt) have been found in several families of deep-sea animals (Gillett et al., 1997; Kelly and Yancey, 1999). For teleosts, the correlation of plasma Na$^+$ concentration, plasma osmotic pressure, and muscle TMAO content is fairly linear with depth of capture, rising from shallow to bathyal (1900 m) to abyssal (2900 m), as shown in Fig. 5. Finally, some (but not all) Antarctic teleosts also have high plasma NaCl contents concomitant with muscle TMAO contents up to 154 mmol/kg wet wt (Raymond and DeVries, 1998); and some northern fishes exhibit elevated blood TMAO and urea levels in winter (Raymond, 1998). For hypotheses regarding these patterns, see Sections IV.C.2, 3, and 4.

As with elasmobranchs, TMAO in teleosts is of concern in the fisheries industry (Hebard et al., 1982). Monitoring of TMAO and its breakdown products— TMA, dimethylamine, and formaldehyde (Fig. 2B)—is important, therefore, in long-term storage of fish muscle for food (Sotelo et al., 1995). With our discovery

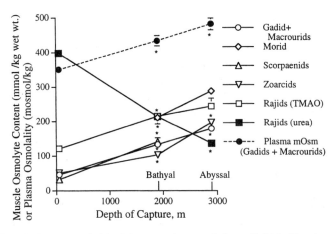

Fig. 5. TMAO (open symbols) of deep-sea teleosts and skates (Rajids). Also shown are skate muscle urea contents (black squares) and osmotic pressure of plasma (black circles) of gadiform (Gadids + Macrourids) teleosts (gadids are shallow-living cods; macrourids are related deep-living grenadiers). Error bars are SD values, most too small to plot (* = $P < 0.05$ compared to lower depth value). Except in the morid, which had high plasma TMAO, plasma osmolalities were higher in deep-sea gadiform fish due to NaCl, approximately balancing the estimated osmotic pressure of higher TMAO in muscles (Data from Gillett *et al.*, 1997, and Kelly and Yancey, 1999.)

of so much TMAO in deep-sea fishes (Fig. 5), this spoilage concern is heightened because of recent efforts to utilize deep-sea fish for food (such as the Russian macrourid fisheries) following depletion of shallow-water stocks.

III. METABOLISM AND REGULATION OF OSMOLYTES

Regulation of the metabolism of nitrogenous solutes is covered in detail in other chapters in this volume; thus, only a few aspects will be reviewed here. Note that little has been done in terms of gene regulation of fish osmolytes. The most detailed studies at this level have been carried out on bacteria, yeast, angiosperm, and mammalian cells. For example, there is evidence in mammals that some osmolyte genes have a common regulatory response element (nGGAAAAnnnC) that may be activated by elevated cellular Na^+ and K^+. Even in these well-studied systems, however, it is not clear how cells sense osmotic change, though signal transduction in yeast, plants, and mammals appears to involve MAP kinase cascades (Burg *et al.*, 1997). These findings provide clues and tools (such as molecular probes) for determining whether similar systems exist in fish.

A. Osmoconformers: Myxini

Little is known of the mechanisms for regulating osmolyte concentrations in hagfish. They are relatively stenohaline and do not experience salinity changes in nature. However, in the laboratory, one species has been shown to tolerate gradual salinity changes from 60 to 150% seawater, acting as a strict osmoconformer. Muscle cells reduce K^+, amino acids, and TMAO levels to maintain cell volume (Cholette and Gagnon, 1973). However, their erythrocytes lack the osmolyte channel found in other fish (Perlman and Goldstein, 1999) (see Sections III.B and III.C.2).

B. Osmoconforming Hypoionic Regulators: Chondrichthyes

A number of elasmobranch species migrate into low-salinity waters where they reduce plasma salt, urea, and TMAO levels; in muscle, urea and TMAO are reduced more than are ions, usually maintaining a similar ratio between the organic solutes (Fig. 6). Urea reduction occurs as the result of either reduced synthesis (Forster and Goldstein, 1976) or a higher renal clearance rate (Goldstein

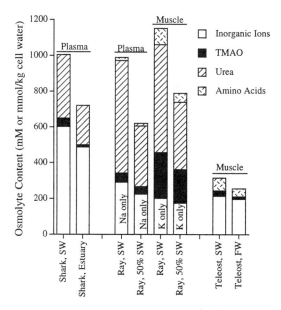

Fig. 6. Changes in osmolytes in fish subjected to different external salinities: plasma in the shark *Carcharinus leucas* (Thorson *et al.*, 1973), plasma and muscle in the ray *Dasyastic americanus* (Forster and Goldstein, 1976), and muscle of the teleost *Platichthys flesus* (Lange and Fugelli, 1965).

and Forster, 1971), but some remains even in euryhaline populations that live full life cycles in freshwater (Piermarini and Evans, 1998).

The organs that regulate urea and TMAO concentrations have been known for some time. The gills are relatively impermeable to both solutes (Shuttleworth, 1988; see also Chapter 8), and the kidneys reabsorb most urea and TMAO from the nephron filtrate (Smith, 1936). The renal nephrons are among the most complex known, probably relating to this reabsorption ability (Lacy and Reale, 1995). The livers of most examined species contain the enzymes to synthesize urea (see Chapter 7) and TMAO. The exception is *Potamotrygon,* a stenohaline Amazonian ray and the only elasmobranch permanently adapted to freshwater. Although it has low levels of some of the enzymes related to urea synthesis (Anderson, 1980), it retains virtually no urea or TMAO *in situ* and cannot accumulate urea in laboratory salinity stress (Thorson *et al.,* 1967; Gerst and Thorson, 1977).

Osmolyte regulation in the coelacanth is not known. Urine reportedly has as much urea as in plasma, but reabsorption from the bladder might occur (Griffith and Pang, 1979).

The TMA oxidase that catalyzes formation of TMAO (Fig. 2B) belongs to the group of flavin-containing monooxygenases (FMOs) that are widespread (but not universal) in prokaryotes and eukaryotes, found in smooth endoplasmic reticulum in the latter (Schlenk, 1998). Some elasmobranchs may not have this enzyme, apparently obtaining TMAO from their diet (Forster and Goldstein, 1969). However, some elasmobranchs originally reported to lack FMO activity have since been found (with improved methods) to contain it, primarily in the liver. Mechanisms of intracellular regulation are not known (Schlenk, 1998).

At the membrane level, mechanisms of osmolyte regulation have been elucidated to some extent *in vitro,* mainly under hypoosmotic stress. Shark rectal gland cells in hypotonic exposure exhibit a 10- to 20-fold increase in taurine, betaine, and TMAO effluxes, but no K^+ efflux increase. F-actin disappears in the initial phases of this response, suggesting a role of the cytoskeleton in signal transduction (Ziyadeh *et al.,* 1992). Skate (*Raja erinacea*) erythrocytes (studied extensively by Goldstein and coworkers) maintain amino acid osmolytes and TMAO at high levels, at least some via Na^+-dependent membrane transporters (Wilson *et al.,* 1999). These erythrocytes reduce organic osmolytes in hypotonic stress via a multispecific membrane channel protein, which releases primarily taurine and also other amino acids, betaine, myo-inositol, and TMAO (Wilson *et al.,* 1999). The channel may be an anion exchanger, such as the so-called Band 3, which is phosphorylated at the time of taurine efflux by a tyrosine kinase activated by cell swelling (Musch *et al.,* 1999). Alternatively, the channel may be activated by an anion exchanger or regulator thereof, such as the pICln protein found throughout the vertebrates (Perlman and Goldstein, 1999). Skate hepatocytes also have a volume-activated taurine release channel (Ballatori *et al.,* 1994), while dogfish erythrocytes have a similar channel that can release TMAO (Wilson *et al.,* 1999).

However, the exact channel and the volume-responsive sensors and transducers remain unknown.

Regulation in marine elasmobranch embryos varies by development type. In the few oviparous species studied, urea and TMAO synthesis appear to arise at the earliest stages (Read, 1968), and egg membranes and embryonic epithelia probably retain them by impermeability properties (Kormanik, 1993; see also Chapter 5). Ontogeny of synthesis is delayed in viviparous species, presumably because maternal intrauterine fluids assist in osmoregulation (Kormanik, 1993).

Little is known about the hormonal regulation of osmolytes in elasmobranchs. Acher *et al.* (1999) reported that cartilaginous fishes have a greater diversity of neurohypophyseal oxytocin-like peptide hormones than do all other vertebrate groups combined. They suggest this is a result of neutral mutations, but might relate to ureotely. They point out that there is no known function of these peptides outside the mammals, despite the occurrence of such peptides in all vertebrates. In contrast to this diversity, the elasmobranchs have a highly conserved vasotocin just like other vertebrates, which may be involved in regulating renal urea transport just as the related hormone vasopressin does in mammals (Acher *et al.*, 1999).

Also, Hazon *et al.* (1999) have found that the renin–angiotensin system occurs in elasmobranchs. Angiotensin II appears to stimulate production of the unique elasmobranch hormone 1α-hydroxycortisol from the interrenal gland; this hormone in turn may affect renal and other sites of ionic regulation (Henderson *et al.*, 1985) as does cortisol in teleosts (see Section III.C). But, none of these studies reveal any clear processes for regulation of urea or TMAO concentration.

C. Hypoosmotic Hypoionic Regulators

1. Agnatha—Cephalaspidomorphi

Osmotic regulation in high salinities in the lampreys is poorly understood, due to the scarcity of captured marine-adapted specimens (Evans, 1993). Little is known about regulation of the low concentrations of organic osmolytes.

2. Actinopterygii

Euryhaline teleosts generally maintain a nearly constant internal osmolality, which changes only slightly in acclimation between seawater and lower salinities. Several studies have shown that, during such acclimations, concentrations of TMAO and amino acids account for most of these modest osmolality changes. For example, in flounder acclimated from seawater to freshwater, intracellular ions in muscle fell only 8% (about 14 mOsm), while TMAO fell 53% (16 mOsm) and amino acids 38% (27 mOsm) (Lange and Fugelli, 1965) (Fig. 6).

At the cellular and tissue level, teleost muscles may use primarily TMAO as an osmolyte (see below), while other tissues such as heart and erythrocytes may rely on taurine (Fig. 2) for such modest osmotic adjustments (King and Goldstein, 1983; Thoroed and Fugelli, 1994). Regulatory studies have focused mainly on the latter; as in elasmobranchs (Section III.B), there is a membrane channel activated by swelling, at least in flounder erythrocytes (Thoroed et al., 1995).

The role of TMAO in teleost osmoregulation is more controversial. On the one hand, some teleosts have FMO activity in liver, kidney, and gills. In flounder and medaka, the levels of FMO activity are regulated appropriately in salinity acclimations (Schlenk, 1998). The tissue contents of TMAO (especially in muscle) are usually higher in marine and seawater-acclimated animals than in freshwater-acclimated animals (Van Waarde, 1988). On the other hand, these patterns do not always occur: for example, TMAO is not higher in muscle of marine-compared to freshwater-adapted salmon (Charest et al., 1988), and TMAO is high in some African freshwater teleosts (Anthoni et al., 1990). Furthermore, FMO activity in the bass *Morone saxatilis* decreased in high-salinity acclimation, and the enzyme is apparently not present in all euryhaline teleosts (Schlenk, 1998). For species without FMO activity, TMAO may be higher in seawater adaptation simply due to diet if the diet contains marine invertebrates that use TMAO as an osmolyte (Forster and Goldstein, 1969).

The hormones involved in osmoregulation are better understood in teleosts than in elasmobranchs. These hormones include prolactin, cortisol, growth hormone, and insulin-like growth factor, all of which modulate gill $Na^+,K^+/ATPase$ activity and hence ionoregulation (Mancera and McCormick, 1998); but it is not clear that these hormones regulate organic osmolyte concentrations. One study showed that sex steroids can modulate hepatic FMO activity in trout (Schlenk, 1998), and another showed that norepinephrine can stimulate taurine release in flounder erythrocytes during hypotonic swelling, via β-receptors and cyclic AMP cascade (Thoroed et al., 1995). However, regulatory processes for the organic osmolytes remain largely unknown (see Chapter 5).

IV. PROPERTIES AND FUNCTIONS OF NITROGEN OSMOLYTES

Though used as osmotic pressure effectors in some fishes, urea and TMAO probably originated as (and are still used for) nitrogenous waste detoxification and excretion (Forster and Goldstein, 1969). Just as urea synthesis detoxifies ammonia, TMAO synthesis can serve to detoxify trimethylamine present in the diet or from choline metabolism (Fig. 2B), especially if water is limiting (Dyer, 1952; Forster and Goldstein, 1969; Van Waarde, 1988). This function also occurs in nonaquatic organisms, including humans (except for "fish-odor syndrome" in individuals unable to convert TMA to TMAO; Schlenk, 1998).

At some point, these wastes and amino acids evolved into osmolytes. To some extent the selection of taurine, urea, and TMAO as osmolytes was likely due to the lack of other metabolic roles for these compounds (unlike α-amino acids, for example). Consideration of these compounds as osmolytes raises several interesting questions. First, why are these solutes and not the more readily available inorganic ions used at high concentrations for cellular osmotic regulation? Second, why do chondrichthyans and the coelacanth exhibit a relatively consistent ratio of urea to other organic osmolytes, especially methylamines (Fig. 3)? Third, why do some osmolyte compositions often vary among tissues and organisms and habitat (Figs. 3, 4, and 5)? The answers to these questions are presumably related to roles and properties of organic osmolytes other than simple exertion of colligative osmotic pressure. Several possibilities have been proposed and tested.

A. Inorganic Ions as Perturbants and Amino Acids as Compatibile Osmolytes

It now seems clear that uptake and release of inorganic ions are used by most cells to adjust cell volume, but usually only over a small range of osmolalities (Lang *et al.,* 1998; Yancey, 1994). In general, in cells capable of surviving long-term or large-scale dehydrational stress, organic osmolytes eventually replace ions for volume regulation. In part this may be to minimize ion effects on membrane potentials. But another key explanation is based on the "compatibility" hypothesis of Brown and Simpson (1972), later extended by Clark, Somero, Wyn Jones, and others (Wyn Jones *et al.,* 1977; Yancey *et al.,* 1982). This hypothesis was initially based on findings that the major organic osmolytes usually do not disrupt macromolecular function *in vitro,* even at concentrations of several molar for some. In contrast, inorganic ions (especially NaCl) at high levels are often quite disruptive. K^+ salts sometimes show compatibility and even stimulation at levels coinciding with those typical of cells, but cause perturbation at higher concentrations. The use of taurine and other amino acids as osmolytes in some fish is generally explained by this hypothesis. Examples of inorganic and organic solute effects are shown in Fig. 7 for tuna and rabbit homologous enzymes.

Numerous studies have shown that exposing living cells to NaCl concentrations higher than normal is deleterious, at least initially. Examples of effects include reduced growth rates of bacteria, mammalian renal cells, and mammalian embryos *in vitro* (reviewed by Somero and Yancey, 1997) and reduced protein synthesis in Atlantic salmon erythrocytes and gill and hepatic tissue *in vitro* (Smith *et al.,* 1999). The latter study also found an upregulation in protein-chaperoning stress proteins in the salmon cells, suggesting that cellular proteins were partly denatured by the exposure to high NaCl concentrations. In contrast, bacterial and mammalian cells and embryos survive and grow remarkably better under osmotic NaCl shock if they are provided key organic osmolytes (reviewed by Yancey, 1994, and Somero and Yancey, 1997).

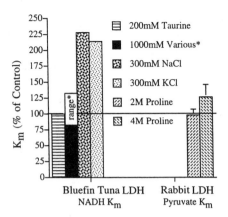

Fig. 7. Osmolyte effects on A_4-lactate dehydrogenases (LDH) from a teleost (bluefin tuna; data from Bowlus and Somero, 1979) and a mammal (data from Wang and Bolen, 1996), showing inhibitory effects of inorganic ions and moderate effects of organic osmolytes even at very high concentrations. *Range for results from individual 1 M solutions of betaine, glycine, proline, alanine, serine, and β–alanine; none were significantly different from the control.

The effects of ions and absence of effects of compatible osmolytes have usually been found to be similar with enzymes from species and tissues with or without high levels of organic osmolytes (Yancey et al., 1982; Yancey, 1994). This has led to an important corollary of the hypothesis: effects of salt and organic osmolytes should be general features of protein–solute–water interactions, not of specific adaptations in protein structure. Use of certain organic osmolytes should maintain cellular functions without significant disruptions over a wide range of external salinities (Brown and Simpson, 1972; Yancey et al., 1982).

B. Urea as Perturbing Osmolyte

Urea seems an odd evolutionary selection for a major osmolyte. At the concentrations found in marine chondrichthyans and mammalian kidneys, urea alters many macromolecular structures and functions (usually disrupting); for example, assembly of collagen (Fessler and Tandberg, 1975) and microtubules from both protozoa (Shigenaka et al., 1971) and mammals (Sackett, 1997) is inhibited by urea concentrations well under 1 M. Enzymes are often inhibited by physiologic concentrations of urea (Fig. 8). More complex systems are also affected: 40% inhibition of rabbit heart contraction in 300 mM urea (Schmidt et al., 1972); 50% mortality of frog embryos in 110 mM urea (McMillan and Battle, 1954); complete mortality of cultured snails in 670 mM urea (Smith et al., 1994); 60–80% decrease in the growth rate of mammalian renal cells *in vitro* in 200 mM urea (Yan-

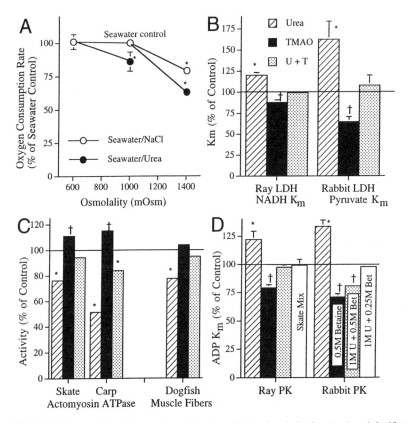

Fig. 8. Inhibitory effects of urea and counteracting effects of methylamines *in vitro;* *significant ($P < 0.05$) functional inhibition relative to control values; †significant enhancement relative to control values. (A) Inhibition of teleost gill respiration by urea, using freshly excized tissue from *Gillichthys mirabilis* (P. H. Yancey, unpublished). (B) K_m values for A_4-lactate dehydrogenases (LDH) with 400 mM urea, 200 mM TMAO, or combination (U+T): from muscle of an elasmobranch (ray, *Rhinobatus productus*) (data from Yancey and Somero, 1980); and from rabbit muscle (data from Baskakov *et al.*, 1998). (C) Activities of actomyosin ATPase with 400 mM urea and/or 200 mM TMAO, from muscle of an elasmobranch (skate, *Raja nevis*) and a carp (*Carassius carassius*) (data from Yancey, 1985) and contractile force of chemically "skinned" shark (*Squalus acanthias*) muscle fibers with 370 mM urea and/or 180 mM TMAO (data from Altringham *et al.*, 1982). Bars as in B. (D) ADP K_m values for pyruvate kinases (PK) from muscle of elasmobranch (*Urolophis haleri*) with 400 mM urea, 200 mM TMAO, and/or "skate mix" (400 mM urea, 65 mM TMAO, 55 mM sarcosine, 30 mM betaine, and 50 mM β-alanine, the osmolyte composition of one skate muscle; King and Goldstein, 1983) (data from Yancey and Somero, 1980) and from rabbit muscle with 1 M urea and betaine as indicated (data from Burg *et al.*, 1996). Bars as in B, except for betaine with rabbit.

cey and Burg, 1990); and a 25–37% inhibition of teleost gill respiration in 300 mM urea (Fig. 8A). Other examples are reviewed by Yancey (1985).

In two studies, teleost fishes (which do not maintain significant levels of urea in their tissues) were placed in urea solutions for several days. In the first study, the fish began to die as the plasma urea levels rose above 250 mM, with 75% mortality as plasma levels reached 370 mM in the ninth day (Griffith *et al.*, 1979). In the second study, freshwater fish fingerlings showed considerable damage (revealed histologically) in melanophores, fins, gills, gut, liver, and kidney after 5 days in 8 mM urea (Sriwastwa and Srivastava, 1982).

Marine cartilaginous fishes clearly survive indefinitely with higher internal levels of urea than that used in these teleost studies. How is this achieved? There are three different adaptation hypotheses, each with supporting evidence.

First, some proteins appear to be *urea insensitive*, which in some ureosmotic regulators may reflect evolutionary adaptation. Elasmobranch actomyosin ATPases, for example, are generally more stable in urea than are the homologs from teleosts (Yancey, 1985; Hasnain and Yasui, 1986). The glycine–cleavage enzyme complex of mitochondria from a skate is urea insensitive. The same complex from a teleost and clam is urea inhibited (Moyes and Moon, 1990). Both elasmobranch and coelacanth cartilage are highly sulfated, making them resistant to urea disruption (Mathews, 1967). The most extensive examples of urea adaptation have been found in hemoglobins from sharks, skates, rays (Bonaventura *et al.*, 1974), and the coelacanth (Mangum, 1991), which unlike amphibian and mammalian hemoglobins are generally insensitive to urea over the physiologic range. However, there are conflicting results for one shark hemoglobin, which in one study (Weber, 1983), but not in another (Mangum, 1991), was found to be sensitive to urea. Finally, membrane properties may also be urea adapted: the membranes of elasmobranch liver mitochondria have a much higher percentage of saturated fatty acids than do those of other fishes, possibly to resist destabilization by urea (Glement and Ballantyne, 1996).

Second, there may be proteins in ureosmotic regulators that have evolved to be *urea requiring*. There are few clear examples. Shark lens protein does not maintain proper structure without at least 250 mM urea (Zigman *et al.*, 1965). Also, pyruvate K_m values for muscle lactate dehydrogenases in the absence of urea have been found to be lower for elasmobranch homologs compared to those of other vertebrates. However, addition of 400 mM urea raises the K_m to values similar to those for this enzyme from other vertebrates, suggesting that the elasmobranch enzymes have evolved higher pyruvate affinities that urea reduces to normal values (Yancey and Somero, 1978). TMAO has only small effects on this parameter (see Section IV.C.1).

Third, the inhibitory effects of urea may be offset by *urea-counteracting osmolytes*. This may be the primary adaptation and is discussed extensively in the next section.

C. Methylamines as Stabilizing and Counteracting Osmolytes

Many studies, often unrelated to osmolyte research, have shown that some osmolyte-type solutes, especially the methylamines, stabilize macromolecular structure *in vitro* in a variety of systems. In doing so, these solutes can offset some destabilizing effects of various perturbants. Such effects are variously termed *counteracting* (Yancey et al., 1982), *compensatory* (Gilles, 1997), or *chemical chaperoning* (Brown et al., 1996).

1. COUNTERACTING UREA

In early studies conducted on elasmobranch and mammalian proteins, Yancey and Somero (1979, 1980) found that methylamines exhibit not just simple compatibility, but often enhancement of protein activity and stability. This latter property was found to be additive with respect to the inhibitory effects of urea such that they counteracted each other, most effectively at about a 2:1 urea:TMAO ratio (similar to physiologic levels, about 400:200 mM in cartilaginous fishes). These discoveries have been extensively confirmed in a variety of systems from many taxa (reviewed by Yancey, 1994) (see examples in Fig. 8). Counteraction between urea and TMAO also occurs on fluidity of elasmobranch erythrocyte membranes, although opposing solute effects may be exerted through integral membrane proteins rather than on the membrane lipids (Barton et al., 1999).

TMAO is usually a better stabilizer than other known osmolytes including betaine (Yancey, 1994, 2000; Göller and Galinski, 1999), perhaps explaining why TMAO is the dominant nonurea osmolyte in most ureosmotic fishes. However, betaine, sarcosine, and β–alanine can also stabilize and counteract (Yancey and Somero, 1979, 1980). Like compatibility, counteraction usually occurs whether a protein is from a urea-accumulating tissue or not (but see Section VII), and thus may reflect universal mechanisms (Yancey et al., 1982).

The importance of the 2:1 ratio is also apparent in elasmobranch development. At least in the oviparous skate *Raja binoculata*, embryos can make urea and TMAO from the very earliest stages of development, and they maintain relatively constant concentrations of both throughout development (Read, 1968).

The hypothesis of counteracting urea-TMAO effects is greatly supported by two cases of possible independent evolution. First is that of the coelacanth (Fig. 3), which is probably more closely related to actinopterygians than to chondrichthyans (Nelson, 1994) and, thus, may have evolved urea-TMAO osmoregulation independently. Second is the discovery of methylamine osmolytes in the mammalian renal inner medulla (Bagnasco et al., 1986). Cells in this tissue, which can have a high urea content due to the urinary concentrating mechanism, have close to a 2:1 ratio of urea to the methylamines glycerophosphorylcholine and betaine (reviewed by Somero and Yancey, 1997).

Among elasmobranchs, two exceptions to the 2:1 urea:methylamine ratio *in vivo* have been reported. In tissue of the skate *R. erinacea*, the 2:1 ratio is between urea and a mixture of nitrogenous solutes of both amino acids (such as taurine and β-alanine) and the methylamines sarcosine (*N*-methylglycine) and TMAO (King and Goldstein, 1983). However, a mixture of these osmolytes is able to counteract urea (skate mix, Fig. 8D). Abyssal *Bathyraja* skates have a reversed 1:2 urea:TMAO ratio (Rajids, Fig. 5), which is part of a recently discovered broader pattern of increasing TMAO concentration with depth in many deep-sea taxa (Section II.C.2.b; Kelly and Yancey, 1999). See Section IV.C.4. for an hypothesis.

Occasionally urea and TMAO have been found to have opposite effects, that is, enhancing and inhibiting effects, respectively, on proteins. Examples include elasmobranch and mammalian LDH activity (Yancey and Somero, 1980) and substrate binding by elasmobranch 5′-monodeiodinase (although urea and TMAO have the more common inhibitory and stimulatory effects, respectively, on that enzyme's V_{max}; Leary *et al.*, 1999). A similar effect was seen with skate mitochondrial respiration (Ballantyne and Moon, 1986). In these cases, perhaps TMAO results in a protein or membrane becoming too rigid for proper function (Yancey *et al.*, 1982; Yancey, 1994). Indeed this concept predicts that too much TMAO relative to urea might be detrimental for this reason (Yancey *et al.*, 1982). Also, some urea-sensitive proteins are not affected by TMAO (see Section VII).

It has long been known that shark hearts *in vitro* require high urea concentration in the perfusing fluids to function properly (Simpson and Ogden, 1932). A recent study showed that this urea requirement is not a simple osmotic effect since replacement of urea with sucrose did not support proper function (Wang *et al.*, 1999). Also, adding or removing TMAO from the perfusant had little effect. However, since the heart cells probably contained impermeable TMAO, while the external urea equilibrated across the membrane, it could be that the heart required a proper urea:TMAO ratio intracellularly.

There appear to be no studies on whether lungfish that estivate also have high stabilizing solutes along with the known high urea levels. However, because these fishes exhibit depressed metabolism during estivation, it has been suggested that urea accumulation in the absence of counteractants might actually facilitate dormancy (Yancey, 1985). Only two relevant studies appear to address this issue. First, estivating terrestrial snails have been found to accumulate up to 300 mM urea without accompanying significant concentrations of methylamines (Rees and Hand, 1993). Second, estivating Australian frogs accumulate up to 220 mM urea but only 35–45 mM total of inositol plus the methylamines TMAO, TMA, betaine, sarcosine, and GPC (Withers and Guppy, 1996). Whether this results in any counteraction, or whether the net effect is a suppression of metabolism by urea, is unknown.

2. COUNTERACTING SALT

Methylamines can sometimes offset perturbing salt effects. This effect was discovered independently by Clark and Zounes (1977) with marine invertebrate enzymes and by Pollard and Wyn Jones (1979) with plant enzymes. Counteraction occurs also with more complex systems; for example, TMAO (more so than betaine) reverses the salt disruption of barnacle muscle fiber architecture (Clark, 1985) and inhibition of force generation in mammalian muscle fibers (Nosek and Andrews, 1998). Whether this occurs in fish with TMAO is not known. Possibilities include hagfish, which apparently have high cellular contents of inorganic ions, at least in muscle (Fig. 3), and those Antarctic fish reported to have both high plasma NaCl and muscle TMAO contents (Section II.C.2), which might, therefore, have high cellular NaCl contents (Raymond and DeVries, 1998).

3. COUNTERACTING HYDROSTATIC PRESSURE

The latest discovery of counteraction involves the deep sea, where high hydrostatic pressure can perturb proteins (Siebenaller and Somero, 1989). As noted above, deep-sea teleosts and skates have up to 300 mmol kg^{-1} TMAO, increasing with depth (Fig. 5), a pattern hypothesized to be related to high pressure (Gillett et al., 1997; Kelly and Yancey, 1999). TMAO, indeed, offsets pressure-induced increases in enzyme K_m values for the two enzymes tested to date (Fig. 9A). And, unlike glycine, TMAO reduces pressure-enhanced instability of several LDH homologs including deep and shallow teleosts and a mammal (e.g., see Fig. 9B) (Yancey and Siebenaller, 1999; Yancey et al., 2000). Similarly, TMAO offsets completely the pressure-inhibited polymerization of actin from a deep-sea grenadier (Yancey et al., 2000). Whether TMAO-pressure counteraction occurs on other cellular systems and properties such as membrane fluidity is unknown.

A corollary to the hypothesis of TMAO-pressure counteraction is that an individual animal migrating to different depths might regulate its TMAO content to match the local pressure (Kelly and Yancey, 1999). This is supported by the results of studies of a gadiform species (the morid cod *Antimora microlepis*) caught at two different depths, 1900 and 2900 m, with the bathyal fish ($n = 5$) having 211 and the abyssal one 288 mmol/kg TMAO (Morid, Fig. 5). However, because the latter value was from one specimen, further studies are needed to test the hypothesis. Interestingly, hagfish live at bathyal depths, but some apparently migrate upward and the specimens that have been analyzed for TMAO content were caught in relatively shallow water (Robertson, 1976). The reported concentrations of TMAO in muscle are higher than in the average shallow-living teleost (Fig. 3) and only slightly less than those in teleosts from a 2000-m depth (Fig. 5). Analysis of hagfish from much greater depths would be a good test of the hypothesis which predicts that the TMAO concentration would be higher.

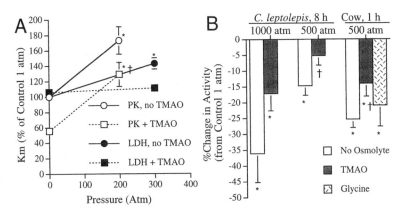

Fig. 9. Effects of osmolytes on enzymes at high hydrostatic pressure. (A) Muscle pyruvate kinase (PK) with 300 mM TMAO from a morid cod (*Antimora microlepis*) from 2000 m (data from Yancey et al., 2000) and A$_4$-lactate dehydrogenase (LDH) with 250 mM TMAO from a macrourid teleost (*Coryphaenoides leptolepis*) from 2900 m (data from Gillett et al., 1997). (B) Loss of activity of LDHs incubated under pressure with and without 250 mM TMAO or glycine for *C. leptolepis*, which lives down to 5000 m (500 atm) (data from Yancey et al., 2000) and for cow (data from Yancey and Siebenaller, 1999). *Significantly ($P < 0.05$) different from control values; †Significantly different from no-osmolyte condition.

Another important observation is the reversed urea:TMAO ratio in deep-sea skates (Section II.C.2), because lower urea contents compensated for the higher TMAO content, apparently maintaining osmoconformation (Rajids, Fig. 5). Thus, the bathyal skates had a 1:1 and the abyssal skates a 1:2 ratio. Again, this suggests that stabilizing properties have a greater selective advantage at greater depth.

The teleosts in these studies also had high plasma sodium levels (Fig. 5), and the high TMAO content might be used for counteracting salt effects. However, the elevated TMAO contents were in muscle, not plasma (for all but one species), and fully accounted for the elevated tissue osmotic pressures, with no evidence for higher intracellular salt contents (Gillett et al., 1997).

4. Counteracting Temperature Extremes

Numerous studies have shown that most osmolytes increase protein thermal stability *in vitro*. Among the zwitterionic nitrogenous osmolytes, methylamines tend to be the best thermostabilizers (reviewed by Yancey, 1994, 2001). Whether this property is physiologically important *in vivo* is unknown. One study (Anthoni et al., 1990) on tropical freshwater teleosts (Nile perch *Lates niloticus* and *Tilapia* sp.) found TMAO levels as high as those of many marine teleosts (up to 27 mM). The authors suggest themostabilization as a possible role, but this has not been tested, and the concentrations seem to be too low to have a significant effect. Another possible role they suggest is that of antioxidation (see Section IV.E).

As noted above, some shallow-water Antarctic and perhaps all deep-sea teleosts (Fig. 5) have unusually high levels of TMAO. A characteristic of both habitats is cold temperatures, suggesting that TMAO content may correlate with temperature. Hand and Somero (1982) reported that modest amounts of TMAO (<100 mM) can prevent cold-denaturation of a mammalian protein. Also, it is possible that some stabilizing properties of osmolytes are enhanced in the cold. For example, the methylated osmolyte β-dimethylsulfonioproprionate has been found to be a more compatible solute at cold than at warm temperatures (Nishiguichi and Somero, 1992). Raymond and DeVries (1998) hypothesize an antifreeze function for the high TMAO contents of Antarctic fish. However, an antifreeze function would not explain the high TMAO content in deep-sea animals (Fig. 5), where temperatures are always above freezing. Raymond and DeVries also note that Antarctic fish have unusually high plasma salt levels (as do the deep-sea fish; Fig. 5), and that perhaps the intracellular salt concentrations are also high, in which case TMAO may serve to protect cell proteins from the deleterious effects of salt.

Stabilization of proteins in the cold would not explain why the TMAO content is not particularly high in many (most?) shallow cold-adapted fish; for example, the Alaskan pollock (*Theragra chalcogramma*) has only 66 mmol TMAO/kg wet wt (Wekell and Barnett, 1991) yet is a gadiform fish, the group having the highest TMAO levels. Also, TMAO contents are roughly linearly correlated with depth (Fig. 5), but due to effects of thermoclines and vertical currents, temperatures are not as linearly correlated (i.e., temperatures are similar at the bathyal and abyssal sites; Fig. 5). These observations are not consistent with the hypothesis that TMAO counteracts temperature extremes. As noted above, counteraction of hydrostatic pressure is a better explanation for the observed increases in TMAO concentration with depth. Because all but one of the Antarctic teleosts with high TMAO content reported by Raymond and DeVries (1998) were notothenids, I speculate that these fish retain TMAO as a remnant of recent deep-sea ancestry. These fishes may have been driven down to 2000 m during periods when the ice sheet covered the continental shelf (J. Eastman, personal communication). However, shallow Antarctic liparids and zoarcids (in families with known deep-water ancestry) have low TMAO contents (Raymond and DeVries, 1998); perhaps their deep-water ancestry is more ancient, or this speculation is incorrect. Other possible explanations for high TMAO contents of Antarctic fish are discussed above and below.

D. Urea and TMAO as Buoyancy Aids

Another important hypothesis is based on calculations that TMAO solutions are less dense (1.000 g/ml for 1 M solution) than common seawater (1.024 g/ml) and common physiologic solutes such as Na^+, K^+, and other osmolytes such as glycine (Fig. 10) (Withers *et al.*, 1994). Betaine, sarcosine, urea, and chloride are

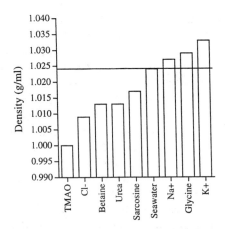

Fig. 10. Densities of 1 M solutions of the indicated solutes (data from Withers *et al.*, 1994). Horizontal line indicates the density of seawater.

also less dense than seawater (Fig. 10). Withers *et al.* propose that urea and TMAO were selected as osmolytes in part because these density properties provide buoyancy for chondrichthyans, which lack swimbladders.

However, this does not explain the 2:1 urea:TMAO ratio because urea is invariably more concentrated than TMAO in shallow-living ureosmotic fishes (Fig. 3), yet urea solutions are less buoyant (denser) than either TMAO (or chloride) solutions (Fig. 10). They speculate that TMAO is not dominant because of its high metabolic energy cost for synthesis and/or low availability in the diet. However, that argument has several problems. First, some teleosts with high TMAO content and the coelacanth have swimbladders, yet can have higher TMAO concentrations than typical for elasmobranchs (Figs. 3 and 5). Second, TMAO levels are present at concentrations estimated at 200–500 mM in cells of deep-sea skates (Fig. 5) in one of the poorest energy habitats. Similarly, many epipelagic invertebrates (which should benefit from buoyancy aids) can make or accumulate TMAO, but apparently only species in the deep sea have high concentrations (Kelly and Yancey, 1999). Third, deep-sea benthic animals (which should not require enhanced buoyancy), including scorpaenid teleosts, skates, crabs, clams, and anemones, have higher TMAO than their shallow-water counterparts (Kelly and Yancey, 1999). Fourth, regardless of its function in chondrichthyans, TMAO levels are high in deep-sea teleosts (Fig. 5) yet would not serve to elevate buoyancy because TMAO solutions have the same buoyancy as pure water, which is effectively what TMAO replaces in these hypoosmotic fishes compared to their shallow-water relatives (Gillett *et al.*, 1997). It is possibile that the density of TMAO solutions is lower relative to pure water at high hydrostatic pressures, but that has not been measured.

Overall, TMAO and urea should provide significant buoyancy for marine chondrichthyans (an estimated 5–6 g/liter of lift), as Withers *et al.* (1994) propose. However, the selection for a 2:1 ratio of urea:TMAO in shallow-living chondrichthyans and the trend of increasing TMAO concentration with depth strongly support the role of other interactions—especially relative perturbing and stabilizing effects—for the particular ratios and concentrations of these osmolytes.

E. Urea, TMAO, and Taurine as Antioxidants

Many osmolytes have been proposed to be antioxidants. Among the fish osmolytes, taurine is the most commonly considered for this function, as a scavenger of hypochlorous ions, for example (Huxtable, 1992). However, this possibility has mainly been studied only in mammalian systems, with inconsistent results (reviewed by Miller *et al.*, 2000).

Urea has recently been found to protect shark and rat heart from oxidative stress induced by both electrolysis and post-ischemic reperfusion. This effect correlates with a concentration-dependent ability of urea (3–300 mM) to scavenge reactive oxygen species generated by electrolysis (Wang *et al.*, 1999). Finally, TMAO has reported antioxidant activity with respect to lipid autooxidation (Ishikawa *et al.*, 1978).

F. Other Functions

Amino acid osmolytes also have other metabolic roles, most of which are beyond the scope of this review (e.g., see Chapter 3). Taurine is perhaps the most complex in terms of possible functions. It is a major solute in many mammalian tissues (Miller *et al.*, 2000) where in addition to demonstrated osmoregulatory and postulated antioxidant functions, it is hypothesized to stabilize membranes by lipid interactions and to modulate ions in excitable tissues (Huxtable, 1992), the latter perhaps explaining why taurine concentration is often highest in mammalian and fish hearts and brains (Fig. 4), for example, about 140 mM in skate heart (King and Goldstein, 1983). In teleosts, it is also conjugated with bile acids to make bile salts (Van Waarde, 1988).

V. MECHANISMS OF COMPATIBILITY AND COUNTERACTION: OSMOLYTES AND WATER STRUCTURE

A key premise of the compatibility and counteracting hypotheses is that osmolyte effects should involve universal water–solute–macromolecule interactions, as originally proposed by Clark, Somero, and colleagues (Yancey *et al.*,

1982) and Wyn Jones and coworkers (1977). It was noted that amino acid and methylamine osmolytes are structurally similar to Hofmeister ions, which form a ranking of anions and cations based on consistent solution effects: *kosmotropes* [such as $N(CH_3)_4^+$] generally salt out, enhance catalysis by, and stabilize macromolecules, whereas *chaotropes* [such as Mg^{2+}] do the opposite (Collins and Washabaugh, 1985). Effects of these ions are additive, leading to counteraction when opposing ion types are paired. Both methylated organic osmolytes such as TMAO and inorganic ions such as $N(CH_3)_4^+$ are usually the best stabilizers.

These universal properties are only partly understood. Inorganic and organic chaotropes are often attracted to protein functional groups; for example, urea appears to unfold proteins by such binding (Wu and Wang, 1999). In some cases, they bind to water less well than water does to itself, resulting in sequestering of solute away from bulk water (since water–water binding is more favorable than water–solute binding) toward surfaces such as membranes and proteins. Both effects lead to *preferential interaction,* which will lead to unfolding of macromolecules because that maximizes the favorable surface interactions (Fig. 11, right).

In contrast, organic kosmotropes usually exhibit a tendency to be excluded from a protein's hydration layer, the shell of bound water molecules around the surface (Timasheff, 1992). This is termed *preferential exclusion,* which creates an entropically unfavorable order of high and low solute concentrations and more

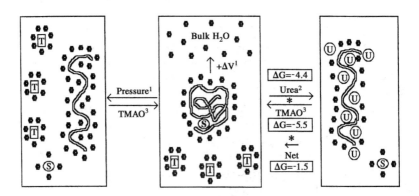

Fig. 11. Model of solute and pressure effects on protein folding, based on models of Low (1985), Siebenaller and Somero (1989), Timasheff (1992), and Noto *et al.* (1995). Small spheres represent water molecules. (1) Pressure can favor unfolding (left box) but only if there is a net expansion ($+\Delta V$, middle box) as water molecules are released into bulk water during folding; (2) addition of urea (U) enhances unfolding (right box) since it maximizes favorable binding sites; and (3) addition of TMAO (T), excluded from the protein hydration layer presumably because of its own structured water layers, favors folding and substrate (S) binding (middle box), since that reduces the total order (higher in left box). *Differences in free energy (kcal/mol) between unfolded and native RNase T1 for the peptide backbone in solutions of 1 M urea, 0.5 M TMAO, or both (Net), showing that urea favors unfolding and TMAO favors folding and can offset the urea effect. (Data from Wang and Bolen, 1997.)

and less ordered water (Fig. 11, left). Proteins reduce this order by minimizing their exposed surface areas by folding more compactly, aggregating, or precipitating (salting out). Binding of ligands to active sites will also be favored if this involves loss of ordered bound water (Fig. 11, middle).

Several hypotheses have been proposed to explain the forces causing preferential exclusion (Low, 1985; Timasheff, 1992). Those relevant to TMAO involve *water structuring*. Zwitterionic organic stabilizers may bind to water molecules better than to other solutes or than water to itself, and enhance water–water hydrogen bonding beyond their immediate hydration layer. Evidence for this hypothesis includes the observed reduction in translational self-diffusion of water in TMAO solutions (Clark, 1985), the tendency of most stabilizers to raise water surface tension (Timasheff, 1992), and molecular dynamic simulations showing that TMAO tightly coordinates water molecules (Noto *et al.*, 1995). Methylamines often show the strongest effects; for example, they elute earlier than predicted from polyacrylamide gel columns, indicating firmly bound hydration shells (Galinski *et al.*, 1997).

Recently Wang and Bolen (1997) showed that unfavorable interactions between TMAO and peptide backbones (but not amino acid side groups) explain the strong exclusion of TMAO and, thus, its enhancement of protein folding (ΔG values; Fig. 11). Finally, Wiggins (1997) has proposed that hydrophobic methyl groups can alter water structure to favor native protein conformations only when part of a molecule with a charged atom: the charge repels these groups from each other to prevent hydrophobic aggregation.

These effects could also explain counteraction of pressure, as recently proposed (Yancey and Siebenaller, 1999). Pressure inhibits release of hydration water from substrates and the folding and assembling proteins in cases where volume expansion occurs (Fig. 11, left and middle) (Siebenaller and Somero, 1989). Again, since TMAO favors the opposite effect (reduction of hydration water), it may counteract pressure simply by additivity of independent effects.

Finally, a few studies have provided evidence that TMAO can induce the formation of S–S bonds in proteins via hydrogen bonding (---) in the reaction SH---ON $\longleftrightarrow$ S$^-$---H$^+$ON. This reaction creates reactive sulfides and may represent an additional mechanism of protein stabilization by TMAO (Brzezinksi and Zundel, 1993).

VI. EVOLUTIONARY CONSIDERATIONS

As has been argued elsewhere (Yancey *et al.*, 1982; Ballantyne *et al.*, 1987; Somero and Yancey, 1997), it seems likely that the acquisition of organic osmolyte regulation should have facilitated adaptation to high and/or variable salinities. This is because the alternative would appear to require changes in most cell macromolecules and membrane-potential regulation to function in a concentrated and/or varying ion (or urea) solution. Possible scenarios for the evolution of fish osmoregulation, from the strictly osmoconforming hagfish to ureosmotic hypoionic regulation to hypoionic hypoosmotic regulation, have been presented in detail elsewhere (Ballantyne *et al.*, 1987; Griffith, 1991) and in this volume (Chapter 1).

Two common evolutionary hypotheses are, first, that both hypoionic systems evolved in estuarine or freshwater environments (Griffith, 1991) and, second, that the hypoosmotic hypoionic pattern is the most metabolically costly for those fish that reinvaded the oceans (Griffith and Pang, 1979). The first is contested by some for ureosmotic fishes and remains to be settled for actinopterygians, though there is suggestive fossil evidence (Ballantyne *et al.*, 1987, Kirschner, 1993). The second hypothesis is based on the need to maintain both ion and water gradients in actinopterygians. In contrast, osmoconformers may use less energy since they do not maintain either gradient, whereas ureosmotic regulators may also use less since they do not maintain a large water gradient (and in fact are often slightly hyperosmotic, with a net influx of water) (Griffith and Pang, 1979). It has been speculated that the (supposedly) costly hypoosmotic hypoionic regulation could not arise before atmospheric oxygen reached a high enough level, perhaps 300 million years ago (Ballantyne *et al.*, 1987).

If these energy costs are correct, elevated TMAO levels in deep-sea teleosts (Fig. 5) might save osmoregulatory energy; indeed, such fish have reduced activities of gill Na^+,K^+/ATPases (the primary ion-regulating enzyme), suggesting lowered osmoregulatory costs (Gibbs, 1997). However, calculations by Kirschner (1993) suggest that hypoosmoregulation is no more metabolically expensive than osmoconforming, since the latter requires synthesis, transport, and retention of organic osmolytes. Also, if a higher osmotic content saves energy, it is not clear why all teleosts have not evolved such a strategy (recall that some polar teleosts are nearly isoosmotic with seawater, so this strategy can arise in this group). The issue remains unresolved.

VII. PRACTICAL APPLICATIONS, EXCEPTIONS, AND UNANSWERED QUESTIONS

Because organic osmolytes are usually universal in their effects on proteins, their properties can be used in biological systems that did not evolve with such solutes. This may have applied medical, agricultural and biotechnological uses. For example, Welch and colleagues have suggested that osmolytes as "chemical chaperones" might rescue misfolded proteins in human diseases. They showed that 75–100 mM TMAO can restore function of one form of cystic fibrosis mutant protein (Brown *et al.*, 1996) and can inhibit the formation of aberrant protein aggregates in scrapie prion disease (Tatzelt *et al.*, 1996). Also, stabilizing properties of betaine have been used to improve the PCR procedure that is central to molecular biology (Henke *et al.*, 1997). Other examples are reviewed by Yancey (2001).

It is important to realize that the universality of osmolyte effects has exceptions and inconsistencies. As universal as preferential exclusion and macromolecular stabilization may be for kosmotropes, these properties do not always correlate with effects on enzyme kinetics (Yancey, 1994; Randall *et al.*, 1998; Burg *et al.*, 1999). A few proteins have been reported to have urea sensitivity with no counteraction by TMAO, for example, several nonelasmobranch enzymes (Mashino and Fridovich, 1987) and the glycine-cleavage system of skate mitochondria (Moyes and Moon, 1990). Another example was reported with dogfish hemoglobin; Weber (1983) speculates that lack of TMAO sensitivity evolved to preserve the conformational changes necessary for the unusually broad Bohr (pH) effect in this species. Also, it was recently found that dogfish 5'-monodeiodinase exhibits urea-TMAO counteraction, but the trout homolog exhibits activity inhibition and K_m enhancement by both (Leary *et al.*, 1999). Such proteins in ureosmotic fishes may have evolved a urea requirement, may receive urea counteraction by other unknown effects *in vivo*, or suffer inhibition *in vivo* that is not selectively important (Yancey, 1994).

Yancey *et al.* (1982) suggest that some urea effects cannot be counteracted by TMAO; for example, competitive inhibition of substrate binding, in which urea binds to the active site. Along these lines, Wiggins (1997) proposes that gross protein stability, which consistently exhibits urea destabilization and TMAO stabilization, may be less important than the (poorly understood) solvation properties of binding sites in explaining osmolyte effects on protein–ligand interactions at physiologic levels. It is this level of osmolyte–water–protein interaction, as well as the cellular and hormonal mechanisms of osmolyte regulation in fish, that needs more study.

REFERENCES

Acher, R., Chauvet, J., Chauvet, M.-T., and Rouille, Y. (1999). Unique evolution of neurohypophysial hormones in cartilaginous fishes: possible implications for urea-based osmoregulation. *J. Exp. Zool.* **284,** 475–484.

Altringham, J. D., Yancey, P. H., and Johnston, I. A. (1982). The effects of osmoregulatory solutes on tension generation by dogfish skinned muscle fibres. *J. Exp. Zool.* **96,** 443–445

Anderson, P. M. (1980). Glutamine- and *N*-acetylglutamate-dependent carbamoyl phosphate synthetase in elasmobranchs. *Science* **208,** 291–293.

Anthoni, U., Børresen, T., Christophersen, C., Gram, L., and Nielsen, P. H. (1990). Is trimethylamine oxide a reliable indicator for the marine origin of fish? *Comp. Physiol. Biochem.* **97B**, 569–571.

Anthoni, U., Christophersen, C., Gram, L., Nielsen, N. H., and Nielsen, P. (1991). Poisonings from flesh of the Greenland shark *Somniosus microcephalus* may be due to trimethylamine. *Toxicon* **29**, 1205–1212.

Bagnasco, S., Balaban, R., Fales, H., Yang, Y.-M., and Burg, M. (1986). Predominant osmotically active organic solutes in rat and rabbit renal medullas. *J. Biol. Chem.* **261**, 5872–5877.

Ballantyne, J. S., and Moon, T. W. (1986). The effects of urea, trimethylamine oxide and ionic strength on the oxidation of acyl carnitines by mitochondria isolated from the liver of the little skate *Raja erinacea*. *J. Comp. Physiol.* **156B**, 845–851.

Ballantyne, J. S., Moyes, C. D., and Moon, T. W. (1987). Compatible and counteracting solutes and the evolution of ion and osmoregulation in fishes. *Can. J. Zool.* **65**, 1883–1888.

Ballatori, N., Simmons, T. W., and Boyer, J. L. (1994). A volume-activated taurine channel in skate hepatocytes: membrane polarity and role of intracellular ATP. *Am. J. Physiol.* **267**, G285–G291.

Barton, K. N., Buhr, M. M., and Ballantyne, J. S. (1999). Effects of urea and trimethylamine N-oxide on fluidity of liposomes and membranes of an elasmobranch. *Am. J. Physiol.* **276**, R397–R406.

Baskakov, I., Wang, A., and Bolen, D. W. (1998). Trimethylamine N-oxide counteracts urea effects on rabbit muscle lactate dehydrogenase function: a test of the counteraction hypothesis. *Biophys. J.* **74**, 2666–2673.

Bedford, J. J., Harper, J. L., Leader, J. P., Yancey, P. H., and Smith, R. A. J. (1998). Tissue composition of the elephant fish, *Callorhyncus milli*: betaine is the principal counteracting osmolyte. *Comp. Biochem. Physiol.* **119B**, 521–526.

Bonaventura, J., Bonaventura, C., and Sullivan, B. (1974). Urea tolerance as a molecular adaptation of elasmobranch hemoglobins. *Science* **186**, 57–59.

Bowlus, R. D., and Somero, G. N. (1979). Solute compatibility with enzyme function and structure: rationales for the selection of osmotic agents and end-products of anaerobic metabolism in marine invertebrates. *J. Exp. Zool.* **208**, 137–152.

Brown, A., and Simpson, J. (1972). Water relations of sugar-tolerant yeasts: the role of intracellular polyols. *J. Gen. Microbiol.* **72**, 589–591.

Brown, C. R., Hong-Brown, L. Q., Biwersi, J., Verkman, A. S., and Welch, W. J. (1996). Chemical chaperones correct the mutant phenotype of the ΔF508 cystic fibrosis transmembrane conductance regulator protein. *Cell Stress Chaper.* **1**, 117–125.

Brzezinksi, B., and Zundel, G. (1993). Formation of disulphide bonds in the reaction of SH group-containing amino acids with trimethylamine N-oxide. *FEBS Lett.* **333**, 331–333.

Burg, M. B., Kwon, E. D., and Peters, E. M. (1996). Glycerophosphorylcholine and betaine counteract the effect of urea on pyruvate kinase. *Kidney Int.* **50** (Suppl. 57), S100–S104.

Burg, M. B., Kwon, E. D., and Kültz, D. (1997). Regulation of gene expression by hypertonicity. *Ann. Rev. Physiol.* **59**, 437–455.

Burg, M. B., Peters, E. M., Bohren, K. M., and Gabbay, K. H. (1999). Factors affecting counteraction by methylamines of urea effects on aldose reductase. *Proc. Natl. Acad. Sci. USA* **96**, 6517–6522.

Carlson, S. R., and Goldstein, L. (1997). Urea transport across the cell membrane of skate erythrocytes. *J. Exp. Zool.* **277**, 275–282.

Carr, W. E. S., Netherton III, J. C., Gleeson, R. A., and Derby, C. D. (1996). Stimulants of feeding behavior in fish: analyses of tissues of diverse marine organisms. *Biol. Bull.* **190**, 149–160.

Charest, R. P., Chenoweth, M., and Dunn, A. (1988). Metabolism of trimethylamines in kelp bass (*Paralabrax clathratus*) and marine and freshwater pink salmon (*Oncorhynchus gorbuscha*). *J. Comp. Physiol.* **158**, 609–619.

Cholette, C., and Gagnon, A. (1973). Isosmotic adaptation in *Myxine glutinosa* L.—II. Variations of the free amino acids, trimethylamine oxide and potassium of the blood and muscle cells. *Comp. Biochem. Physiol.* **45A**, 1009–1021.

Clark, M. E. (1985). The osmotic role of amino acids: discovery and function. In "Transport Processes, Iono- and Osmoregulation" (R. Gilles and M. Gilles-Ballien, eds.), pp. 412–423. Springer-Verlag, Berlin.

Clark, M. E., and Zounes, M. (1977). The effects of selected cell osmolytes on the activity of lactate dehydrogenase from the euryhaline polychaete, *Nereis succinea. Biol. Bull.* **153**, 468–484.

Collins, K. D., and Washabaugh, M. W. (1985). The Hofmeister effect and the behavior of water at interfaces. *Qu. Rev. Biophys.* **18**, 323–422.

Dyer, W. J. (1952). Amines in fish muscle. VI. Trimethylamine oxide content of fish and marine invertebrates. *J. Fish. Res. Bd. Canada* **8**, 314–324.

Evans, D. H. (1993). Osmotic and ionic regulation. In "The Physiology of Fishes" (D. H. Evans, ed.), pp. 315–342. CRC Press, Boca Raton, FL.

Fessler, J. H., and Tandberg, W. D. (1975). Interactions between collagen chains and fiber formation. *J. Supramol. Struct.* **3**, 17–23.

Forster, R. P., and Goldstein, L. (1969). Formation of excretory products. In "Fish Physiology" (W. S. Hoar and D. J. Randall, eds.), Vol. 1, pp. 313–350. Academic Press, New York.

Forster, R. P., and Goldstein, L. (1976). Intracellular osmoregulatory role of amino acids and urea in marine elasmobranchs. *Am. J. Physiol.* **230**, 925–931.

Galinski, E. A., Stein, M., Amendt, B., and Kinder, M. (1997). The kosmotropic (structure-forming) effect of compensatory solutes. *Comp. Biochem. Physiol.* **117A**, 357–365.

Gerst, J. W., and Thorson, T. B. (1977). Effects of saline acclimation on plasma electrolytes, urea excretion, and hepatic urea biosynthesis in a freshwater stingray, *Potamotrygon* sp. Garman, 1877. *Comp. Biochem. Physiol.* **56A**, 87–93.

Gibbs, A. G. (1997). Biochemistry at depth. In "Fish Physiology, Volume 16, Deep-Sea Fishes" (D. J. Randall and A. P Farrell, eds.), pp. 239–275. Academic Press, San Diego.

Gilles, R. (1997). "Compensatory" organic osmolytes in high osmolarity and dehydration stresses: history and perspectives. *Comp. Biochem. Physiol.* **117A**, 279–290.

Gillett, M. B., Suko, J. R., Santoso, F. O., and Yancey, P. H. (1997). Elevated levels of trimethylamine oxide in muscles of deep-sea gadiform teleosts: a high-pressure adaptation? *J. Exp. Zool.* **279**, 386–391.

Glemet, H. C., and Ballantyne, J. S. (1996). Comparison of liver mitochondrial membranes from an agnathan (*Myxine glutinosa*), an elasmobranch (*Raja erinacea*) and a teleost fish (*Pleuronectes americanus*). *Mar. Biol.* **124**, 509–518.

Goldstein, L., and Forster, R. P. (1971). Osmoregulation and urea metabolism in the little skate *Raja erinacea. Am. J. Physiol.* **220**, 742–746.

Goldstein, L., and Kleinzeller, A. (1987). Cell volume regulation in lower vertebrates. *Curr. Top. Membr. Transp.* **30**, 181–204.

Göller, K., and Galinski, E. A. (1999). Protection of a model enzyme (lactate dehydrogenase) against heat, urea and freeze–thaw treatment by compatible solute additives. *J. Mol. Catal. B.* **7**, 37–45.

Griffith, R. W. (1991). Guppies, toadfish, lungfish, coelacanths and frogs: a scenario for the evolution of urea retention in fishes. *Environ. Biol. Fishes* **32**, 199–218.

Griffith, R. W., and Pang, P. K. T. (1979). Mechanisms of osmoregulation of the coelacanth: evolutionary implications. *Occas. Papers Calif. Acad. Sci.*, **134**, 79–92.

Griffith, R. W., Pang, R. K. T., and Benedetto, L. A. (1979). Urea tolerance in the killifish, *Fundulus heteroclitus. Comp. Biochem. Physiol.* **62A**, 327–330.

Hand, S. C., and Somero, G. N. (1982). Urea and methylamine effects on rabbit muscle phosphofructokinase. *J. Biol. Chem.* **257**, 734–741.

Hasnain, A., and Yasui, T. (1986). Urea tolerance of myofibrillar proteins of two elasmobranchs: *Squalus acanthias* and *Raja tengu. Arch. Int. Physiol. Biochim.* **94**, 233–237.

Hazon, N., Tierney, M. L., and Takei, Y. (1999). Renin-angiotensin system in elasmobranch fish: a review. *J. Exp. Zool.* **284**, 526–534.

Hebard, C. E., Flick, G. J., and Martin, R. E. (1982). Occurrence and significance of trimethylamine oxide and its derivatives in fish and shellfish. *In* "Chemistry and Biochemistry of Marine Food Products" (R. E. Martin, ed.), pp. 149–175. AVI Publishing, Westport, CT.

Henderson, I. W., Hazon, N., and Hughes, K. (1985). Hormones, ionic regulation and kidney function in fishes. *In* "Physiological Adaptations of Marine Animals" (M. S. Laverack, ed.), pp. 245–265. The Company of Biologists Ltd., Cambridge.

Henke, W., Herdel, K., Jung, K., Schnorr, D., and Loening, S. A. (1997). Betaine improves the PCR amplification of GC-rich DNA sequences. *Nucleic Acids Res.* **25,** 3957–3958.

Huxtable, R. J. (1992). The physiological actions of taurine. *Physiol. Rev.* **72,** 101–164.

Ishikawa, Y., Yuki, E., Kato, H., and Fujimaki, M. (1978). The mechanism of synergism between tocopherols and trimethylamine oxide in the inhibition of the autooxidation of methyl linoleate. *Agric. Biol. Chem.* **42,** 711–716.

Ito, Y., Suzuki, T., Shirai, T., and Hirano, T. (1994). Presence of cyclic betaines in fish. *Comp. Biochem. Physiol.* **109B,** 115–124.

Kelly, R. H., and Yancey, P. H. (1999). High contents of trimethylamine oxide correlating with depth in deep-sea teleost fishes, skates, and decapod crustaceans. *Biol. Bull.* **196,** 18–25.

King, P., and Goldstein, L. (1983). Organic osmolytes and cell volume regulation in fish. *Mol. Physiol.* **4,** 53–66.

Kirschner, L. B. (1991). Water and ions. *In* "Environmental and Metabolic Animal Physiology" (C. L. Prosser, ed.), pp. 13–108. Wiley-Liss, New York.

Kirschner, L. B. (1993). The energetics of osmotic regulation in ureotelic and hypoosmotic fishes. *J. Exp. Zool.*, **267,** 19–26.

Kormanik, G. A. (1993). Ionic and osmotic environment of developing elasmobranch embryos. *Environ. Biol. Fishes* **38,** 233–240.

Lacy, E. R., and Reale, E. (1995). Functional morphology of the elasmobranch nephron and retention of urea. *In* "Cellular and Molecular Approaches to Fish Ionic Regulation" (C. M. Wood and T. J. Shuttleworth, eds.), pp. 106–146. Academic Press, New York.

Lang, F., Busch, G. L., and Völkl, H. (1998). The diversity of volume regulatory mechanisms. *Cell Physiol. Biochem.* **8,** 1–45.

Lange, R., and Fugelli, K. (1965). The osmotic adjustment in the euryhaline teleosts, the flounder *Platichthys flesus* L., and the three-spined stickleback, *Gatterosteus aculeatus* L. *Comp. Biochem. Physiol.* **15,** 283–292.

Leary, S. C., Ballantyne, J. S., and Leatherland, J. F. (1999). Evaluation of thyroid hormone economy in elasmobranch fishes, with measurements of hepatic 5′-monodeiodinase activity in wild dogfish. *J. Exp. Zool.* **284,** 492–499.

Low, P. S. (1985). Molecular basis of the biological compatibility of nature's osmolytes. *In* "Transport Processes, Iono- and Osmoregulation" (R. Gilles and M. Gilles-Ballien, eds.), pp. 469–477. Springer-Verlag, Berlin.

Lutz, P. L., and Robertson, J. D. (1971). Osmotic constituents of the coelacanth, *Latimeria chalumnae* Smith. *Biol. Bull.* **141,** 553–560.

Mancera, J. M., and McCormick, S. D. (1998). Osmoregulatory actions of the GH/IGF axis in non-salmonid teleosts. *Comp. Biochem. Physiol.* **121B,** 43–48.

Mangum, C. P. (1991). Urea and chloride sensitivities of coelacanth hemoglobin. *Environ. Biol. Fishes* **32,** 219–222.

Mashino, T., and Fridovich, I. (1987). Effects of urea and trimethylamine-N-oxide on enzyme activity and stability. *Arch. Biochem. Biophys.* **258,** 356–360.

Mathews, M. B. (1967). Macromolecular evolution of connective tissue. *Biol. Rev.* **42,** 499–551.

McMillan, D. B., and Battle, H. I. (1954). Effects of ethyl carbamate and related compounds on early developmental processes in the leopard frog *Rana pipiens. Cancer R*es. **14,** 319–323.

Miller, T. J., Hanson, R. D., and Yancey, P. H. (2000). Developmental changes in organic osmolytes in prenatal and postnatal rat tissues. *Comp. Biochem. Physiol.* **125,** 45–56.

Moyes, C. D., and Moon, T. W. (1990). Solute effects on the glycine cleavage system of two osmoconformers *Raja erinacea* and *Mya arenaria* and an osmoregulator *Pseudopleuronectes americanus*. *J. Exp. Zool.* **242**, 1–8.

Musch, M. W., Hubert, E. M., and Goldstein, L. (1999). Volume expansion stimulates p72^{syk} and p56^{lyn} in skate erythrocytes. *J. Biol. Chem.* **274**, 7923–7928.

Nelson, J. S. (1994). "Fishes of the World." Wiley & Sons, New York.

Nishiguichi, M. K., and Somero, G. N. (1992). Temperature- and concentration-dependence of compatibility of the organic osmolyte ß-dimethylsulfoniopropionate. *Cryobiology* **29**, 118–123.

Nosek, T. M., and Andrews, M. A. W. (1998). Ion-specific protein destabilization of the contractile proteins of cardiac muscle fibers. *Plügers Arch. Eur. J. Physiol.* **435**, 394–401.

Noto, R., Martorana, V., Emanuele, A., and Fornili, S. L. (1995). Comparison of the water perturbations induced by two small organic solutes: *ab initio* calculations and molecular dynamics simulation. *J. Chem. Soc. Faraday Trans.* **91**, 3803–3808.

Perlman, D. F., and Goldstein, L. (1999). Organic osmolyte channels in cell volume regulation in vertebrates. *J. Exp. Zool.* **283**, 725–733.

Piermarini, P. M., and Evans, D. H. (1998). Osmoregulation of the Atlantic stingray (*Dasyastis sabina*) from the freshwater Lake Jesup of the St. Johns River, Florida. *Physiol. Zool.* **71**, 553–560.

Pollard, A., and Wyn Jones, R. G. (1979). Enzyme activities in concentrated solutions of glycinebetaine and other solutes. *Planta* **144**, 291–298.

Randall, K., Lever, M., and Chambers, S. T. (1998). Counteraction of urea denaturation by glycine betaine and other kosmotropes. *J. Biochem. Mol. Biol. Biophys.* **2**, 51–58.

Raymond, J. A. (1992). Glycerol is a colligative antifreeze in some northern fishes. *J. Exp. Zool.* **262**, 347–352.

Raymond, J. A. (1998). Trimethylamine oxide and urea synthesis in rainbow smelt and some other northern fishes. *Physiol. Zool.* **71**, 515–523.

Raymond, J. A., and DeVries, A. L. (1998). Elevated concentrations and synthetic pathways of trimethylamine oxide and urea in some teleost fishes of McMurdo Sound, Antarctica. *Fish Physiol. Biochem.* **18**, 387–398.

Read, L. J. (1968). Urea and trimethylamine oxide levels in elasmobranch embryos. *Biol. Bull.* **135**, 537–547.

Rees, B. B., and Hand, S. C. (1993). Biochemical correlates of estivation tolerance in the mountainsnail *Oreohelix* (Pulmonata: Orehelicidae). *Biol. Bull.* **184**, 230–242.

Robertson, J. D. (1976). Chemical composition of the hagfish *Myxine glutinosa* and the rabbit-fish *Chimaera monstrosa*. *J. Zool. Lond.* **178**, 261–277.

Robertson, J. D. (1984). The composition of blood plasma and parietal muscle of Oslo Fjord eels [*Anguilla anguilla* (L.)] and the river lamprey [*Lampetra fluviatilis* (L.)]. *Comp. Biochem. Physiol.* **77A**, 431–439.

Robertson, J. D. (1989). Osmotic constituents of the blood plasma and parietal muscle of *Scyliorhinus canicula* (L.). *Comp. Biochem. Physiol.* **93A**, 799–805.

Sackett, D. L. (1997). Natural osmolyte trimethylamine *N*-oxide stimulates tubulin polymerization and reverses urea inhibition. *Am. J. Physiol.* **273**, R669–R676.

Schlenk, D. (1998). Occurrence of flavin-containing monooxygenases in nonmammalian eukaryotic organisms. *Comp. Biochem. Physiol.* **121C**, 185–195.

Schmidt, E., Wilkes, A. B., and Holland, W. C. (1972). Effects of various glycerol or urea concentrations and incubation times on atrial contraction and ultrastructure. *J. Mol. Cell. Cardiol.* **4**, 113–120.

Shigenaka, J. M., Roth, L. E., and Pihlaja, D. J. (1971). Microtubules in the heliozoan axopodium. III. Degradation and reformation after dilute urea treatment. *J. Cell Sci.* **8**, 127–151.

Shuttleworth, T. J. (1988). Salt and water balance—extrarenal mechanisms. *In* "Physiology of Elasmobranch Fishes" (T. J. Shuttleworth, ed.), pp. 171–199. Springer-Verlag, Berlin.

Siebenaller, J. F., and Somero, G. N. (1989). Biochemical adaptation to the deep sea. *CRC Crit. Rev. Aquat. Sci.* **1,** 1–25.

Simpson, W. W., and Ogden, E. (1932). The physiological significance of urea. I. The elasmobranch heart. *J. Exp. Biol.* **9,** 1–5.

Smith, H. J. (1936). The retention and physiological role of urea in the elasmobranchii. *Biol. Rev.* **11,** 49–82.

Smith, M. E., Steiner, S. A., and Isseroff, H. (1994). Urea: inhibitor of growth and reproduction in *Bulinus truncatus*. *Comp. Biochem. Physiol.* **108A,** 569–577.

Smith, T. R., Tremblay, G. C., and Bradley, T. M. (1999). Hsp70 and a 54 kDa protein (Osp54) are induced in salmon (*Salmo salar*) in response to hyperosmotic stress. *J. Exp. Zool.* **284,** 286–298.

Somero, G. N., and Yancey, P. H. (1997). Osmolytes and cell volume regulation: physiological and evolutionary principles. *In* "Handbook of Physiology" (J. F. Hoffman and J. D. Jamieson, eds.), Sec. 14, pp. 441–484. Oxford University Press, Oxford.

Sotelo, C. G., Gallardo, J. M., Piñeiro, C., and Pérez-Martin, R. (1995). Trimethylamine oxide and derived compounds' changes during frozen storage of hake (*Merluccius merluccius*). *Food Chem.* **53,** 61–65.

Sriwastwa, V. M. S., and Srivastava, D. K. (1982). Histopathological changes in *Cirrhinus mrigala* (Hamilton) fingerlings exposed to urea. *J. Environ. Biol.* **3,** 171–174.

Stark, G. R. (1972). Modifications of proteins with cyanate. *In* "Methods in Enzymology" (C. H. W. Hirs and S. N. Timasheff, eds.), Vol. XXV, pp. 579–584. Academic Press, New York.

Suyama, M., and Tokuhiro, T. (1954). Urea content and ammonia formation of the muscle of cartilaginous fishes. II. The distribution of urea and trimethylamine oxide in different parts of the body. *Bull. Jap. Soc. Sci. Fisher.* **19,** 1003–1006.

Tatzelt, J., Prusiner, S. B., and Welch, W. J. (1996). Chemical chaperones interfere with the formation of scrapie prion protein. *EMBO J.* **15,** 6363–6373.

Thoroed, S. M., and Fugelli, K. (1994). Free amino acid compounds and cell volume regulation in erythrocytes from different marine fish species under hypoosmotic conditions: the role of a taurine channel. *J. Comp. Physiol.* **164,** 1–10.

Thoroed, S. M., Soergaard, M., Cragoe, E. J. Jr., and Fugelli, K. (1995). The osmolality-sensitive taurine channel in flounder erythrocytes is strongly stimulated by noradrenaline under hypoosmotic conditions. *J. Exp. Biol.* **198,** 311–324.

Thorson, T. B., Cowan, C. M., and Watson, D. E. (1967). *Potamotrygon* spp.: elasmobranchs with low urea content. *Science* **158,** 375–377.

Thorson, T. B., Cowan, C. M., and Watson, D. E. (1973). Body fluid solutes of juveniles and adults of the euryhaline bull shark *Carcharhinus leucas* from freshwater and saline environments. *Physiol. Zool.* **46,** 29–42.

Timasheff, S. N. (1992). A physicochemical basis for the selection of osmolytes by nature. *In* "Water and Life: A Comparative Analysis of Water Relationships at the Organismic, Cellular, and Molecular Levels" (G. N. Somero, C. B. Osmond, and C. L. Bolis, eds.), pp. 70–84. Springer-Verlag, Berlin.

Van Waarde, A. (1988). Biochemistry of nonprotein nitrogenous compounds in fish including the use of amino acids for anaerobic energy production. *Comp. Biochem. Physiol.* **91B,** 207–228.

Wang, A., and Bolen, D. W. (1996). Effect of proline on lactate dehydrogenase activity: testing the generality and scope of the compatibility paradigm. *Biophys. J.* **71,** 2117–2122.

Wang, A., and Bolen, D. W. (1997). A naturally occurring protective system in urea-rich cells: mechanism of osmolyte protection of proteins against urea denaturation. *Biochemistry* **36,** 9101–9108.

Wang, X. T., Wu, L. Y., Aouffen, M., Mateescu, M. A., Nadeau, R., and Wang, R. (1999). Novel protective effects of urea: from shark to rat. *Br. J. Pharmacol.* **128,** 1477–1484.

Weber, R. E. (1983). TMAO (trimethylamine oxide)-independence of oxygen affinity and its urea and ATP sensitivities in an elasmobranch hemoglobin. *J. Exp. Zool.* **228,** 551–554.

Wekell, J. C., and Barnett, H. (1991). New method for analysis of trimethylamine oxide using ferrous sulfate and EDTA. *J. Food Sci.* **56,** 132–138.

Wiggins, P. M. (1997). Hydrophobic hydration, hydrophobic forces and protein folding. *Physica A* **238,** 113–128.

Wilkie, M. P., Wang, Y., Walsh, P. J., and Youson, J. H. (1999). Nitrogenous waste excretion by the larvae of a phylogenetically ancient vertebrate: the sea lamprey (*Petromyzon marinus*). *Can. J. Zool.* **77,** 707–715.

Wilson, E. D., McGuinn, M. R., and Goldstein, L. (1999). Trimethylamine oxide transport across plasma membranes of elasmobranch erythrocytes. *J. Exp. Zool.* **284,** 605–609.

Withers, P. C., and Guppy, M. (1996). Do Australian desert frogs co-accumulate counteracting solutes with urea during aestivation? *J. Exp. Biol.* **199,** 1809–1816.

Withers, P. C., Morrison, G., Hefter, G. T., and Pang, T.-S. (1994). Role of urea and methylamines in buoyancy of elasmobranchs. *J. Exp. Biol.* **188,** 175–189.

Wood, C. M. (1993). Ammonia and urea metabolism and excretion. *In* "The Physiology of Fishes" (D. H. Evans, ed.), pp. 379–426. CRC Press, Boca Raton, FL.

Wu, J.-W., and Wang, Z.-X. (1999). New evidence for the denaturant binding model. *Prot. Sci.* **8,** 2090–2097.

Wyn Jones, R. G., Storey, R., Leigh, R. A., Ahmad, N., and Pollard, A. (1977). A hypothesis on cytoplasmic osmoregulation. *In* "Regulation of Cell Membrane Activities in Plants" (E. Marre and O. Ciferri, eds.), pp. 121–136. Elsevier Press, Amsterdam.

Yancey, P. H. (1985). Organic osmotic effectors in cartilaginous fishes. *In* "Transport Processes, Iono- and Osmoregulation" (R. Gilles and M. Gilles-Ballien, eds.), pp. 424–436. Springer-Verlag, Berlin.

Yancey, P. H. (1994). Compatible and counteracting solutes. *In* "Cellular and Molecular Physiology of Cell Volume Regulation" (K. Strange, ed.), pp. 81–109. CRC Press, Boca Raton, FL.

Yancey, P. H. (2001). Water stress, osmolytes and proteins. *Am. Zool.* (in press).

Yancey, P. H., and Burg, M. B. (1990). Counteracting effects of urea and betaine on colony-forming efficiency of mammalian cells in culture. *Am. J. Physiol.* **258,** R198–204.

Yancey, P. H., and Siebenaller, J. F. (1999). Trimethylamine oxide stabilizes teleost and mammalian lactate dehydrogenases against inactivation by hydrostatic pressure and trypsinolysis. *J. Exp. Biol.* **202,** 3597–3603.

Yancey, P. H., and Somero, G. N. (1978). Temperature dependence of intracellular pH: its role in the conservation of pyruvate apparent K_m values of vertebrate lactate dehydrogenases. *J. Comp. Physiol.* **125,** 129–134.

Yancey, P. H., and Somero, G. N. (1979). Counteraction of urea destabilization of protein structure by methylamine osmoregulatory compounds of elasmobranch fishes. *Biochem. J.* **182,** 317–323

Yancey, P. H., and Somero, G. N. (1980). Methylamine osmoregulatory compounds in elasmobranch fishes reverse urea inhibition of enzymes. *J. Exp. Zool.* **212,** 205–213

Yancey, P. H., Clark, M. E., Hand, S. C., Bowlus, R. D., and Somero, G. N. (1982). Living with water stress: evolution of osmolyte systems. *Science* **217,** 1214–1222.

Yancey, P. H., Kelly, R. H., Fyfe-Johnson, A. L., Auñón, M. T., Walker, V. P., and Siebenaller, J. F. (2000). Effects of osmolytes of deep-sea animals on enzyme function and stability under high hydrostatic pressure. *In* "Science and Technology of High Pressure: Proceedings of AIRAPT-17" (M. H. Manghnani, Nellis, W. J., and Nicol, M. T., eds.), pp 328–330. Universities Press, Hyderabad, India.

Zigman, S. Munro, J., and Lerman, S. (1965). Effect of urea on the cold precipitation of proteins in the lens of dogfish. *Nature* **207,** 414–415.

Ziyadeh, F. N., Mills, J. W., and Kleinzeller, A. (1992). Hypotonicity and cell volume regulation in shark rectal gland: role of organic osmolytes and F-actin. *Am. J. Physiol.* **262,** F468–F479.

INDEX

A

1α-Hydroxycortisol, 319
5-hydroxytryptamine, 83
Acanthuridae, 154
Acipenser, amino acid metabolism, 80, 82, 95
Acipenseriformes, 19
Actin, 327
Actinopterygean fishes, osmolytes and osmoregulation, 311, 314–315, 319–320
Actinopterygians, 163
Actinopterygii, 19
Actomyosin ATPase, 323
Adenyl succinate lyase, 213
Adenyl succinate synthetase, 213
Adenylate breakdown, 210, 212–215
Agnatha, 5, 150
  osmolytes and osmoregulation, 311–312
Agnatha-Class Cephalaspidomorphi, osmoregulation, 314
Agnatha-Class Myxini, osmoregulation, 311
Air exposure
  responses of air breathing fish, 265–266
  ureotelism in *H. fossilis,* 263–266
Alanine, 23, 79, 83–86, 90–92, 94–96, 98, 123, 126–128, 140, 226
  beta, 311, 313, 322, 325
Albumin, 81, 83
Alcolapia grahami, 12, 14, 20, 298
Alevin, 150–151, 173, 179, 182
Alkaline lake-adapted tilapia, *see* Tilapia, Lake Magadi
Alkalosis, 225
Allantoate, 22
Allantoicase, 9, 243
Allantoin, 22

Allantoinase, 243
Alpha-helices, 282
Alpha-ketoglutarate, 213
Amia calva, 19, 132, 260
  amino acid metabolism, 89
Amino acid
  composition of protein, 39, 52
  free pool, 36–38
  requirements, 52–57
Amino acid catabolism, 123–128
Amino acid metabolism
  diurnal changes, 93
  effects of anoxia, 97–98
  effects of diet and starvation, 94–95
  effects of salinity, 96–97
  effects of temperature, 95–96
  hormonal regulation
    cortisol, 91–92
    insulin, glucagon, 92–93
    thyroid hormones, 93
  in gill, 86
  in kidney, 86–87
Amino acids
  aerobic metabolism, 165–166
  catabolism during exercise, 91
  demersal egg, 154, 161
  digestion and uptake, 79
  dispensable, 153, 156
  egg size, 153
  essential, 206–207, 212
  excretion, 223
  free, 123, 152–162, 164–166, 168–169, 185
    in development, 151–164
  freshwater, 155

Amino acids (*continued*)
  growth, 153, 157, 164
  in liver
    catabolism, 83–84
    gluconeogenesis, 85
    lipogenesis, 85–86
  indispensable, 153, 155–156
  intracellular levels, 215
  levels in plasma, 79–83
  membrane transporters, 79
  metabolism
    in blood cells, 87–88
    in brain, 90
    in intestine, 90–91
    in muscle, 88–90
    sleep, 83
  osmoregulation, 154, 168
  pelagic egg, 152–156, 158–161
  phosphoserine, 156
  plasma levels, 206–207, 209–210, 215, 223, 226
  salinity, 152, 154
  synthesis, 164
  taurine, 154–157, 159, 161
  yolk, 149, 152–153, 155–157, 160, 166, 177, 185
Aminoacyl tRNA synthetases, 78
Aminooxyacetate, 84, 89–91
Ammocoete, 5, 15, 181, 183
Ammonia
  active transport, 123, 134–139, 140
  biochemistry, 2–4
  branchial permeability, 223–226
  cell membrane permeability, 215
  chemistry, 169
  concentration
    in embryos and yolk, 171
    pH, 111
    pressure, 111
    temperature, 111
  detoxification, 123–140
    ATP requirement, 130, 138
    glutamate dehydrogenase, 128–129
    glutamine production, 128–132, 139
    glutamine synthetase, 128, 131
    urea production, 132
  diffusion, 112–113, 132, 139
  diffusity, 114
  distribution ratio, 215
  environment, 183
  environmental sources, 109–110
  factors affecting toxicity, 116–121
  hatching, 162, 171, 176, 185
  impermeable membrane, 113, 138
  intracellular levels, 215
  metabolic basis for toxicity, 115–116
  passive retention in muscle, 213–215
  permeability, 113–114, 280
    in embryos, 169–171
  plasma levels, 214–215, 224–226
  production and excretion, 112–114, 124–126
  properties, 110–113
  retention, 280
  solubility, 113
  synthesis, 173, 176
  toxicity levels, 110
  toxicity, 110–122, 139, 183
    avoidance
      alanine formation, 126–128
      glutamine formation, 128–132
      reduced rates of production, 124–126
      urea formation, 132
    criteria, 119–122
    criterion continuous concentration, 119–121
    criterion maximum concentration, 119–121
    environmental factors
      hardness of water, 117
      oxygen level, 118
      pH, 116
      seawater, 118–119
      temperature, 116
  transport, 3
  trapping, 112, 133–134
  trapping, glutamine synthetase, 289, 294
  un-ionized, 3
  volatilization, 133, 140
  water quality criteria, 110, 139
Ammonia detoxification, 210
  synthesis and storage of glutamine and other amino acids, 244
  urea synthesis, 239–244
Ammonia effects
  alanine production, 126–128, 139
  amino acid catabolism, 126, 139
  ATP production, 127
  brain, 115–116, 130
  feeding, 115
  gill, 115

INDEX

glycolysis, 115
ion transport, 115
ionic balance, 113, 115
liver, 129, 131
membrane potential, 115, 122
NMDA receptor, 116
proteloysis, 124–126, 140
proton excretion, 117, 136
swimming, 122
TCA cycle, 115, 132
total free amino acid, 124–126, 140
uncoupling mitochondria, 114
Ammonia excretion, 36, 112–113, 123, 132–140, 162, 179, 182–185
air-breathing, 20
alkaline environment, 20
ATP requirement, 138
branchial, 223–226
exercise effects, 209–216, 225–228
feeding effects, 202–210, 212, 216–219, 225–227
in elasmobranchs, 225–228
in teleosts, 201–225
K^+/NH_4^+ antiporter, 135–136
Na^+,K^+-ATPase, 114–115, 135–138
Na^+/H^+ exchanger, 114–115, 135–138
$Na^+/K^+/2Cl^-$ cotransport, 135–136
NH_3 volatilization, 133, 140
temperature effects, 216–219, 228
urinary, 223–224
Ammonia exposure, tolerance, 269
Ammonia metabolism, embryos, 176
Ammonia retention, 213–215
Ammoniagenesis, 124
Ammonia-producing reactions, 2
Ammonium ion, active transport, 134–139
Ammonotelic, 241
Amphibians, 180
Amphioxus, 6
Anarhichas lupus, 51, 58
Anchovy, 222
Angiotensin II, elasmobranchs, 319
Angiotensin, 319
Anguilla anguilla, 312
amino acid metabolism, 96–97
Anguilla japonica, 160
Anoplopoma fimbria, 162
Anoxia, amino acid metabolism, 76, 85, 97–98
Anserine, 315
Anthropogenic nitrogen, 25

Antimora microlepis, 327–328
Antioxidant, 331
Antiproteolytic response, 128
AQP, *see* Aquaporin water channels
Aquaculture, 32, 57
Aquaporin water channels, 283, 287
Areolated grouper, 216
Arginase, 7, 9, 92–94, 177, 179, 243
Arginine, 79, 81, 92–94
dietary requirement, 11
metabolic hub, 9
metabolism, 7–13
role in urea excretion by teleosts, 243
synthesis, 12–13
transport, 11
Arginine:glycine amidinotransferase, 24
Argininolyase, 177
Argininosuccinate, 177
Arginolysis, 7–8, 132, 217
Artemia, 169
Asparaginase, 84
Asparagine, 84–85
Aspartate, 79, 81, 84, 86–87, 98, 213
Autotrophic, 2

B

Band-3 anion exchanger, 318
Bass
kelp, 223
largemouth, *see Micropterus salmoides*
smallmouth, 117
striped, 118, 153, 158
Bathyraja, 326
Beoplasia, 18
Betaine, 311, 313, 318, 322–323, 325, 329–330
Betaine aldehyde dehydrogenase, 311
Betaine homocysteine transmethylase, 311
Beta-sheet, 282
Blastopore, 151
Blennidae, 154
Blennius pavo, 174
Blenny, 174
Blood, amino acid metabolism, 81, 87–88, 92, 97
Bluegill, 118
Boleophthalmus boddaerti, 125–127, 131–135
Bone, 7
Bostrichthyes sinensis, 130

Bowfin, 260
Brain, amino acid metabolism, 75, 83, 90, 92, 98
Bromofuroate, 90
Buoyancy, 5, 16, 329–331
　amino acids, 162
　NH_4^+, 162

C

Calpain, 90
Calpastatin, 90
Carassius auratus, 260
　amino acid metabolism, 97, 80, 82
Carassius carassius, 323
　amino acid metabolism, 98
Carbamoyl phosphate synthetase, 12–14, 19–20, 23
　evolution, 13
　types, 12, 244–245
Carbamoyl phosphate synthetase I, 20
Carbamoyl phosphate synthetase III, 20, 23, 177, 179–181, 220
　adapted properties in alkaline lake tilapia, 261–262
　amino acid sequences, 246–250
　evolution, 249–250
　gene sequences and regulation, 255–256
　in gulf toadfish, 267–269
　in *H. fossilis,* 264–265
　in largemouth bass, 249, 256–258
　in muscle, 247, 257–262, 264, 269
　in rainbow trout, 258–259
　in teleosts, 247
　properties and distribution, 245–250
　regulation by acetyl glutamate, 250
Carbonic anhydrase, 224
Carboxypeptidase A, 90
Carcharhinus amblyrhynchos, amino acids, 80, 82
Carcharhinus leucas, 296, 317
Carcharhinus melanopterus, amino acids, 80, 82
Carnosine, 315
Carp, 9, 155, 174, 182, 206, 216, 222
　amino acid metabolism, 86, 90, 94–96, 98
　common, 54, 64–66, 130, 260
　crucian, 44, 59
　grass, 47
Carrassius carassius, 59
Cartilaginous fishes, osmolytes, 310, 312–313, 330–331

Catfish, 157, 182–183
　African, 61, 157, 169, 171–172, 174, 176, 181–182
　amino acid metabolism, 84
　channel, 117–118, 260
　Indian, *see Heteropneustes fossilis*
　Madurai, 209
　Singhi, 263
　walking, *see Clarias batrachus*
Cathepsin D, 90
Cathepsin, 90
Catostomus, 120
Cell volume adaptation, strategies, 309
Centroscyllium fabricius, amino acids, 80, 82
Cephalospidomorphi, 6
Ceriodaphnia, 120
Chaenocephalus aceratus, 40
Chaetodontidae, 154
Channa asiatica, 128
Chaotropes, 332
Char
　Arctic, 206, 210, 222
　brook, 221
　lake, 221
Chimaera monstrosa, 312
Chimaera, osmolytes, 312
Chironomus, 117
Choline dehydrogenase, 311
Chondrichthyan fishes, osmolytes, 310, 317–318
Chondrichthyes, 150
Chondrostei, 19
　osmolytes and osmoregulation, 314
Chorion, 151, 162, 170, 172, 176, 182, 185
Chymotrypsin, 79
Ciliates, 168
Circadian rhythm, amino acid metabolism, 93
Clarias batrachus, 20, 265
Clarias gariepinus, 157, 172, 174
Cod, 153–155, 160, 162, 173, 177, 179, 182
　Atlantic, 39–40, 43–44, 47, 49, 51, 58, 153–155, 158–163, 165, 168, 171–172, 174, 177, 182, 220
　Murray, 155
　Pacific black, 162
Coelacanth, 5, 7, 18–19
　osmolytes, 312, 314, 318, 324–325
Collecting duct, 284
Conodonts, 6
Copepods, 168–169
Copper rock fish, 131
Coregonus gariepinus, 174

INDEX

Coregonus lavaretus, 174
Coregonus schinzii, 62
Cortisol, 215, 267, 319–320
 amino acid metabolism, 75–76, 91–92
 induction of GSase mRNA expression, 267
Coryphaenoides leptolepis, 328
CPSase, *see* Carbamoyl phosphate synthetase
CPSase I, *see* Carbamoyl phosphate synthetase I
CPSase III, *see* Carbamoyl phosphate synthetase III
Creatine phosphate, 23
Creatine, 20, 23, 222, 315
Creatinine, 20, 222
Cyclopterus lumpus, 155, 174, 176
Cymatogaster aggregata, 182
Cyprinis carpio, 21, 54, 80, 82, 96–97, 130, 155, 174, 260
Cysteine, 79, 90

D

Dab, 45, 59
 Long rough, 153
Daphnia, 120
Daphnia sp., 117
Dasyastis americanus, 317
Demersal egg, 154, 161
Development, stages, 150
Dexamethasone, 92
D-glycerate, 85
Dicentrarchus labrax, 154–155
Diet, amino acid metabolism, 94–95
Digestion, amino acid metabolism, 75, 79
Dimethylglycine oxidase, 311
Dimethylsulonioproprionate, beta, 329
Dipeptides, 79
Dipnoi, 19, 314
Dogfish, see also *Squalus acanthias*
 larger spotted, 228
 lesser spotted, 226
 spiny, *see Squalus acanthias*

E

Eel, 128
 European, 47, 224
 Japanese, 160
Eelpout, 183
Eeutheroembryo, 150–151, 171, 178–180, 184

Egg, 149–151, 153–156, 158–166, 170–171, 176, 181, 185
Elasmobranchs, 7, 13, 15, 18, 131, 150, 177, 225–228
 amino acid metabolism, 76, 80–82, 88, 90, 96–97
 low incidence of neoplasms, 254–255
 osmolytes, 312, 315, 319, 325–326
 osmoregulation, 312
 parenchymal hepatocytes, 8
 red blood cells, 8
 urea-utilizing bacteria, 25
Embryogenesis, 150, 171–172, 176–177, 179, 185
 urea cycle, 269
Embryos, 14, 24
 developmental changes in nitrogen excretion, 181–183
 elasmobranch fishes, osmolytes, 319, 325
 nutritional demand, 164–169
 tolerance to elevated ammonia and pH, 183
Entosphenus tridentatus, 183
Epinephelus coioides, 180
Erythrocytes
 amino acid metabolism, 81, 87, 92, 97
 osmolytes, 312, 318, 321
Esox lucius, amino acids, 80, 82
Ethanol, 98
Euryhaline, 310
Evolution, osmoregulation, 334
Exercise, 209–216, 227–228
 amino acid metabolism, 85, 90–91, 96
Exogenous fraction, 202–207

F

Faciocranial bone structure, 17
Fathead minnow, 117–118
Feeding, 149–150, 152, 155, 157, 162, 164, 166, 168–169, 182, 185, 202–210, 212, 216, 219, 222–223, 226–227, 289
Flavin-containing monooxygenase, 318, 320
Flounder, 36, 42, 161, 180
 Barfin, 154
 Japanese, 168, 180
 starry, 131
 winter, 161
 yellowtail, 159–160

FMO, see Flavin-containing monooxygenase
Frogs, estivating, 326
Fuel usage
  carbohydrate, 207–210, 212
  compositional approach, 207
  instantaneous approach, 208–210, 212, 219, 223
  lipid, 207–210, 212
  protein, 207–210, 212, 216, 219, 223
Fumarate, 213
Fundulus grandis, 40

G

Gadiform fish, 316
Gadus morhua, 40, 163, 172, 174
Gamma-aminobutyric acid, 98
Gamma-guanidine urea, hydrolase pathway, 7–8
GDH, see Glutamate dehydrogenase
Gill, amino acid metabolism, 75, 86, 93, 96
Gillichthys mirabilis, 323
  amino acid metabolism, 97
Gilthead sea bream, 154–155, 160, 172, 174
Global warming, 217
Glomerulus, 284
Glucagon, 287
  amino acid metabolism, 75, 92–93
Glucocorticoid response element, 267
  glutamine synthetase gene, 267
Glucocorticoids, 287
Gluconeogenesis, amino acids, 75–76, 83, 85–86, 90–92, 94
Glucose-alanine cycle, 23
Glutamate dehydrogenase, 83–84, 86–95, 97–99, 128–129, 176
Glutamate oxaloacetate transaminase, 87, 92–95, 176
Glutamate pyruvate transaminase, 176
Glutamate, 23, 79, 81, 83–84, 86–87, 89–90, 96, 98, 213
Glutaminase, 84, 86–87, 89, 93, 130
Glutamine, 22–23, 76, 78, 81, 84, 86–91, 93, 95, 98–99, 123, 128–132, 140, 227
  detoxification role, 23
  intertissue nitrogen carrier, 23
Glutamine synthetase, 13, 23, 86–88, 90, 92–94, 128, 131, 177, 179, 181, 226
  assay, 251

essential role in urea synthesis in elasmobranchs, 251–252
gene regulation, 252–253, 267
mitochondrial and cytosolic isozymes, 252
mitochondrial localization in liver of elasmobranchs, 251
role in ammonia detoxification, 250–253
subcellular localization in teleosts, 252–253
Glycerate dehydrogenase, 85
Glycerol, osmoregulation, 315
Glycerophosphorylcholine, 325–326
Glycine, 79, 85, 96–98, 310–311, 322
Gnathostomata-Class Actinopterygii, osmoregulation, 314
Gnathostomata-Class Chondrichthyes, osmoregulation, 312
Gnathostomata-Class Sarcopterygii, osmoregulation, 314
Gnathostomes, 5
Gobiidae, 154
Goldfish, 130–131, 220, 260
  amino acid metabolism, 83–84, 86–87, 89–90, 93, 98
GOT, see Glutamate oxaloacetate transaminase
Green sunfish, 117
Grouper, 180
Growth hormone, 64
Grunion, 173
GS, see Glutamine synthetase
GSase, see Glutamine synthetase
Guanidino-acetate, 23
Gulf toadfish, see *Opsanus beta*
Guppy, 117, 177, 180

H

H^+ ATPase, 3
Haddock, 159–161
Hagfish, 6, 138, 311
  amino acid metabolism, 97
  osmolytes, 312, 317, 327
Halibut, 61
  Atlantic, 152–160, 162, 165–168, 170–172, 174, 176–177, 260
Heart, shark, 326
Helicobacter pylori, 4
Hemitripterus americanus, amino acid metabolism, 92
Hemoglobin, 25, 324

Hepatocytes, 83–85, 92, 94, 97
  osmolytes, 318
Hepatosomatic index, 95
Heteropneustes fossilis, 20, 132, 219
  liver perfusion studies, effects of ammonia, 264–265
  tolerance to high ammonia levels, 264
  ureogenic nature, 264
  ureotelic response to air exposure, 263, 266
Heterotrophy, 2
High respiratory quotient, 208
Hippoglossoides platessoides limandoides, 153
Hippoglossus hippoglossus, 152, 172, 174
Hippoglossus hippoglossus L., 260
Histidase, 84
Histidine, 86, 92, 94, 96, 315
Homarine, 315
Hormones, amino acid metabolism, 91–93
Hyalella, 117, 120
Hypomesus olidus, 155
Hypoosmotic, 17
Hypoosmotic hypoionic regulation, 311
Hypoosmotic hypoionic regulators, 314–316
Hypoosmotic strategy, 17

I

Icefish, 40, 43, 60
Ictalurus, 120
Ictalurus nebulosus, amino acids, 80, 82
Ictalurus punctatus, 131, 260
IMCD$_t$, *see* Inner medullary collecting duct
Immune system, amino acid metabolism, 88
Indian catfish, *see Heteropneustes fossilis*
Inner medullar collecting duct, 281, 286
Inositol, myo-, 313, 318, 326
Insulin, amino acid metabolism, 75, 81, 92–93
Intestine, amino acid metabolism, 75, 79, 88, 90–91
Isolated perfused head preparation, 295, 299

J

Jack, 45
Jk null, 287
Juvenile, 151, 184

K

K^+/NH_4^+ antiporter, 135–136
Ketone bodies, 16
Kidney
  amino acid metabolism, 75, 86–88, 94, 96–97
  anatomy, 284–285
Killifish, 40, 220
Kjeldahl digestion, 222
Kosmotropes
  preferential exclusion, 332–333
  preferential interaction, 332

L

Labridae, 154
Lactate, 83, 85, 87
Lactate dehydrogenase, 326–328, 322–324
Lake Magadi tilapia, *see* Tilapia, Lake Magadi
Lampetra fluviatilis, 312
Lamprey, 5, 15, 311
  osmolytes and osmoregulation, 312, 314, 319
  Pacific, 183
  Sea, 179
    amino acid metabolism, 80, 82
Largemouth bass, *see Micropterus salmoides*
Larvae, 150, 152–157, 161–162, 164–166, 168–169, 171, 177, 182, 184–185
  ammonia and urea transport, 184–185
  nutritional demand, 164–169
Larval fish, 38–39, 60–62
Lates calcarifer, 154–155
Lates niloticus, 328
Latimeria chalumnae, 19, 312
Latimeria menadoensis, 19
Leiostomus xanthurus, 154
Lepisosteus osseus, amino acids, 80, 82
Lepomis, 120
Leucine, 78–79, 83, 86–87, 92, 93, 96–97
Leucocytes, amino acid metabolism, 87
Leuresthes tenuis, 173
Limanda limanda, 59
Liparid, 329
Lipogenesis, amino acids, 75–76, 85–86, 95
Liver, amino acid metabolism, 83–86
Lophius piscatorias, 301
Lumbriculus, 117
Lumpsucker, 155, 174, 176

Lungfish, 19
  amino acid metabolism, 84
  osmolytes, 314, 326
  osmoregulation, 314
Lysine, 79, 81, 92, 94

M

Maccullochella macquariensis, 155
Maccullochella peelii peelii, 155
Macrourid fish, 316
Malate, 213
Malic enzyme, 88
Mammals, 152, 156, 180
Mangrove snapper, 216
Marble goby, *see Oxyeleotris marmoratus*
Medaka, 156
Melanogrammus aeglefinus, 159
Membrane potential, 215
Metabolic acidosis, 223, 225
Metabolic fuels, 207–209, 219, 223
Metabolic rate, 203–206, 208, 210, 216–217, 219, 221, 223, 228
Methemoglobin, 25
Methionine, 81, 94
Methylamines, 310, 312–313, 321, 325–331
  counteracting hydrostatic pressure effects, 327–328
  counteracting salt effects, 327
  counteracting temperature effects, 328–331
  counteracting urea effects, 325–326
Methylamines, osmoregulation, 312–322
Micropterus salmoides, 118, 120, 220, 224, 256
  CPSase III and OCTase in muscle, 257
  exposure to ammonia, 257
  urea cycle, 256–258
  ureogenic and ammonotelic nature, 257
Microstomus kitt, 153, 172, 174
Midshipman, see also *Porichthys notatus*
  Pacific, 223
Misgurnus anguillicaudatus, 133
Mitochondria, respiration, 326
Monodeiodinase, 5', osmolyte effects, 326, 335
Monopterus albus, 128
Moonfish, 45
Morid fish, 316
Morone saxatilis, 153, 320
Mucous, 85
Mucus, 223

Mudskipper, 124–128, 131–135, 140
  amino acid metabolism, 85
Mullet, 208–210
Muscle, amino acid metabolism, 75–76, 78, 88–90, 92, 93, 95–98
Musculium, 120
Mustelus canis, 171, 290
Mustelus manazo, 313
Myosin, 38
Myxine glutinosa, 312
Myxine, amino acid metabolism, 97
Myxini, 6
  osmolytes, 311–312, 317

N

Na,K/2 Cl$^-$ cotransporter, 224
Na,K-ATPase, 224
Na/H exchange, 224–225
Na/NH$_4$ exchange, 224–225
Na$^+$,K$^+$/2 Cl$^-$ cotransporter, 25
Na$^+$,K$^+$-ATPase, 91, 96, 114–115, 135–138, 288, 292, 320, 334
Na$^+$/H$^+$ exchange, 3
Na$^+$/H$^+$ exchanger, 114–115, 135–138
Na$^+$/K$^+$/2 Cl$^-$ cotransport, 135–136
Na$^+$-coupled urea cotransporter, 292
Na$^+$-urea cotransport, 288
N-acetylglutamate, 13
Nase, 61
Nebrius ferrugineus, amino acids, 80, 82
Neopterygii, 19
  osmolytes and osmoregulation, 314
Neurotransmitters, amino acids, 90, 98
NH$_3$ diffusion/permeability, 213–216, 224, 226
Nitrate, 25
Nitrogen budget, 36
Nitrogen excretion
  changes during early development, 181–183
  effect of exercise, 209–216
  effect of feeding, exercise and temperature in elasmobranchs, 225–228
  effect of temperature, 216–219
  endogenous fraction, 202–207
  exercise effects, 209–216, 227–228
  exogenous fraction, 202–207, 212, 227
  fasting versus fed, 202–207
  feeding effects, 202–210, 212, 216–219, 222–223, 226–227

INDEX 351

in elasmobranchs, 225–228
in teleosts, 201–225
mechanisms, 223–225
other compounds, 220–223
other than urea or ammonia, 221–223
temperature effects, 216–219, 228
urinary, 203, 223–224
Nitrogen metabolism, 149, 176
  exercise effects, 209–216, 227–228
  feeding effects, 202–210, 212, 216–219, 222–223, 226–227
  in elasmobranchs, 225–228
  in teleosts, 201–225
  temperature effects, 216–219, 228
Nitrogen oxidizer, 222
Nitrogen quotient, 208, 210–211
Nitrogen retention, 203, 215, 226
Non-salmonlids, 120–121
Norepinephrine, 320
Notothenia corriceps, 43, 46, 60
Nototheniid, 329
N-trimethylamine oxide, *see* Trimethylamine oxide
Nutrient requirements, 65–66

O

Octopine, 15
Octopine dehydrogenase, 15
Oligocottus maculosus, 298
Oncorhynchus, 120
  amino acid metabolism, 80, 82, 97
Oncorhynchus clarki henshawi, 117
Oncorhynchus gorbuscha, 183
Oncorhynchus keta, 176, 183
Oncorhynchus kisutch, 11, 59, 122
Oncorhynchus mossambicus, 184–185
Oncorhynchus mykiss, 11, 34–35, 38–39, 43–47, 49–50, 52–58, 61, 65, 113, 117–118, 124, 130, 169–170, 172–174, 176–180, 182, 184–185, 203–204, 206, 208, 215–217, 220–222, 224, 256
  amino acid metabolism, 78, 84–87, 91–97
  CPSase III and OCTase in muscle, 258
  possible role of CPSase III and OCTase in muscle during exercise, 259
Oncorhynchus tshawytscha, 11
Ontogeny, 150, 185

Opsanus, 299
  amino acid metabolism, 92
Opsanus beta, 12, 20, 22, 33, 40, 44, 130, 137, 219, 224, 298–299, 301
  CPSase III in muscle, 269
  induction of GSase expression by environmental changes, 266–267
  urea cycle capability, 266
  ureogenic and ammonotelic nature, 266
Opsanus tau, 33, 301
Oreochromis alcalicus grahami, 132, 260
  adaptation to high pH, 260–263
  adapted properties of CPSase III, 261–262
  high levels of urea cycle enzymes in muscle, 261
Oreochromis mossambicus, 94, 97
Oreochromis niloticus, 155
Organic osmolytes, 310
Ornithine, 11
Ornithine carbamoyltransferase, 92, 177, 181
Ornithine decarboxylase, 11
Ornithine transcarbamylase, *see* Ornithine carbamoyltransferase
Ornithine-urea cycle, *see* Urea cycle
Oryzias latipes, 156
Osmoconformer, 309–312, 317, 334
Osmoconforming hypoionic regulation, 310
Osmoconforming hypoionic regulators, 312–314
Osmolyte genes, regulatory response element, 316
Osmolytes, 5
  antioxidant, 330
  buoyancy, 329–331
  compatibile, 320–324
  counteracting, 325–329, 335
  evolution, 333–334
  genes, 316
  metabolism and regulation, 316–320
  practical use, 334–335
  properties and functions, 320–333
  types of organic nitrogenous, 310–311
  water structure, 333
Osmoregulation, 96–97, 241
  energy costs, 334
  evolution, 334
  hyperosmotic, 311
  hypoosmotic hypoionic, 311, 314–315, 319–320, 334
  osmoconforming hypoionic, 310, 312–314, 317–319, 334

Osteichthyes, 19, 150
Oubain, 288
Oxaloacetate, 213
Oxyeleotris marmoratus, 126, 129–130, 259
   CPSase III and OCTase activities in muscle, 259

P

Pacific staghorn sculpin, 222
Pagrus major, 155
Paralichthys olivaceus, 180
PDG, *see* Phosphate-dependent glutaminase
Pedogenesis, 15
Pelagic eggs, 151–156, 158–161, 164, 185
PEPCK, *see* Phosphoenolpyruvate carboxykinase
Pepsin, 79
Periophthalmodon schlosseir, 124–125, 127–128, 131–135, 138, 140
Periophthalmus, amino acid metabolism, 90
Periophthalmus cantonensis, 124, 126, 131
Perivitelline
   fluid, 151, 170
   membrane, 151, 156
   space, 151, 153
Perturbation of macromolecular structure
   amino acids, 321–322
   inorganic ions, 321–322
   methylamines as stabilizing agents, 325–331
   urea, 322–324
Petromyzon fluviatilis, 312
Petromyzon marinus, 179, 181, 183
   amino acid metabolism, 80, 82
Phenylalanine, 79, 81, 92
Phloretin, 281, 288, 290, 292
Phosphate-dependent glutaminase, 86–89, 95
Phosphoenolpyruvate carboxykinase, 85
Phosphofructokinase, 212
   activation by ammonia, 213
Photoperiod, amino acid metabolism, 93
Pimephales, 120
PKA, *see* Protein kinase A
PKC, *see* Protein kinase C
Plaice, 155, 158, 160–162, 174, 208, 216
Platichthys flesus, 317
Platichthys stellatus, 24, 131
Plecoglossus altivelis, 155

Pleuronectes, amino acid metabolism, 96
Pleuronectes americanus, 161
Pleuronectes ferrugineus, 159
Pleuronectes flesus, 36
Pleuronectes platessa, 174
Poecilia reticulata, 177
Polyamine biosynthesis, 11
Polypteriformes, 19
Pomacentridae, 154
Porichthys notatus, 259
   CPSase III and OCTase activities in muscle, 259
   ureogenic but ammonotelic nature, 259
Potamotrygon, 318
Potomotrygonidae, 226
Prawn larvae, 117
Pressure
   hydrostatic, osmolyte counteraction, 327–328, 332–333
   osmotic, 309
Prolactin, 320
Proline, 79, 85, 322
Protease, 90
Protein as a fuel
   burst anaerobic swimming, 212–216
   fed state, 209
   measurement, instantaneous method, 208
   resting state, 208
   sustainable aerobic swimming, 209–212
Protein degradation, 35, 210
   lysosomes, 36
   proteasome ubiquitin pathway, 36
Protein kinase A, 283
Protein kinase C, 283
Protein synthesis, 77, 206–207, 210, 212, 216–217, 219
   amino acid requirements, 52–57
   anabolic stimulation, 54–57
   and amino acid flux, 37–38
   and body weight, 62
   and daily cycles, 38–40
   and diet, 66
   and exercise, 63
   and humans, 64
   and life-history stage, 60–62
   capacity for, 33, 37–38, 52–53, 58–59, 62, 64, 66
   effect of feed intake, 64
   effect of oxygen, 59
   effect of pollutants, 59–60

effect of salinity, 59
effect of starvation, 64
effect of temperature, 58–59
energetic cost, 67
fractional rate, 33
influence of diet, 52–57, 65–67
influence of feeding, 34–41, 64–65
influence of temperature, 52, 57–59
influence of weight, 52, 62–63
integrated model, 36–37
measurement, 33–35
   end-product, 33–34, 64
   flooding dose, 34
   stable isotope, 34
mechanism, 32–33
nutrient requirements, 65
rates within organs and cells, 40–52
retention efficiency, 50, 54–57
species differences, 60
Protein turnover, 33
Proteolysis, 79, 90, 123–126
Pseudopleuronectes americanus, 301
Puffer fish, 155
Purine degradation, 7–8
Purine nucleotide biosynthetic pathway, scheme, 241
Purine nucleotide cycle, 83, 89, 213
  during exercise, 213
Purines, 22
Pyrimidine nucleotide biosynthesis, absent in liver of elasmobranchs, 254
Pyruvate carboxylase, 85
Pyruvate kinase, 323, 328

Q

Q10, 216, 228

R

Rainbow trout, *see Oncorhynchus mykiss*
Raja, amino acid metabolism, 96, 97
Raja binoculata, 177, 325
Raja eglanteria, 297
Raja erinacea, 138, 172, 290, 296, 326
Raja nevis, 323
Raja ocellata, 172

Rays, freshwater, 226
RBC, *see* Red blood cells
Red blood cells, 281, 287–288, 296
Red drum, 174, 182
Red sea bream, 155
Renin, 319
Respiration
  cutaneous, 184
  gill, 170, 184
Respiratory quotient, 208
Rhinobatus productus, 323
Ribosome, 32
RNA
  activity, 33, 52, 58–59, 62, 64, 66
  messenger, 32
  ribosomal, 52, 67
  transfer, 32, 67
Rotifers, 168
Ruminant, 5

S

Salinity, amino acid metabolism, 76, 96–97
Salmo salar, 40, 55, 155
  amino acid metabolism, 94, 97
Salmon
  Atlantic, 35, 39–40, 43, 45–46, 55, 61–62, 155, 158, 206, 208, 222
  chinook, 131, 220
  chum, 176, 183
  coho, 59, 117, 122, 221
  osmoregulation, 315, 320
  Pacific, 203
  pink, 183
  sockeye, 205, 208
Salmonids, 121, 150, 164, 183
Salt, osmolyte counteraction, 327
Salvelinus alpinus, amino acid metabolism, 80, 82
Salvelinus fontinalis, amino acid metabolism, 93
Salvelinus namaycush, 132
  amino acid metabolism, 89
Sarcopterygian fishes, osmolytes, 314
Sarcopterygii, 19
Sarcosine oxidase, 311
Sarcosine, 313, 326, 329–330
Scaridae, 154
Sciaenops ocellatus, 174, 182, 301

Scophthalmus maximus, 163, 172, 174
Scyliorhinus canicula, 171, 296, 312
Sea bass, 61, 66, 206
  Asian, 154–155, 166
  European, 154–155, 160
Sea ravens, amino acid metabolism, 92
Serine, 79, 84–86, 94
Serine dehydratase, 85
Serine hydroxymethyltransferase, 84
Serine pyruvate transaminase, 84–85
Shark
  osmolytes, 312, 316
  oviparous, 171
  pyjama, 226–227
  viviparous, 171
ShUT, 291
Skate, 138
  osmolytes, 312, 316, 323, 326, 328, 330
Sleep, amino acid metabolism, 83
Smelt
  pond, 155
  sweet, 155
Smolt, 151
Snakehead, 128
Sole,
  Dover, 115
  lemon, 153, 155, 158, 160, 162, 173, 174
  Senegal, 155, 168
Solea senegalensis, 155
Sparus aurata, 154–155, 172, 174
Specific dynamic action, 206, 212
Spiny dogfish, *see Squalus acanthias*
Spot, 154
Squalus acanthias, 137, 177, 226, 289–290, 294–296, 313, 323
  amino acid metabolism, 80, 82, 90
  ureotelism, 289
Squalus suckleyi, 177, 296
Starvation, 64–65
  amino acid metabolism, 75, 94–95
Stenohaline, 310
Stickleback, 118
Stizostedion vitreum, 168
Sturgeon, 206, 212
  amino acid metabolism, 78, 80, 82, 94–95
Surf steenbras, 216, 222
Swimming
  aerobic, 209–212, 227–228
  anaerobic, 212–216, 227–228

T

T_3, amino acid metabolism, 92–93
T_4, amino acid metabolism, 93
Tadpoles, 180
Taeniura lymma, amino acids, 80, 82
Takifugu niphobles, 155
Taurine, 81, 90, 94, 96–97, 315
  as antioxidant, 331
  compatibile osmolyte, 322
  osmolyte usage, 313, 315, 318, 320
  structure of, 311
Teleost, 151, 153–158, 161–165, 169, 176, 179, 181, 184
Teleostean, 18–19
Teleostei, 19
  Arctic and Antarctic, 315, 327, 329
  deep sea, 315–316, 327–328
  osmolytes and osmoregulation, 312, 314–315, 319–320
Teleosts, 201–225; *see also* Teleostei
Temperature
  amino acid metabolism, 76, 95–96
  osmolyte counteraction, 328–329
Theragra chalcogramma, 329
Threonine dehydratase, 84
Thrombocytes, amino acid metabolism, 87
Thyroid, amino acid metabolism, 92–93
Thyroxine, amino acid metabolism, 93
Tilapia mossambica, amino acid metabolism, 96
Tilapia, 61, 63, 328
  amino acid metabolism, 91, 96
  Lake Magadi, 131, 134, 219, 260, 298
  Mozambique, 209
  Nile, 155, 206, 208, 216
TMA, *see* Trimethylamine
TMAO, *see* Trimethylamine oxide
Toadfish, *see Opsanus beta*
Transcription, 32
Translation
  elongation, 32
  initiation, 32
  termination, 32
Transport, amino acids, 78–79, 81, 83–84, 87, 91–92
Triggerfish, 44–45
Triiodothyronine, amino acid metabolism, 92
Trimethylamine oxidase, 311, 318
Trimethylamine oxide, 16–17, 222, 226, 289, 312–322

INDEX

antioxidant, 331
breakdown, 313–316
buoyancy, 16, 329–331
deep sea animals, 316
in detoxification, 320
osmolyte usage, 313, 315, 317–320
osmoregulation, 312–322
regulation, 317–319
stabilizing solute, 16
stabilizing/counteracting, 323, 325–329, 333, 335
structure of, 311
Trimethylamine, 222, 314–315, 320
Trout
  Lahontan cutthroat, 117, 219
  Rainbow, see Oncorhynchus mykiss
Trypsin inhibitor, 90
Trypsin, 79
Tryptophan, 81, 83
Tuna, amino acid metabolism, 83, 85
Turbot, 61–62, 154–156, 162–163, 165, 167, 171–172, 174
tUT, 298
Typhlogobius californiensis, amino acid metabolism, 97
Tyrosine, 81, 92–94
Tyrosine aminotransferase, 92

U

Unstirred layer, 170, 185
Urea, 123, 132, 140
  active transporter, mammalian kidney, 288
  alternative roles, 5
  as antioxidant, 331
  as osmolyte, 220, 225–227, 312–325
  back transporter, gill, 294–295
  branchial permeability, 226
  breakdown, 314
  buoyancy 16, 329–331
  concentration in embryos and yolk, 171–173
  excretion, 36, 168, 171, 173, 179–180, 182–183, 185, 219–221
    branchial, 223–226
    elasmobranch, euryhaline adaptation, 296–297
    elasmobranch, kidney, 290–294
    elevated, 220–221
    embryogenesis, 220
  exercise effects, 209–216, 225–228
  feeding effects, 202–209, 212, 216–219, 222–223, 225–227
  gill
    pulsatile, 298–300
    toadfish, 298–300
    in elasmobranchs, 225–228
    in teleosts 201–225
    predominating, 219–221
    pulsatile, 268
    sources of urea in teleosts, 242–244
    temperature effects, 216–219, 228
    urinary, 224–226
  in detoxification, 320
  insensitive proteins, 324
  movement, 286
  osmolyte usage, 312–315, 317–319
  osmoregulation, 312
  permeability, 15, 171, 280
    elasmobranch, gill, 294
  perturbation of macromolecular structure, 321–324
  perturbing osmolyte, 322–324, 335
  plasma levels, 224, 226–227
  ratio to TMAO and methylamines, 325–326, 330–331
  reabsorption, 290
  regulation, 317–319
  requiring proteins, 324
  resistance to inhibitory effects, 324
  retention, 15–16, 173, 176–177, 226, 279
  structure of, 311
  synthesis, 7–15, 173, 176–177, 179–182
  tolerance, 6–17
  transport, 4, 24, 173, 184–185, 279–280
    elasmobranch kidney, active, passive, 291–293
    elasmobranch
      hepatocytes, 296
      red blood cells, 296
    evolution, 301
    hepatocytes, red blood cells, 301
    mammalian systems, 281–288
    Na$^+$,K$^+$-ATPase, 292
    Na$^+$-coupled urea cotransporter, 292
  transporter
    gills
      distribution in teleosts, 300
      toadfish, 298–300
    kidney, 300–301

Urea (continued)
  transporters, 16, 280
    mammalian kidney, 284–288
    membrane-spanning structure, 281–283
    protein kinase A site, 283
    protein kinase B site, 283
    species distribution, 288
    UT-A and UT-B1, 281–283
    UTA-A and UTP-B
      expression, 283
      specificity, 283
Urea cycle, 7–13, 210, 213, 219–220, 226
  adaptations, 256–269
  ammonia detoxification, 183
  ammonia-dependent, scheme, 240
  during embryogenesis, 269
  early development, 176–181
    localization, 180–181
    regulation, 180
  embryo, 177, 179–180
  embryogenesis, 14
  evolution, 13
  glutamine-dependent, scheme, 242
  in elasmobranchs, 253–256
  in non-ureotelic teleosts, 256–260
  in teleosts, 256–269
  N-acetyl-L-glutamate, 179
  pathway, elasmobranchs, 242
  phylogenetic survey, 18
  regulation, 177, 180
  role of glutamine synthetase in regulation, 266–268
  role of muscle, in teleosts, 257–262
  zonation, 25
Urease, 4
Ureogenic, 239
Ureoosmotic, 239, 241, 312
Ureoosmotic regulation
  adaptation, 7
  evolution, 5–7
Ureoosmotic strategy, 17
Ureotelic, 239
Uric acid
  antioxidant, 176
  chemistry, 173
  embryos, 173
  excretion, 22

Uricase, 93, 95, 243
Uricolysis pathway, scheme, 241
Uricolysis, 132, 179, 181, 206, 210, 213, 220
  rate dependent on flux through purine biosynthesis and degradation pathway, 243
Urolophis haleri, 323

V

Valine, 85
Vasopressin, 281, 286–288, 319
Vasopressin receptor, 286
Vasotocin, 319
Verasper moseri, 154
Vitellin, 14
Vitellogenin, 5, 11–12

W

Walleye, 168
Walking catfish, *see Clarias batrachus*
Water, structure, 333
Weather loach, 133
White perch, 117
Whitefish, 62, 174, 182
Wolffish, 51, 58, 216

X

Xenopus oocytes, 281, 283

Y

Yolk
  lipovitellins, 160–161
  phosvitins, 156, 160
Yolk sac stage, 150, 169–171, 173, 182, 184

Z

Zoarces viviparus, 183
Zoarcid, 316, 329

OTHER VOLUMES IN THE FISH PHYSIOLOGY SERIES

VOLUME 1	Excretion, Ionic Regulation, and Metabolism
	Edited by W. S. Hoar and D. J. Randall
VOLUME 2	The Endocrine System
	Edited by W. S. Hoar and D. J. Randall
VOLUME 3	Reproduction and Growth: Bioluminescence, Pigments, and Poisons
	Edited by W. S. Hoar and D. J. Randall
VOLUME 4	The Nervous System, Circulation, and Respiration
	Edited by W. S. Hoar and D. J. Randall
VOLUME 5	Sensory Systems and Electric Organs
	Edited by W. S. Hoar and D. J. Randall
VOLUME 6	Environmental Relations and Behavior
	Edited by W. S. Hoar and D. J. Randall
VOLUME 7	Locomotion
	Edited by W. S. Hoar and D. J. Randall
VOLUME 8	Bioenergetics and Growth
	Edited by W. S. Hoar, D. J. Randall, and J. R. Brett
VOLUME 9A	Reproduction: Endocrine Tissues and Hormones
	Edited by W. S. Hoar, D. J. Randall, and E. M. Donaldson
VOLUME 9B	Reproduction: Behavior and Fertility Control
	Edited by W. S. Hoar, D. J. Randall, and E. M. Donaldson
VOLUME 10A	Gills: Anatomy, Gas Transfer, and Acid-Base Regulation
	Edited by W. S. Hoar and D. J. Randall
VOLUME 10B	Gills: Ion and Water Transfer
	Edited by W. S. Hoar and D. J. Randall
VOLUME 11A	The Physiology of Developing Fish: Eggs and Larvae
	Edited by W. S. Hoar and D. J. Randall

VOLUME 11B	The Physiology of Developing Fish: Viviparity and Posthatching Juveniles *Edited by W. S. Hoar and D. J. Randall*
VOLUME 12A	The Cardiovascular System *Edited by W. S. Hoar, D. J. Randall, and A. P. Farrell*
VOLUME 12B	The Cardiovascular System *Edited by W. S. Hoar, D. J. Randall, and A. P. Farrell*
VOLUME 13	Molecular Endocrinology of Fish *Edited by N. M. Sherwood and C. L. Hew*
VOLUME 14	Cellular and Molecular Approaches to Fish Ionic Regulation *Edited by Chris M. Wood and Trevor J. Shuttleworth*
VOLUME 15	The Fish Immune System: Organism, Pathogen, and Environment *Edited by George Iwama and Teruyuki Nakanishi*
VOLUME 16	Deep Sea Fishes *Edited by D. J. Randall and A. P. Farrell*
VOLUME 17	Fish Respiration *Edited by Steve F. Perry and Bruce Tufts*
VOLUME 18	Muscle Growth and Development *Edited by Ian A. Johnston*
VOLUME 19	Tuna *Edited by Barbara Block and E. Donald Stevens*